Chemistry from First Principles

Jan C. A. Boeyens

Chemistry from First Principles

Springer

Jan C. A. Boeyens
Unit for Advanced Study
University of Pretoria
Pretoria 0002
South Africa

ISBN: 978-90-481-7907-7 e-ISBN: 978-1-4020-8546-8

Softcover reprint of the hardcover 1st edition 2008

Printed on acid-free paper.

9 8 7 6 5 4 3 2 1

springer.com

Preface

The events of 1925/26 that revolutionized physics held out the promise of solving all problems in chemistry. For physics these events represented the fastest paradigm shift on record. Many great ideas in science meet with scepticism and conservative resistance which can delay their acceptance, even by centuries, as in the case of Copernicus and Galileo. The announcement in 1925 that the old quantum theory had been decisively swept away by a fundamentally different profound new understanding of the atomic world was accepted with acclaim, not within decades or years, but within a few months. A notable exception was Albert Einstein, who wrote in a letter of September 1925 [1](page 225):

> In Göttingen they believe it (I don't).

He remained unconvinced for all his life.

The rest of the physics world was dazzled by the mathematical wizardry and the stature of Niels Bohr who championed the new theory from its inception. In retrospect some of the claims about the new theory as "the end of the road" for theoretical physics appear bizarre, making the universal uncritical acceptance of the new theory all the more remarkable. The further claim that the new development represented a total break with classical physics, although equally bizarre, was enthusiastically hailed as the biggest single advance ever achieved in physics.

The extravagent claims by which the new *Quantum Mechanics* was announced, are now largely forgotten, but not the belief that a new world order was established in science, free of concepts such as reality, causality, objectivity, certainty, predictability and many other notions based on classical views of the macroscopic world; all of these to be replaced by statistical probabilities.

The new theory developed from two independent publications – a purely mathematical model and the Schrödinger alternative with a clear physical foundation. The latter was immediately branded as a futile attempt to revive the concepts of classical physics, already refuted by the new paradigm.

All of this and the subsequent attacks to discredit Einstein and re-interpret Schrödinger's results are historically documented facts, to be frequently referenced in the following.

The interminable discussions on the interpretation of quantum theory that followed the pioneering events are now considered to be of interest only to philosophers and historians, but not to physicists. In their view, finality had been reached on acceptance of the Copenhagen interpretation and the mathematical demonstration by John von Neumann of the impossibility of any alternative interpretation. The fact that theoretical chemists still have not managed to realize the initial promise of solving all chemical problems by quantum mechanics probably only means some lack of insight on the their part.

The chemical literature bristles with failed attempts to find a quantum-mechanical model that accounts for all aspects of chemistry, including chemical bonding, molecular structure, molecular rearrangement, stereochemistry, photochemistry, chirality, reactivity, electronegativity, the valence state and too many more to mention. A small group of enthusiasts still believe that it's all a question of computing power, but that hope is also fading fast.

The present volume is a final attempt, after fifty years of probing, to retrace the steps that produced the theories of physics and to identify the point at which chemistry missed the boat. It is well known that in the days of the old quantum theory chemists and physicists could speak with one voice, which produced the solution to the Balmer numbers, the development of the Bohr-Sommerfeld model of the hydrogen atom and explained the periodic table of the elements. After that the paths of chemistry and physics have diverged. The definition of the periodic table and the tetrahedral carbon atom is no longer as convincing as before and electronic orbital angular momentum has been replaced by the ill-defined concept of *atomic orbitals.* There is no theoretical guidance to the understanding of chemistry's empirical truths.

The historical record shows that the success and failure of the first structural model of the atom resulted from a correct assumption made by Bohr for the wrong reasons. It was correct to assume that orbital angular momentum is quantized, but the assumed value in the hydrogen ground state was wrong. Apart from this understandable error, the Bohr model is shown to contain all the necessary ingredients that could have led directly to the mathematical structure of quantum mechanics discovered more than ten years later. In retrospect, it was the wrong decision not to concentrate on the mathematical formalism, rather than trying to improve the physical Kepler model, along with Sommerfeld.

It is interesting to note that the Göttingen school, who later developed matrix mechanics, followed the mathematical route, while Schrödinger linked his wave mechanics to a physical picture. Despite their mathematical equivalence as Sturm-Liouville problems, the two approaches have never been reconciled. It will be argued that Schrödinger's physical model had no room for classical particles, as later assumed in the Copenhagen interpretation of quantum mechanics. Rather than contemplate the wave alternative the Copenhagen orthodoxy preferred to disperse their point particles in a probability density and to dress up their interpretation with the uncertainty principle and a quantum measurement problem to avoid any wave structure.

The weird properties that came to be associated with quantum systems, because of the probability doctrine, obscured the simple mathematical relationship that exists between classical and quantum mechanics. The lenghthy discussion of this aspect may be of less interest to chemical readers, but it may dispel the myth that a revolution in scientific thinking occured in 1925. Actually there is no break between classical and non-classical systems apart from the relative importance of Planck's action constant in macroscopic and microscopic systems respectively. Along with this argument goes the realization that even in classical mechanics, as in optics, there is a wave-like aspect associated with all forms of motion, which becomes more apparent, at the expense of particle behaviour, in the microscopic domain.

These comments will undoubtedly lead to the criticism that here is just another attempt to return to classical physics. As already explained, this assessment will not be entirely wrong and not entirely right. In order to recognize the distinctive new features of quantum theories it is necessary to examine some alternative interpretations, which have failed to enter mainstream physics, and having sensed that: "Something is rotten in the state of Denmark". The truly novel feature of quantum theory, its non-locality, has been lost in the arguments over completeness and uncertainty.

The book consists of two parts: A summary and critical examination of chemical theory as it developed from early beginnings through the dramatic events of the twentieth century, and a reconstruction based on a reinterpretation of the three seminal theories of periodicity, relativity and quantum mechanics in chemical context.

Anticipating the final conclusion that matter and energy are special configurations of space-time, the investigation starts with the topic of relativity, the only theory that has a direct bearing on the topology of space-time and which demonstrates the equivalence of energy and matter and a reciprocal relationship between matter and the curvature of space.

Re-examination of the first quantitative model of the atom, proposed by Bohr, reveals that this theory was abandoned before it had received the attention it deserved. It provided a natural explanation of the Balmer formula that firmly established number as a fundamental parameter in science, rationalized the interaction between radiation and matter, defined the unit of electronic magnetism and produced the fine-structure constant. These are not accidental achievements and in reworking the model it is shown, after all, to be compatible with the theory of angular momentum, on the basis of which it was first rejected with unbecoming haste.

The Sommerfeld extension of the Bohr model was based on more general quantization rules and, although more successful at the time, is demonstrated to have introduced the red herring of tetrahedrally directed elliptic orbits, which still haunts most models of chemical bonding.

The gestation period between Bohr and the formulation of quantum mechanics was dominated by the discovery and recognition of wave phenomena in theories of matter, to the extent that all formulations of the quantum theory developed from the same classical-mechanical background and the Hamiltonian description of multiply-periodic systems. The reasons for the fierce debates on the interpretation of phenomena such as quantum jumps and wave models of the atom are discussed in the context of later developments. The successful, but unreasonable, suppression of the Schrödinger, Madelung and Bohm interpretations of quantum theory is shown not to have served chemistry well. The inflated claims about uniqueness of quantum systems created a mystique that continues to frighten students of chemistry. Unreasonable models of electrons, atoms and molecules have alienated chemists from their roots, paying lip service to borrowed concepts such as measurement problems, quantum uncertainty, lack of reality, quantum logic, probability density and other ghostlike phenomena without any relevance in chemistry. In fact, classical and non-classical sytems are closely linked through concepts such as wave motion, quantum potential and dynamic variables.

The second part of the book re-examines the traditional concepts of chemistry against the background of physical theories adapted for chemistry. An alternative theory is formulated from the recognition that the processes of chemistry happen in crowded environments that promote activated states of matter. Compressive activation, modelled by the methods of Hartree-Fock-Slater atomic structure simulation, leads to an understanding of elemental periodicity, the electronegativity function and covalence as a manifestation of space-time structure and the golden ratio.

The cover drawing shows the set of calculated general covalence curves, in dimensionless units, with an empirical reconstruction, as circular segments, within a golden rectangle. The absolute limit to covalent interaction is reached at values of interatomic distance and binding energy conditioned by the golden ratio τ. The turning point occurs where the maximum concentration of valence density, allowed by the Pauli exclusion principle, is reached between interacting nuclei. By this interpretation the exclusion principle is

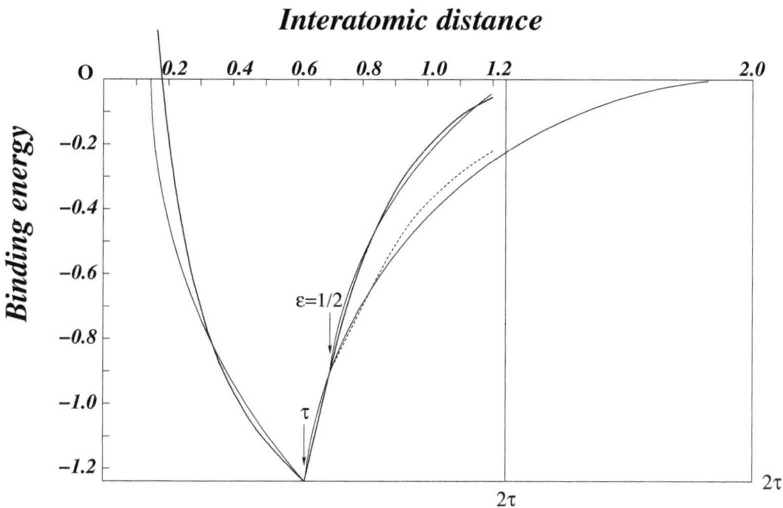

also defined as a property of space-time geometry. This makes good scientific sense as a fundamental basis of the principle has not been recognized before.

Molecular structure and shape are related to orbital angular momentum and chemical change is shown to be dictated by the quantum potential. The empirical parameters used in computer simulations such as molecular mechanics and dynamics are shown to derive in a fundamental way from the relationship between covalence and the golden ratio.

Reconstruction of the periodic properties of all forms of atomic matter, in terms of the same number-theoretic concepts that give meaning to intramolecular interaction, points at a universal self-similarity, which may extend through biological systems to cosmic proportions. The importance of the golden ratio is already known from botanical Fibonacci phyllotaxis and the same principles are now recognized in the structure of the solar system and galactic images. Differences in detail are brought about by special properties that emerge at each new level of organization. The emergent properties at the chemical level are the exclusion principle, molecular structure and the second law of thermodynamics – concepts not predicted by the more fundamental laws of physics. Self-similarity at the cosmic scale has important implications for cosmology and several discrepancies with the standard theories are identified.

These ideas have matured over many years, been recorded in scattered publications and discussed with countless colleagues. I appreciate their honest criticism, which made me aware of some general reluctance, akin to a mental block, to argue against established authority. Once a scientific contribution has been recognized by the award of a prize and trivialized by popular science writers, it turns into dogma – no longer subject to scrutiny, analysis or understanding. This respect for authority has been the bane of twentieth century theoretical chemistry. Should this book therefore stir up nothing but healthy scepticism among a next generation of chemists, the effort will be considered worth while.

I owe the courage to proceed with the project to the enthusiasm of many graduate students and the intellectual support, over the years, of several fellow scientists, in alphabetical rather than chronological order: Peter Comba, Rob Hancock, Demetrius Levendis, Casper Schutte, Pete Wedepohl and the two prematurely deceased, Amatz Meyer and Carl Pistorius. I acknowledge the helpful interest of Robin Crewe, Director of the Unit for Advanced Study at the University of Pretoria.

Jan Boeyens, Pretoria, June 2008

Abbreviations

BCC,FCC	Body(Face)-centred cubic
BDE	Bond dissociation energy
BO	Born-Oppenheimer
CCP,HCP	Cubic (hexagonal) close-packed
DFT	Density Functional Theory
esu	Electrostatic units
eV	Electron volt
FF	Force field
COT	Cyclo-octatetraene
GT	General Relativity
HF	Hellmann-Feynman
HF(S)	Hartree-Fock-(Slater)
HJ	Hamilton-Jacobi
HL	Heitler-London
JT	Jahn-Teller
LCAO	Linear Combination of Atomic Orbitals
MM	Molecular Mechanics
MO	Molecular orbital
NMR	Nuclear Magnetic Resonance
o-a-m	Orbital angular momentum
RDF	Radial Distribution Function
SCF	Self consistent field
SI	International Scientific Units
SR	Special Relativity
UV	Ultra Violet
VSEPR	Valence Shell Electron Pair Repulsion

Important Constants:

Avogadro's number	$L = 6.0221 \times 10^{23}\text{mol}^{-1}$
Bohr radius	$a_0 = 0.5292 \times 10^{-10}\text{m}$
Boltzmann's constant	$k = 1.3807 \times 10^{-23}\text{JK}^{-1}$
Compton wavelength	$\lambda_C = 2.4263 \times 10^{-12}\text{m}$
Electron charge	$e = 1.6022 \times 10^{-19}\text{C}$
Electron mass	$m = 9.1095 \times 10^{-31}\text{kg}$
Fine structure constant	$\alpha = 7.297 \times 10^{-3}$
Gas constant	$R = kL = 8.3145\text{Jmol}^{-1}\text{K}^{-1}$
Permeability constant	$\mu_0 = 4\pi \times 10^{-7}\text{Hm}^{-1}$
Permittivity constant	$\epsilon_0 = 8.8542 \times 10^{-12}\text{Fm}^{-1}$
Planck's constant	$h = 6.6268 \times 10^{-34}\text{Js}$
Speed of light	$c = 2.9979 \times 10^{8}\text{ms}^{-1}$

II Alternative Theory 127

Part I

A New Look at Old Theories

Chapter 1

Historical Perspective

Any scientific pursuit starts out with an examination of objects and phenomena of interest and proceeds by the accumulation of relevant data. As regularities emerge, classification of related facts inevitably leads to the formulation of laws and hypotheses that stimulate experimental design, until better understanding culminates in a general theory. The wider the field of enquiry the more cumbersome the development of theoretical understanding would be. In a subject like chemistry with so many facets it is even more difficult to recognize the central issues to feature in a comprehensive theory.

Chemistry has its roots in alchemy, best described as the most extensive project in applied research of all time. It pursued a single-minded search for the philosopher's stone and the elixir of life for more than a thousand years, through the middle ages and into the modern era. It relied with dogmatic certainty on a given theory that clearly specified the powers of the philosopher's stone and its hidden existence. No room was left for improvement or falsification of the theory and failed experiments were documented with the sole purpose of avoiding the same mistakes in future. Claims of successful production of alchemical gold were fiercely protected secrets and the only visible benefits were in the isolation, purification and characterization of theoretically irrelevant chemical substances. Alchemy, in this sense, is the exact antithesis of scientific endeavour. In science there is no authority or infallible theory. Any theory that claims final validity stifles further progress.

The current chemical version of quantum theory is in danger of assuming such stature. As with alchemy, too many resources are committed under its assumed infallibility and the presumed reward awaiting ultimate success is simply too alluring to ignore. A fresh look at the uncritical use of quantum mechanics in chemistry could therefore be a rewarding exercise. For example, the exact meaning of a familiar concept, such as orbital hybridization, as a working model, may well appear not to be of crucial theoretical importance, except that it also features in the interpretation of a large number of important secondary phenomena. Any unresolved primary ambiguity could easily become inflated to produce serious conceptual problems down the line. It may, for instance, not lead to the anticipated quantum-mechanical resolution of a problem and, unawares conceptualize chemical interactions in terms of the classical Lewis electron-pair description of molecules.

One of the problems faced by quantum chemistry is that it is based on a theory borrowed from physics. It therefore is important to note that a given variable or concept may be interpreted very differently in physical and chemical context, respectively. The physicist who is interested in the motion of a molecule in a force-free environment, treats it as a mass point, without any loss of generality. Such a molecule is of no interest to the chemist who studies the interaction of a molecule with its environment. In chemical context the size and shape of the molecule, left undefined in theoretical physics, must be taken into account. The interaction between mass points can simply not account for the observed behaviour of chemical substances.

Most theoretical concepts in chemistry are in fact borrowed from other disciplines and should properly be re-examined to ensure their use in an appropriate sense. Such an exercise demands a clear understanding of which systems are of interest to chemistry. In its broadest sense, chemistry deals with the interaction between substances and transformations between different forms of matter. This definition is akin to describing thermodynamics as the study of interconversions between different forms of energy. Conversion between matter and energy is considered irrelevant in both chemical and thermodynamic contexts. The minimum requirement for sensible study in each of the separate fields is a *conservation law.* Thermodynamics is based, in the first instance, on the conservation of *energy*, chemistry is based on the conservation of *energy and mass*, and nuclear physics on the conservation of *mass-energy.* It is important to note that recognition of mass conservation initiated the final break between chemistry and alchemy. During the transition period an interesting controversy arose around the theory of combustion. The prevailing phlogiston theory, despite many attractive features, failed to obey the law of mass conservation and eventually had to make way for a matter-based theory. It is an ironic fact that the modern electronic theory of oxidation and reduction is a virtual carbon copy of the phlogiston theory.

The first strides after recognition of mass conservation led to the formulation of several phenomenological laws of *chemical composition*, such as the laws of *constant proportions*, *multiple proportions* and *equivalent proportions*, found to be obeyed during interaction between chemical substances. These laws served to catalogue and systematize a large body of empirical data without providing a logical framework to rationalize the observations. Such a framework was provided by Dalton's atomic theory, borrowed from the ancient concept of an indivisible unit of matter, with the added new proposition that each chemical *element* is made up of identical atoms, different from those of any other element. A necessary by-product of the theory was the introduction of the concept *molecule.* The initial confusion between atom and molecule was cleared up by the work of Avogadro who analyzed volume relationships among interacting gases. The distinction between elements and simple compounds was an even harder experimental nut to crack before the next significant theoretical advance, based on accurate atomic weights, became possible.

Even before the experimental techniques of 19th century chemistry succeeded in isolating all elements in their pure form, a brilliant regularity that links all atoms and their properties together in a single scheme, was recognized. Construction of the periodic table of the elements still shines as the highest achievement of theoretical chemistry. The regular increase in atomic mass, which becomes evident when the elements are arranged in the correct numerical order, indicated the build up of all atoms by a common mechanism from common constituents. The quest to identify this mechanism originated with an anonymously announced hypothesis, later credited to Prout, and continues to this day. Prout's hypothesis based on atomic hydrogen as the building block, failed to account for the formation of atoms with fractional, rather than integral atomic weights, such as chlorine and copper. It found a new lease of life only when atomic theory had developed far enough to explain the existence of isotopes, but that was too late.

The theory of elemental periodicity reached maturity at the same time as quantum mechanics and general relativity, the great theories of physics. The mistaken assumption that quantum theory explained periodicity in detail caused the emphasis in chemistry to shift into the new paradigm of *quantum chemistry*, which has now remained sterile for more than half a century. Evidence is emerging that the periodicity of matter reaches way beyond the electronic quantum theory of chemistry and that many important answers have been missed during the 20th century search for the structure of matter and the nature of chemical interaction. The time is ripe to re-examine the theories, either prematurely rejected and forgotten or too hastily adopted into chemical thinking, during the heyday of pioneering quantum physics. The first question to face is whether chemistry needs quantum mechanics and the theory of relativity at all. Anticipating an affirmative answer, the next question is whether these theories can be reformulated to address the problems of chemistry directly. Is it realistic to expect that a three-dimensionally structured molecule can be analyzed meaningfully in terms of zero-dimensional point particles, with no extension, apart from an association with waves in multidimensional configuration space? If not, all conceivable alternatives, including those rejected by the founding fathers, should be explored.

Schrödinger's fertile mind spawned many concepts that failed to meet with the approval of his less imaginative, but more vociferous contemporaries. Many of these ideas have been forgotten and deserve to be reconsidered. One of his most exciting proposals, which has been dormant since 1921, links the quantum variables that characterize the hydrogen atom, to the same principle of gauge invariance that generates the electromagnetic field, and hence to the properties of space-time, or aether.

In 1952 David Bohm rediscovered aspects of earlier proposals by de Broglie and Madelung, which had been rejected years before, and established the concept of non-local interaction *via* the *quantum potential.* It appears to provide fundamental answers for the understanding of chemistry, but remains on the fringes, while awaiting recognition by the establishment.

For theoretical chemistry to succeed it must develop the power to elucidate the behaviour of chemical substances to the satisfaction of experimental chemists, known to operate at many different levels. Understanding is not promoted by the generation of numbers, however accurate or numerous, without a simple picture that tells the story. It is inevitable that the chain of reasoning must reduce the problem of understanding the behaviour of substances, to the understanding of molecules, atoms, electrons, and eventually the aether. Again, this ladder of understanding should not be obscured by complicated mathematical relationships that cannot be projected into a simple picture. Small wonder that the planetary model of the atom, inspired by Kepler, and discredited almost a hundred years ago, is still the preferred icon to represent nuclear installations and activity in the commercial world. Theoretical chemistry should also communicate with the predominantly non-scientist population of the world, but in order to tell a story it is first of all necessary to know the story.

A programme to develop a theory of chemistry, not dictated by theoretical physics and free of unnecessary mathematical complications, is not supposed to be a paradigm in isolation. It should respect the discoveries of related disciplines, but not necessarily all of their *interpretations*. The implications of relativity and quantum theory are as important for the understanding of chemical phenomena as for physics, particularly in so far as these theories elucidate the structure of matter. This aspect is of vital importance to chemistry, but only a philosophical curiosity in physics. In the orthodox view of physics it is the outcome of experimental measurements which has theoretical significance – the chemist needs insight into the nature of elementary substances to understand and manipulate their systems of interest. Without relating experimental readings to some structural basis, chemistry comes to a standstill. As quantum physics has no need to define the structure of matter, a quantum-mechanical model of molecular structure has never been developed, apart from the classical models of chemistry. As the basis for a quantum theory of chemistry these models are all but useless. To address this problem it will be necessary to examine the nature of electrons, atoms and molecules in a way that physicists have neglected.

The theories of special (SR) and general relativity (GR) appear to be even more remote from chemistry, but no less important. GR is the only theory that provides direct insight into the origin and the nature of matter. It must obviously be the basis on which any theory of chemistry can develop, but it features only rarely and peripherally in any chemistry curriculum. In the present instance it is the first topic to be considered in some detail. It could certainly be argued that relativity is pure physics and not a topic with which to burden already overcommitted students of chemistry. On the other hand, the implications of the theory stretches way beyond the reaches of physics, and if not recognized by the chemist, fundamental insight into the origin and structure of matter will be lost. Without that insight the basis of chemistry remains hearsay.

Chapter 2

The Important Concepts

The universally accepted theory of chemistry as a synthesis of the 19th century notions of chemical affinity, molecular structure and thermodynamics, with the theories of physics, which developed in the early 20th century, has gained almost universal acceptance as a closed set of concepts under the heading of *Physical Chemistry*. For more than fifty years textbooks on the subject have been revised and reorganized with the addition of preciously little new material. Today, these treatises are standardized over the world and translated into all relevant languages, emulating the standard models of particle physics and cosmology.

The seminal theories are respected as received wisdom, all flaws have been rationalized and the only remaining challenge is to dress up the old material with electronic wizardry, as if theoretical innovation ceased to operate in 1950. If indeed, there is nothing new or controversial in theoretical chemistry, with everything securely locked up in computer software at different levels of theory, the excitement is gone and dissident views are taboo. However, the nature of knowledge and of science is different. There are no closed books, not even on Euclidean geometry, and certainly not on chemistry. The standard models neglect to tell us how matter originates, what limits the variety of atomic matter, what is a chemical bond, and why is it necessary to assume the most fundamental concept that dictates the stability of matter – the exclusion principle – on faith? Even if these questions cannot be answered, they should be asked continually, maybe from a point of view overlooked by the founding theorists. It is in this spirit that the important concepts, fundamental to chemistry, will be re-examined in Part I of this work.

2.1 The Principle of Relativity

The principle of relativity, even more so than quantum theory, has acquired an aura of almost mythical inscrutability. The volume of popular literature that refutes the conclusions of the theory probably outweighs the published efforts to elucidate the principle. Students of chemistry, quite understandably, are probably reading these lines with trepidation, debating the prospects of continuing with this effort if it requires them to attempt the impossible. One way around the dilemma would be to skip this chapter and continue with the next topic without serious interruption of the central argument. Fully aware of the fact that this brash introduction of material, considered by most as largely irrelevant and past comprehension, could scare off many prospective readers, the author also seriously contemplated a soothing rearrangement of the chapters.

The conscious final decision to take the risk, with the current sequence, should be read as a personal conviction that the beauty of chemistry can never be fully appreciated unless viewed against the background in which all matter originates – space-time, or the vacuum. Not only matter, but all modes of interaction are shaped by the geometry of space, which at the moment remains a matter of conjecture. However, the theory of general relativity points the way by firmly demonstrating that the known material world can only exist in curved space-time. The theory of special relativity affirms that space-time has a minimum of four dimensions. Again, spaces of more dimensions are conjectural at present.

Most students of science must be aware of those arguments that relativity is illogical and unnecessary, that time is immutable and that gravity is adequately explained by Newton's laws. All of these statements are flawed. One of the monumental achievements of 19th century science was the study of electricity and magnetism, which culminated in the recognition of an electromagnetic field, which is described by Maxwell's equations. Nobody doubts the reality of electromagnetic phenomena and everybody should be aware of the fact that electromagnetic signals propagate through the vacuum at constant speed. This observation however, is totally incomprehensible in terms of the equally respected mechanical laws of relative motion, known as *Galilean* relativity. Resolution of this fundamental discrepancy in the laws of physics was achieved by the formulation of a new principle of relativity that applies to both mechanical and electromagnetic systems.

2.1.1 Relative Motion

The idea of relative motion is readily understood in terms of an observer who measures the position of an object on a riverboat that floats by at a constant speed, v, as in Figure 2.1. In the coordinate system (S) defined on shore the

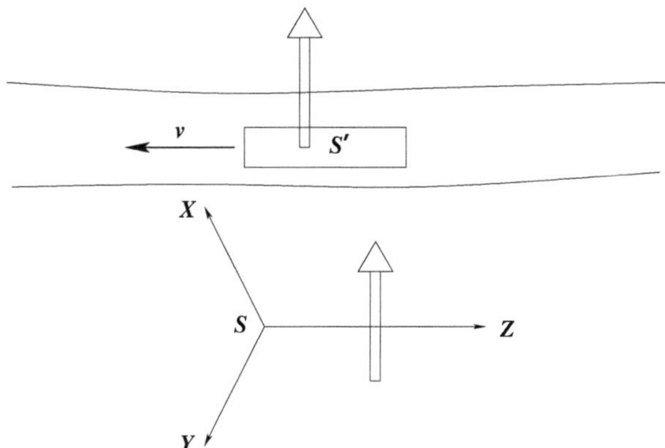

Figure 2.1: *Diagram to illustrate relative motion.*

object is observed to move at a velocity v downstream, covering a distance $-vt$ in the Z direction during a time t. In the parallel coordinate system (S') defined on the boat the object remains stationary at z' =constant. The two measurements are, in general, proportional to each other, such that

$$z - vt = \alpha z' \tag{2.1}$$

Seen from the boat a stationary object on shore appears to move at velocity $-v$ upstream, covering a distance vt' during time t'. Again the two measurements of z are proportional to each other, but now

$$z' + vt' = \alpha z \tag{2.2}$$

This (Galilean) description of relative motion had been accepted as universally valid, with proportionality constant $\alpha = 1$, until it was discovered by Maxwell that the electromagnetic field was carried through the vacuum at a constant velocity, c, which is also the velocity of light. Whereas c is not affected by the motion of a light source, the simple formulae that describe relative mechanical motion are no longer adequate when applied to photons. In this case the proportionality constant $\alpha \neq 1$.

2.1.2 Lorentz Transformation

To allow for constant c in terms of the previous equations it is necessary to define the velocities of a light signal as measured in the two coordinate systems to be equal and constant, *i.e.*

$$u = \frac{z}{t} = u' = \frac{z'}{t'} = c$$

From eqns (2.1) and (2.2) it follows that

$$\begin{aligned} vt' &= \alpha z - z' \\ &= \alpha z - \left(\frac{z - vt}{\alpha}\right) \\ &= \frac{1}{\alpha}\left\{\left(\alpha^2 - 1\right) z + vt\right\} \end{aligned} \tag{2.3}$$

Also from

$$\frac{\alpha z'}{v} + \alpha t' = \frac{\alpha^2 z}{v}$$

follows

$$\begin{aligned} \alpha t' &= \frac{\alpha^2 z}{v} - \frac{\alpha z'}{v} \\ &= \frac{\alpha^2 z}{v} - \frac{z - vt}{v} \\ &= \frac{z}{v}(\alpha^2 - 1) + t \end{aligned} \tag{2.4}$$

Finally

$$u' = \frac{z'}{t'} = \frac{\alpha z'}{\alpha t'} = \frac{z - vt}{(z/v)(\alpha^2 - 1) + t}$$

Hence

$$u' = \frac{(z/t) - v}{[(\alpha^2 - 1)/v][(z/t) + 1]} = \frac{z - vt}{[(\alpha^2 - 1)/v]u + 1}$$

On setting $u' = u = c$ this expression,

$$c = \frac{c - v}{[(\alpha^2 - 1)/v]c + 1}$$

reduces to

$$\alpha^2 - 1 = -\frac{v^2}{c^2} \quad , \quad \alpha = \sqrt{1 - v^2/c^2}$$

Substitution of this value into (2.1) gives

$$z' = \frac{z - vt}{\sqrt{1 - v^2/c^2}} \tag{2.5}$$

For $v << c$, $\alpha \rightarrow 1$ and $z' = z - vt$.

Equation (2.5) redefines the transformation between coordinate systems in relative motion, allowing for a signal with constant velocity c. Compared to the known velocity of light $c = 3 \times 10^8 \text{ms}^{-1}$, the ratio v^2/c^2 for the fastest common objects (*e.g.* an aircraft moving at 1000 km/h=2.8×10^2m/s) is only $10^{-8} \simeq 0$, and $\alpha = 1$. For such objects use of the simple (Galilean) transformation formula ($\alpha = 1$, $c >> v$) is more appropriate than the *Lorentzian* transformation (2.5). A surprising new feature of the Lorentz transformation is that the intuitively valid Galilean condition $t' = t$ no longer applies. From (2.4) follows instead:

$$\begin{aligned} \alpha t' &= \frac{z}{v}\left(-\frac{v^2}{c^2}\right) + t \\ t' &= \frac{t - vz/c^2}{\sqrt{1 - v^2/c^2}} \end{aligned} \tag{2.6}$$

Equally surprising is that the three-dimensional line element of Pythagoras,

$$(\Delta r)^2 = \left((\Delta x)^2 + (\Delta y)^2 + (\Delta z)^2\right)^{\frac{1}{2}}$$

is not invariant under Lorentz transformation, as under the Galilean,

$$(\Delta z')_G = (z_1 - vt) - (z_2 - vt) = \Delta z$$

$$\text{but} \qquad (\Delta z')_L = \Delta z / \sqrt{1 - v^2/c^2}$$

Under Lorentz transformation a 3D line element appears to be contracted in the direction of motion. A time interval as measured in two relatively moving coordinate system is likewise, not invariant under Lorentz transformation,

$$\begin{aligned} \Delta t' &= \frac{(t_1 - vz/c^2) - (t_2 - vz/c^2)}{\alpha} \\ &= \frac{\Delta t}{\sqrt{1 - v^2/c^2}} \end{aligned} \tag{2.7}$$

Relative motion in this case causes a *time dilation.*

To identify the invariant quantity under Lorentz transformation it is noted that a light wave emitted from a point source at time $t = 0$ spreads to the surface of a sphere, radius r, such that

$$r^2 = x^2 + y^2 + z^2 = (ct)^2 \tag{2.8}$$

at time t. The transformed wave front as observed in a moving frame is described by

$$\begin{aligned}(r')^2 &= (x')^2 + (y')^2 + (z')^2 \\ &= x^2 + y^2 + \frac{(z - vt)^2}{\alpha^2} \\ &= (ct)^2 - z^2 + \frac{(z - vt)^2}{\alpha^2} \qquad \text{, using 2.8}\end{aligned}$$

This equation reduces to

$$\begin{aligned}(r')^2 &= \frac{(ct)^2 + (vz/c)^2 - 2zvt}{\alpha^2} \\ &= \frac{(ct - vz/c)^2}{\alpha^2} = (ct')^2\end{aligned}$$

The transformed equation

$$(x')^2 + (y')^2 + (z')^2 = (ct')^2$$

once again describes a spherical wave with velocity of propagation c when viewed in the moving system [2]. The result

$$(r')^2 - (ct')^2 = r^2 - (ct)^2 = 0$$

shows that the quantity

$$\sigma^2 = x^2 + y^2 + z^2 + (ict)^2$$

defines the invariant interval σ, known as the *proper time* between two nearby events at $(\boldsymbol{r},t)$ and $(\boldsymbol{r}+\mathrm{d}\boldsymbol{r},t+\mathrm{d}t)$. The interval Δt of eqn. (2.7) describes the time interval on a clock that travels with the moving object. It represents an interval in *proper time*, or the *world time* τ of the moving object. Like σ, τ is also an invariant and hence has absolute meaning,independent of any observer.

The important conclusion to be drawn from the foregoing discussion is that space and time coordinates are relativistically linked together in a way that compensates for apparent length contraction and time dilation due to uniform motion. Considered together as the coordinates of a four-dimensional space-time they define a single space-time interval in four dimensions. This effect is entirely equivalent to the perception which is gained on observing three-dimensional events in two dimensions.

During two-dimensional rotation the line element $s_2 = \sqrt{x^2 + y^2}$ is an invariant, as shown on the left in Figure 2.2. However, should this line

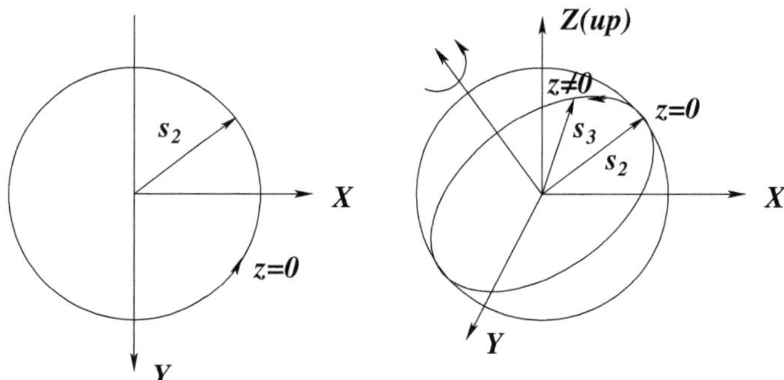

Figure 2.2: *A two-dimensional vector s_2 remains constant in length on rotation in a plane ($z = 0$). On rotation about an axis which is not perpendicular to the $X - Y$ plane, s_2 moves out of this plane. Its length, s_3, remains invariant in 3 dimensions, but its projection along Z, into the $X - Y$ plane, appears to be contracted.*

element be rotated about an axis which is not perpendicular to the $X - Y$ plane, its projection in this plane shows an apparent contraction of s_2, which is no longer invariant. Instead, the line element $s_3 = \sqrt{x^2 + y^2 + z^2}$ becomes invariant. The contraction of s_2 is compensated for by dilation of z. These virtual effects, easily interpreted by three-dimensional observation, are very real in flatland. As real as the deformation of an electron moving at relativistic speed in three dimensions [4].

Not only is c constant, but it also represents a limiting velocity. It is noted that any object moving at a velocity that approaches c, *i.e.* $v \to c$, must contract to zero thickness and suffer an infinite time dilation. Certain combinations of space and time coordinates are apparently disallowed by these effects, as shown in Figure 2.3. The x-axis, in this diagram, represents all space coordinates, perpendicular to the time axis. Any stationary point therefore traces out a *world line*, perpendicular to x. An object that moves at velocity v follows a world line at an angle θ with respect to t. The faster the object moves the larger is θ. It reaches a limiting value at $v = c$. The possible directions of c sweep out a *light* cone. An object starting from O,

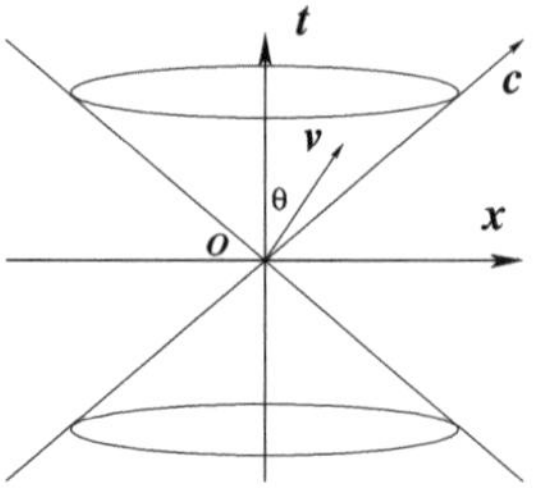

Figure 2.3: *The Minkowski diagram.*

with velocity $v < c$ can never move out of the light cone, and its behaviour is called *time-like*. *Space-like* behaviour, which implies moving out of the light cone at superluminal speed, is commonly considered impossible, although there is no physical or mathematical reason for this conclusion.

For points in the surface of the light cone

$$c = \frac{\sqrt{x^2 + y^2 + z^2}}{t} = \frac{\boldsymbol{r}}{t}$$

If the squared world vector $\boldsymbol{X}_\mu^2 = x^2 + y^2 + z^2 - (ct)^2 \leq 0$ it lies within the light cone and $\boldsymbol{X}_\mu$ is called *time-like*. Otherwise $\boldsymbol{X}_\mu$ is space-like. A space-like four-vector can always be transformed such that the fourth component vanishes. A time-like vector must always have the fourth component, but it can be transformed such that the first three vanish. The difference between two world points

$$(\Delta \boldsymbol{X})^2 = |\boldsymbol{r_1} - \boldsymbol{r_2}|^2 - c^2(t_1 - t_2)^2$$

can also be either time-like or space-like. In the special case of relative velocity c, $\Delta \boldsymbol{X} = 0$. $\Delta \boldsymbol{X}$ is the distance between two points in four-dimensional space-time. $\Delta \boldsymbol{X} = 0$ at all points along the singular line Oc in the Minkowski diagram. As first pointed out by the chemist Gilbert Lewis [5], the emitter and absorber of a light signal therefore remain in virtual contact, which indicates that c should not be interpreted as a velocity at all. Transmission of signals along the world line of a photon does not correspond to the type of motion normally associated with massive particles and represents a situation in which time and space coordinates coincide.

The conservation of momentum is one of the basic principles of mechanics and it is to be expected that momentum remains invariant under Lorentz transformation. The fact that the velocity of a moving body is observed to be different in relatively moving coordinate systems therefore implies that, in order to keep $p = p'$, the transformed mass $m' \neq m$, as measured for the stationary object.

To estimate the variation in mass consider a particle moving along the y-direction. The y-coordinate of the particle is not affected by the relative motion, *i.e.* $y_1' = y_1$ and $y_2' = y_2$, as measured at times t_1 and t_2 in relatively moving frames. From (2.7) the time difference $t_2 - t_1$, measured in S' will be $\Delta t/\sqrt{1 - v^2/c^2}$, so that the particle velocity in S' becomes

$$v_y' = \frac{y_2' - y_1'}{t_2' - t_1'} = \frac{(y_2 - y_1)(\sqrt{1 - v^2/c^2})}{t_2 - t_1} = v_y\sqrt{1 - v^2/c^2}$$

To ensure the invariance of momentum it is required that $mv_y = m'v_y'$. Hence

$$m' = \frac{m}{\sqrt{1 - v^2/c^2}} \tag{2.9}$$

This conclusion is subject to the assumption that $v_y << v$, the relative velocity. However, more detailed analysis shows this result to be generally valid. If the rest mass of a stationary particle is defined as m_0, its mass increases to $m = m_0/\sqrt{1 - v^2/c^2}$ when moving at velocity v, becoming infinite at $v = c$.

The relativistic energy of a mass point is computed by differentiation of the relativistic momentum, to yield the relativistic force

$$\boldsymbol{F} = \frac{\mathrm{d}}{\mathrm{d}t}(m\boldsymbol{v}) = m\frac{\mathrm{d}\boldsymbol{v}}{\mathrm{d}t} + \boldsymbol{v}\frac{\mathrm{d}m}{\mathrm{d}t}$$

The work done by the force in moving the mass point by $\mathrm{d}\boldsymbol{l}$, defines the energy

$$\begin{aligned}
\mathrm{d}E = \boldsymbol{F}\cdot\mathrm{d}\boldsymbol{l} &= \frac{\mathrm{d}}{\mathrm{d}t}(m\boldsymbol{v})\cdot\mathrm{d}\boldsymbol{l} \\
&= \mathrm{d}(m\boldsymbol{v})\frac{\mathrm{d}\boldsymbol{l}}{\mathrm{d}t} = \mathrm{d}(m\boldsymbol{v})\cdot\boldsymbol{v} \\
&= m\boldsymbol{v}\cdot\mathrm{d}\boldsymbol{v} + \boldsymbol{v}\cdot\boldsymbol{v}\mathrm{d}m \\
&= \frac{1}{2}m\mathrm{d}(v^2) + v^2\mathrm{d}m
\end{aligned}$$

Using (2.9) the first term on the right can be expressed as a function of mass,

$$\begin{aligned}
\frac{1}{2}m\mathrm{d}(v^2) &= \frac{1}{2}m\mathrm{d}[c^2 - (m_0c)^2m^{-2}] \\
&= \frac{1}{2}m[2(m_0c)^2m^{-3}]\mathrm{d}m
\end{aligned}$$

Hence

$$\begin{aligned}\Delta E &= \int_m^{m+\Delta m} \left[\left(\frac{m_0 c}{m}\right)^2 + v^2\right] \mathrm{d}m \\ &= \int_m^{m+\Delta m} [(1 - v^2/c^2)c^2 + v^2]\mathrm{d}m = \int_m^{m+\Delta m} c^2 \mathrm{d}m \\ &= c^2 \Delta m\end{aligned}$$

This recognition that energy and mass are equivalent is the most important conclusion of special relativity. Classically there are separate conservation laws for mass and energy, which is now replaced by a conservation law for total mass-energy. Any closed system that suffers a change in mass shows an increase in kinetic energy which may be written

$$\Delta E = mc^2 - m_0 c^2 \tag{2.10}$$

The relativistic momentum of a particle moving with velocity v can now be written as $p = m_0 v/\sqrt{1 - v^2/c^2}$, which rearrranges into

$$v^2 = \frac{p^2}{m_0^2}(1 - v^2/c^2)$$

i.e.

$$\left(\frac{p}{m_0}\right)^2 = v^2 \left[1 + \left(\frac{p}{m_0 c}\right)^2\right]$$

or $p = m_0 v\sqrt{1 + p^2/(m_0 c)^2}$, which shows that

$$\frac{1}{\sqrt{1 - v^2/c^2}} = \sqrt{1 + p^2/(m_0 c)^2}$$

Hence the total energy can be expressed as a function of momentum:

$$E = mc^2 = \frac{m_0 c^2}{\sqrt{1 - v^2/c^2}} = m_0 c^2 \sqrt{1 + p^2/(m_0 c)^2}$$

i.e.

$$E^2 = p^2 c^2 + m_0^2 c^4$$

For a photon with zero rest mass it follows that $E = pc$.

2.1.3 General Relativity

Relative motion according to Lorentz transformation refers specifically to unaccelerated uniform motion and is therefore known as *special relativity* (SR). The theory which developed to also take acceleration into account is known as general relativity (TGR). Based on the demonstration, by Eötvös and others, that there is no difference between the inertial and the gravitational mass of an object, TGR also became the theory of the gravitational field.

The world line of an accelerated object appears curved in a Minkowski

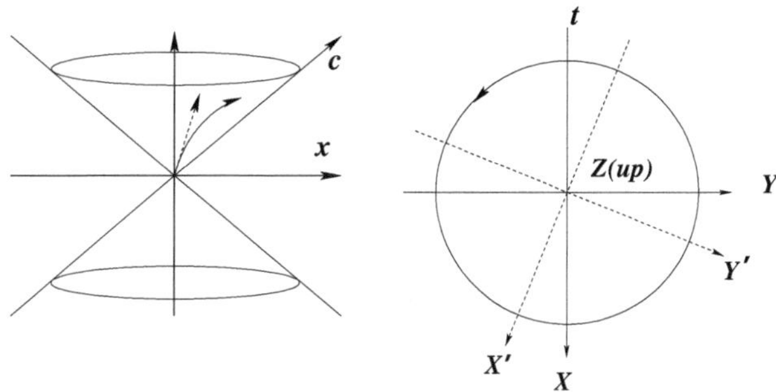

Figure 2.4: *(Left): Accelerated motion in Minkowski space. (Right): Two coordinate systems in relative rotational motion.*

diagram as shown in Figure 2.4. Because of the equivalence of acceleration and gravity the world line of a photon in a gravitational field is inferred to be curved as well, which implies a velocity that exceeds the constant c of SR. This contradiction is avoided if the geometry of space-time in a gravitational field is no longer euclidean. What appears to be the curved path in euclidean space could then be interpreted as a *geodesic* in non-euclidean space, which is the equivalent of a straight euclidean line. Because of the time dilation in the gravitational field the photon has the same constant velocity as before. The apparent displacement of a star as observed near the surface of the sun during an eclipse provides experimental proof of this effect, described as the curving of space in a gravitational field.

Einstein [6] illustrated the curvature of space-time by considering two coordinate systems, K and K', with a common origin, one of them stationary and the other rotating (accelerated) about the common Z-axis, in a space free of gravitational fields, shown in Figure 2.4 on the right. A circle around the origin in the $X - Y$ plane of K is also a circle in the $X' - Y'$ plane of K'. Measurement of the circumference S and diameter $2R$ in the stationary system must yield

$$\frac{S}{2R} = \pi$$

When measured in the rotating system however,

$$\frac{S'}{2R'} > \pi$$

The reason is that the measuring rod that moves with the rotating circumference suffers Lorentz contraction, but not for measurement along the radius. The only interpretation is that euclidean geometry does not apply to K'.

In SR the invariant quadratic form, in differential notation,

$$\mathrm{d}s^2 = \mathrm{d}x^2 + \mathrm{d}y^2 + \mathrm{d}z^2 - c^2\mathrm{d}t^2$$

is a special case of the more general expression for non-euclidean or Riemannian geometry,

$$\mathrm{d}s^2 = \sum_{\mu,\nu=1}^{4} \eta_{\mu\nu}\mathrm{d}x^\nu \mathrm{d}x^\mu \tag{2.11}$$

with $x^1 = x$, $x^2 = y$, $x^3 = z$, $x^4 = ict$ and the *metric tensor*,

$$\eta = \begin{pmatrix} 1 & 0 & 0 & 0 \\ 0 & 1 & 0 & 0 \\ 0 & 0 & 1 & 0 \\ 0 & 0 & 0 & -c^2 \end{pmatrix}$$

Equation (2.11) with variable metric tensor describes the invariance in the gravitational case which is characterized by curved space-time. The summation extends over all values of μ and ν, so that the sum consists of 4×4 terms, of which 12 are equal in pairs, hence 10 independent functions. The motion of a free material point in this field will take the form of curvilinear non-uniform motion. If the matrix of the metric tensor can be diagonalized it is independent of position and the corresponding geometry is said to be flat, which is the special case of SR.

The mathematical detail of TGR depends on complicated tensor analysis which will not be considered here. The important result for purposes of the present discussion is the relationship, which is found to exist between two fundamental tensors[1]: The symmetric Riemann *curvature tensor* $R_{\mu\nu}$ (with

[1] In an isotropic medium, vectors such as stress $\boldsymbol{S}$ and strain $\boldsymbol{X}$ are directly proportional,

ten independent components), which describes the geometry of space, and $T_{\mu\nu}$, the *energy-momentum* (or stress) tensor of the matter field, which describes the density of energy (including matter) in space. The proportionality factor contains Newton's gravitational constant.

The symmetry between curvature and matter is the most important result of Einstein's gravitational field equations. Both of these tensors vanish in empty euclidean space and the symmetry implies that whereas the presence of matter causes space to curve, curvature of space generates matter. This reciprocity has the important consequence that, because the stress tensor never vanishes in the real world, a non-vanishing curvature tensor must exist everywhere. The simplifying assumption of effective euclidean space-time therefore is a delusion and the simplification it effects is outweighed by the contradiction with reality. Flat space, by definition, is void.

All solutions of Einstein's equations are conditioned by the need of some *ad hoc* assumption about the geometry of space-time. The only indisputably valid assumption is that space-time is of *absolute* non-euclidean geometry. It is interesting to note that chiral space-time, probably demanded by the existence of antimatter and other chiral forms of matter, rules out the possibility of affine geometry, the standard assumption of modern TGR [7].

An even bigger dilemma for TGR is how to avoid the infinite gravity at vanishing distances between mass points that interact according to an inverse-square law. Theories that address the problem are known as quantum gravity, but to date it has not produced any convincing solution. The complicating fact is that no experimental measurement is possible within the domain of quantum gravity. Candidate theories such as higher-dimensional string theories, the current favourites, can therefore never be elevated beyond the level of untestable conjecture. It is almost comical to note that quantum theory has its own infinity problem because of the inverse-square Coulomb law. The accepted solution, provided by renormalization in quantum field theory, is by no means universally accepted, and like quantum gravity remains an essentially unsolved problem.

$\boldsymbol{S}= k\boldsymbol{X}$, provided they point in the same direction. If not, the quotient $\boldsymbol{S}/\boldsymbol{X}$ is of more complicated form. Like the quotient of two integers, which is not always another integer, the quotient of two non-aligned vectors is not necessarily a vector itself, but something else, called a *tensor*. The scalar k for co-aligned vectors is a tensor of *rank* zero. A tensor of the first rank has three components, *e.g.* $T_i' = \sum_j a_{ij}T_j$ and is equivalent to a vector. A second rank tensor has the form of a square matrix with nine elements in three dimensions *e.g.* $T_{ij} = \sum_{k,l} a_{ik}a_{jl}T_{kl}$, and so forth.

2.2 The Old Quantum Theory

2.2.1 The Bohr Model

The first quantitative atomic model appeared early in the previous century, based on the pioneering work of Lord Rutherford and the Danish physicist Niels Bohr. It was devised in simple analogy with Kepler's model of the solar system and, despite a number of known fatal defects, it has such intuitive appeal that, even today, scientists and non-scientists alike accept it as the most reasonable working model for understanding the distribution of electrons in atoms. Formulation of the model was guided by three important experimental observations which had no obvious explanation in terms of 19th century physics.

Most puzzling was the discovery by Balmer of a simple relationship between the wavelength distribution in the light emitted by hydrogen gas in an incandescent state, and the natural numbers. Rather than having a continuous variation of wavelength (λ) this observed *emission spectrum* is found to consist of a series of sharp maxima, called *lines*, at specific wavelengths that obey the simple formula:

$$\frac{1}{\lambda} = R\left(\frac{1}{2^2} - \frac{1}{n^2}\right) \text{ , } n = 3, 4, 5, \ldots$$

This formula is reminiscent of the famous relationship that was discovered by Pythagoras to exist between the pitch of sound produced by the plucked string of a lyre and the effective length of the string. Based on this observation he proposed, not only a theory of music, but also the structure of the cosmos.

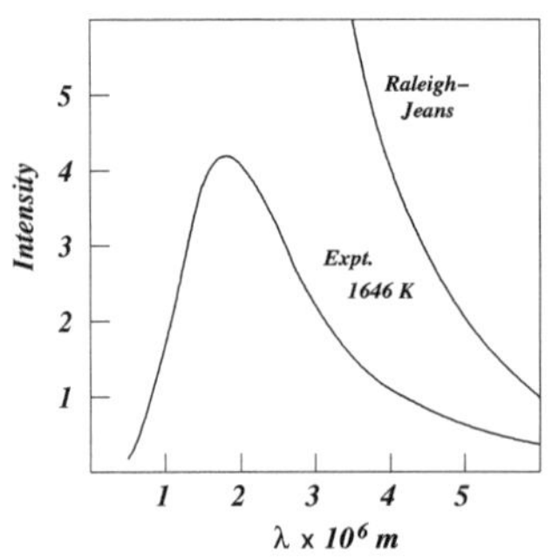

Figure 2.5: *The intensity distribution of radiation trapped in a closed cavity.*

The next crucial observation came from a thermodynamic study of the radiation which is emitted through an aperture in the wall of a heated and otherwise closed oven. Once more, it was the intensity distribution of the radiation emitted at different wavelengths that defied analysis. Presented in graphical form the observed distribution is as shown in the Figure 2.5. The distribution predicted by the laws of thermodynamics is shown as the Raleigh-Jeans curve. It disagrees disastrously with the observed. In hindsight it is obvious that, in order to explain the observed distribution, the intensity had to be assumed to depend, not only on the laws of thermodynamics, but also on the wavelength of the emitted light. In other words, the shorter the wavelength, the more energy (heat) is required to cause emission of light at that wavelength or frequency. Frequency (ν) is inversely proportional to wavelength, $\nu = c/\lambda$. The constant c is identified as the speed of light. The formula that correctly describes the intensity distribution of, what became known as *black-body radiation*, was first proposed by Max Planck, based on the assumption that relates energy to frequency, according to the equation:

$$E = h\nu$$

The proportionality constant h is now known as *Planck's constant* and it has the dimensions of mechanical action, Js in SI units. Another revolutionary aspect of Planck's model, which is needed to reproduce the experimental result, was the assumption that, instead of a continuous flow, energy is transmitted in discrete units of $h\nu$.

The third factor that inspired Bohr's atomic theory was the observation by Rutherford that high-energy particles (the so-called α-particles emitted by radioactive substances) directed to impinge on metal foils are scattered as if all mass in the metal foil is concentrated in high density at regularly spaced points, relatively far apart, thus allowing most α-particles to suffer minimum deviation and causing a small percentage to rebound at high scattering angles.

The Bohr Conjecture

Synthesis of the three observations led to Bohr's proposal of a planetary atom consisting of a heavy small stationary heavy nucleus and a number of orbiting electrons. Each electron, like a planet, had its own stable orbit centred at the atomic nucleus. The simplest atom, that of hydrogen, with atomic number 1 could therefore be described as a single electron orbiting a proton at a fixed, relatively large, distance. The mechanical requirement to stabilize the orbit is a balance between electrostatic and mechanical forces, expressed in simple electrostatic units, and particle momentum $p = mv$, as:

$$\frac{e^2}{r^2} = \frac{mv^2}{r} = \frac{p^2}{mr}$$

These expressions are easily rearranged to produce expressions for total energy (using esu),

$$E = T + V = \frac{1}{2}\frac{e^2}{r} - \frac{e^2}{r} = -\frac{1}{2}\frac{e^2}{r}$$

and orbital radius,

$$r = \frac{(pr)^2}{me^2} \quad i.e. \quad E = -\frac{1}{2}\frac{me^4}{(pr)^2}$$

The quantity in parentheses is recognized as the angular momentum of the orbiting particle, which, like Planck's constant has the units of action. This observation sets the stage for Bohr's conjecture[2] that if the angular momentum of the orbiting electron, like energy, is also restricted to occur as multiples of an elementary unit, such that $pr = nh/2\pi = (n\hbar)$, the electron in orbit on a hydrogen atom has the energy,

$$E = -\frac{2\pi^2 me^4}{n^2h^2}$$

The immediate success of this conjecture is how it leads directly to the Balmer formula on the assumption that a *quantum* of emitted radiation corresponds to the energy difference between two allowed orbits with $n = n_1, n_2$ respectively, *i.e.*

$$h\nu = \Delta E = \frac{2\pi^2 me^4}{h^2}\left(\frac{1}{n_1^2} - \frac{1}{n_2^2}\right)$$

Not only does it lead to the correct form of the Balmer formula with the correct value of the *Rydberg constant* $R = 2\pi^2 me^4/h^3c$, but also predicts more spectral series of the same type with different values of $n_1 \neq 2$. These predictions were soon to be confirmed by experiment.

Apart from the assumed quantization of orbital angular momentum the Bohr model predicted the quantization of electronic energy, radius, velocity and magnetic moment of atoms:

$$\begin{aligned} E_n &= -\frac{Rhc}{n^2} \\ r_n &= \frac{n^2h^2}{4\pi^2 me^2} \quad , \quad v_n = \sqrt{e^2/r_n m} = \frac{2\pi e^2}{nh} \\ \mu_B &= \frac{eh}{4\pi mc} \quad , \quad \alpha = v_1/c = \frac{2\pi e^2}{hc} \approx \frac{1}{137} \end{aligned}$$

The orbital frequency $\omega_n = v_n/2\pi r_n = (2/nh)W_n$, where W_n is the ionization energy of an electron on the nth orbit. At the lowest level, $n = 1$, the orbital

[2]This conjecture did not feature explicitly in Bohr's original argument, which he based on a correspondence principle, and only emerged in later work.

energy can be written as $W_1 = \frac{1}{2}h\omega_0$, which is the *zero-point energy* of an electron on a hydrogen atom. The dimensionless *fine-structure constant* α is still considered one of the most fundamental numbers in science.

Although the Bohr model gave quantitatively correct values for many measurable atomic properties it seemed to violate the fundamental rule of electrodynamics that requires an accelerated charge to radiate energy. This problem was partially overcome by postulating that the orbiting electron is in a *stationary state*, subject to new quantum laws. The angular-momentum conjecture was substantiated in 1921 by Stern and Gerlach who measured the predicted magnetic moment $(h/2\pi)$ of atoms such as Ag and the alkali metals, characterized by single valence electrons in an s-state, as in H.

Following the wave-mechanical reformulation of the quantum atomic model it became evident that the observed angular momentum of an s-state was not the result of orbital rotation of charge. As a result, the Bohr model was finally rejected within twenty years of publication and replaced by a whole succession of more refined atomic models. Closer examination will show however, that even the most refined contemporary model is still beset by conceptual problems. It could therefore be argued that some other hidden assumption, rather than Bohr's quantization rule, is responsible for the failure of the entire family of quantum-mechanical atomic models. Not only should the Bohr model be re-examined for some fatal flaw, but also for the valid assumptions that led on to the successful features of the quantum approach.

Reinterpretation

The assumptions of the Bohr model will now be re-examined without reference to the hydrogen atom. The conjecture that orbital angular momentum of an electron cannot change by an amount less than $h/2\pi = \hbar$, is certainly a valid assumption, without implying this to be the angular momentum in the hydrogen ground state. In the same way, the Bohr magneton μ_B only defines the smallest measurable quantum of orbital, as well as spin, magnetic moment of an electron, again without reference to the hydrogen atom. On the other hand, the Bohr model predicts the correct zero-point energy for hydrogen and the radius of the smallest orbit agrees with the accepted size of a hydrogen atom inferred from gas kinetics. However, the likely shape of the hydrogen atom is spherical rather than disc-like as implied by the Bohr model.

The sound part of Bohr's atomic model, and its successors, appears to be the assumed quantization of electronic angular momentum and energy, as well as atomic size. Had Bohr gone one step further the proposed quantiza-

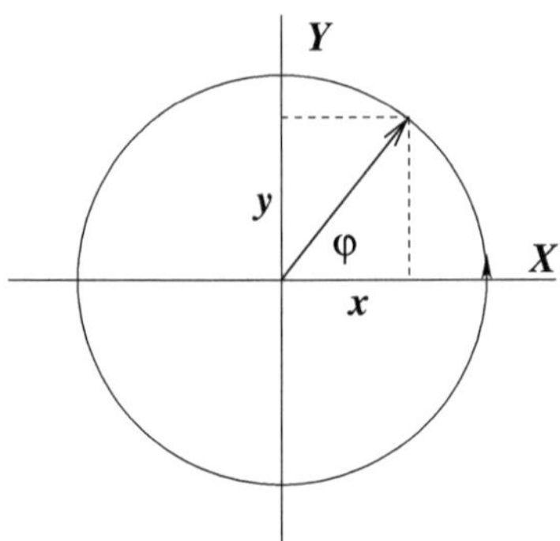

Figure 2.6: *Diagram to describe the angular momentum associated with a rotating vector.*

tion of angular momentum could have been recognized as a Sturm-Liouville problem[3]:

$$L(x)y(x) = \lambda_i y(x)$$

in which $L(x)$ is a differential operator and the *eigenvalues* λ_i are the allowed solutions under suitable boundary conditions. Angular momentum λ, as a vector directed along Z, (Figure 2.6)is a function of the rotational angle φ, *i.e.*

$$\frac{\mathrm{d}f(\varphi)}{\mathrm{d}\varphi} \propto \lambda f(\varphi)$$

The solution to this equation is a complex exponential function. Hence

$$f(\varphi) = \Phi = e^{i\alpha\lambda\varphi}$$

is a solution, *i.e.*

$$\frac{\mathrm{d}\Phi}{\mathrm{d}\varphi} = i\alpha\lambda\Phi$$

which rearranges into the Sturm-Liouville eigenvalue equation

$$L(\varphi)\Phi = -\frac{i}{\alpha}\frac{\mathrm{d}\Phi}{\mathrm{d}\varphi} = \lambda\Phi$$

[3]Sturm-Liouville differential equations, resulting from a separation of variables have been known since the middle of the 19th century. Separation constants, subject to boundary conditions, yield sets of characteristic, or eigenvalue, solutions.

For Φ to be single valued it should return to the same value after each rotation of 2π, *i.e.* $\Phi = \Phi + 2\pi$, which implies $\exp(2\pi i\alpha\lambda) = 1$, *i.e.* $\lambda\alpha = m$, an integer. But, by Bohr's conjecture $\lambda = m\hbar$, *i.e* $\alpha = 1/\hbar$, and hence

$$-i\hbar\frac{\mathrm{d}\Phi}{\mathrm{d}\varphi} = m\hbar\Phi \ , \ \Phi = e^{im\Phi}$$

This result is interpreted to mean that angular momentum is described by the operator $L(\varphi) = i\hbar\partial/\partial\varphi$, which is equivalent to the postulate that linear momentum is quantum-mechanically represented by the operator $-i\hbar\nabla$. The Bohr operator for angular momentum is converted into cartesian coordinates by writing

$$\frac{\mathrm{d}\Phi}{\mathrm{d}\varphi} = \frac{\partial\Phi}{\partial x}\cdot\frac{\partial x}{\partial\varphi} + \frac{\partial\Phi}{\partial y}\cdot\frac{\partial y}{\partial\varphi}$$

From

$$x = r\cos\varphi \quad , \quad \frac{\partial x}{\partial\varphi} = -r\sin\varphi$$

$$y = r\sin\varphi \quad , \quad \frac{\partial y}{\partial\varphi} = r\cos\varphi$$

it follows that

$$\frac{\mathrm{d}\Phi}{\mathrm{d}\varphi} = -r\sin\varphi\frac{\partial\Phi}{\partial x} + r\cos\varphi\frac{\partial\Phi}{\partial y}$$

The r.h.s. resembles the expression for the z-component of angular momentum $l_z = xp_y - yp_x$, and the expressions become identical provided

$$p_x \rightarrow i\hbar\frac{\partial}{\partial x} \ , \ etc.$$

The problematic part of the Bohr model is that quantization goes together with the accelerated orbital motion of an electronic particle. In order to avoid this problem an alternative explanation of orbital quantization would be required, which means discarding the particle model of the electron.

2.2.2 The Sommerfeld Model

Bohr's atomic model was accepted in physics, with some reservation and received even less enthusiastically in chemistry, as there was no visible prospect of extending the treatment to other atoms, more complex than hydrogen. Chemical models of the era were all conditioned by the need to account for chemical interactions that bind atoms together into molecules. One of the more successful, due to Lewis, Langmuir and others, proposed a static

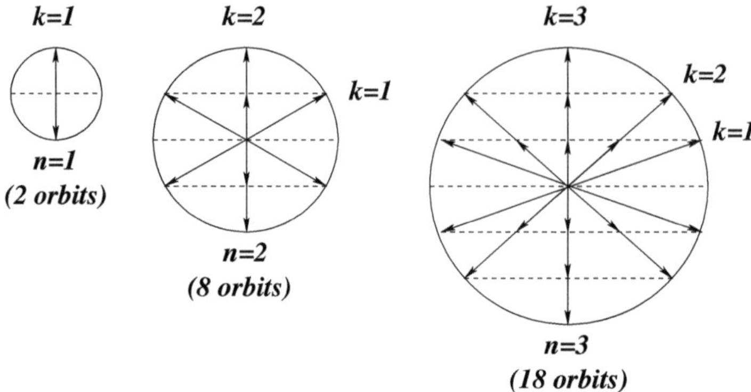

Figure 2.7: *Sommerfeld space quantization*

distribution of electrons which allowed the formation of *electron-pair bonds* between neighbouring pairs of atoms.

Prompted by the structure of the periodic table of the elements, electrons were assumed to occur in concentric shells around the nucleus with a positive charge of Z units, equal to the number of extranuclear electrons. In any period of 8 elements, arranged in order of increasing Z, electrons are postulated to occupy an increasing number of sites (from 1 to 8) at the corners of a cube centred at the nucleus. Any vacancy in the shell of eight enables the relevant atom to share an electron with a neighbouring atom to form a covalent bond and to complete the octet of electrons for that shell. This view has now endured for almost hundred years and still forms the basis for teaching elementary chemistry. The simple planetary model, proposed by Bohr, allows for only one electron per orbit and has little in common with the Lewis model.

An obvious possible improvement of the Bohr model was to bring it better into line with Kepler's model of the solar sxstem, which placed the planets in elliptical, rather than circular, orbits. Sommerfeld managed to solve this problem by the introduction of two extra *quantum numbers* in addition to the *principal* quantum number (n) of the Bohr model, and the formulation of general quantization rules for periodic systems, which contained the Bohr conjecture as a special case.

The *radial* quantum number (n') was introduced to specify the eccentricity of elliptic orbits and an *azimuthal* quantum number (k) to specify the orientation of orbits in space. The three quantum numbers are related by

$$n = k + n'$$

Circular orbits are defined by $n' = 0$. The principal quantum number specifies energy shells. For $n = 1$ the only solution is $n' = 0$, $k = 1$, which specifies two orbits with angular momentum vectors in opposite directions. The solutions $n' = 0$, $k = 2$ and $n' = 1$, $k = 1$ define 8 possible orbits, 4 circular and 4 elliptic. The angular momentum vectors of each set are directed in four tetrahedral directions to define zero angular momentum when fully occupied. Taken together, these tetrahedra define a cubic arrangement, closely related to the Lewis model for the Ne atom.

In Figure 2.7 the arrows indicate possible orientations of the total angular momentum vector such that the component in the line-of-force direction is always a rational fraction of the total measure. The possible vectors are identified by their projection on the radius of the unit circle as fractions k/n. The quantum number $k = 0$ is considered meaningless. In Sommerfeld's words [8]:

> (This) space quantization of the Kepler orbits is without doubt the most surprising result of the quantum theory. The simplicity of the results and their derivation is almost like magic.

The Sommerfeld model for Ne is shown in figure 2.8. The He atom presented a special problem as the quantum numbers restrict the two electrons to the same circular orbit, on a collision course. One way to overcome this dilemma was by assuming an azimuthal quantum number $k = \frac{1}{2}$ for each electron, confining them to coplanar elliptic orbits with a common focal point. To avoid interference they need to stay precisely out of phase. This postulate, which antedates the discovery of electron spin was never seen as an acceptable solution to the problem which eventually led to the demise of the Sommerfeld model.

However, long before the postulate of electron spin (1925), Sommerfeld gave the correct interpretation of the Stern-Gerlach measurement of the angular momentum of the valence electron of Ag ([8] – p. 653) without calling it spin, but by writing the total angular momentum of s-states as $j = \frac{1}{2}$.

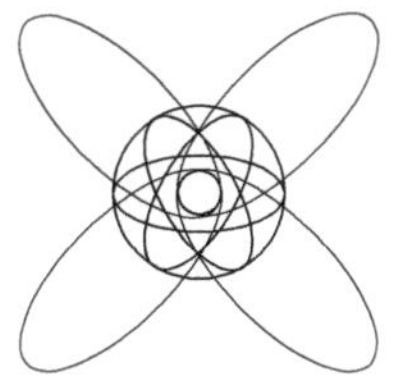

Figure 2.8: *Sommerfeld model of the Ne atom.*

For a brief period the Sommerfeld model enjoyed general acceptance in chemistry. The most powerful argument in its favour was the obvious agreement with the periodic law, *e.g.* by providing for 2 and 8 electrons in the first two shells, respectively. The predicted number of orbits for higher values of n are also in agreement with the periodic law.

Another appealing feature of the Sommerfeld model, still unintentionally in general use among chemists is the prediction of tetrahedral carbon, in line with van't Hoff's model. With only four electrons at the second energy level it was argued that these should occupy the degenerate set of four elliptic orbits. To quench the resulting angular momentum the plane normals of the four orbits should be arranged tetrahedrally with respect to each other. This arrangement implies that the four elliptic orbits extend towards the corners of a regular tetrahedron centred on the carbon nucleus. The result is complete agreement among the Sommerfeld, Lewis and van't Hoff models and perfectly in line with the theory of covalent bonding.

The more general quantization rules formulated in terms of periodic action integrals of the type

$$\oint p_k \mathrm{d}q_k = n_k h \,,\; k = 1, 2, \ldots,\; n = \text{integer}$$

lent further credibility to the Sommerfeld model. These rules could be used to derive correct quantum rules for molecules and crystals. All molecular spectra, including electronic, vibrational and rotational spectra could be explained in detail and Bragg's diffraction law for periodic crystals could be derived directly. The assignment of electronic spectra in terms of sharp, principal, diffuse and fundamental series, is still recognizable in the s, p, d, f distinction between electrons with angular momentum quantum numbers $l = 0, 1, 2, 3$.

Theoretical chemistry reached its pinnacle during the Sommerfeld era, before the advent of wave mechanics. The theoretically superior new theory, although it eliminated the paradoxes of zero angular momentum of the hydrogen ground-state, the orbital motion in helium and the nature of stationary states, it defined the periodic table less well and confused the simple picture of chemical bonding. Theoretical chemistry still suffers from that body blow.

There are several reasons for this unsatisfactory state of affairs. Most important is perhars the different conceptual demands on theories of chemistry and physics respectively. In this instance there has been no effort to re-interpret mathematical quantum theory to satisfy the needs of chemistry. The physical, or Copenhagen, interpretation, which is essentially an ensemble theory, is simply not able to handle the individual elementary units needed to formulate a successful theory of chemical cohesion and interaction. Computational dexterity without some mechanistic basis does not constitute a theory. Equally unfortunate has been the dogmatic insistence of theoretical chemists to drag their outdated phenomenological notions into the formulation of a hybrid theory, neither classical nor quantum; even to the point of discarding the fundamental law of conservation of angular momentum. The ironic fact is that the Sommerfeld tetrahedral model of carbon is dictated explicitly by the conservation of angular momentum, but retention of the concept within wave mechanics is only possible at the expense of the conservation law.

2.3 Wave-Particle Duality

The main objection against the Bohr and Sommerfeld atomic models was the *ad hoc* definition of stationary states. Simply declaring these as quantum states offers no explanation for the failure of an accelerated charge to radiate energy. The quantization of neither energy nor angular momentum implies such an effect.

The first convincing description of stationary quantum states was provided by the assumed wave nature of matter proposed by Louis de Broglie. The proposal had its roots in Einstein's explanation of the photoelectric effect and Compton's analysis of X-ray scattering.

2.3.1 Photoelectric Effect

The emission of electrons from irradiated metal surfaces, although well documented experimentally, is notoriously difficult to envisage as caused by the transfer of energy carried by electromagnetic waves. To understand the effect it is important to note that, irrespective of light intensity, no photoelectrons are produced by radiation of frequency less than some minimum, characteristic of the metal concerned. This observation has the reasonable implication that some minimum energy, W_0 say, is required to dislodge an electron from the metal surface, but fails to explain why the required energy cannot be accumulated during prolonged low-frequency irradiation. The only conclusion seems to be that, in line with Planck's earlier proposal, radiant energy only occurs in discrete packets (quanta) of magnitude $h\nu$, and that electrons can absorb energy only in multiples thereof. The metal *work function* hence corresponds to some minimum frequency, such that $W_0 = h\nu_0$. More energetic quanta of light dislodge electrons with kinetic energy $T = h\nu - W_0$, which is readily measurable and confirmed by experiment. It appears that, although the propagation of radiant energy is correctly described in wave formalism, the transfer of energy resembles a collision between particles of energy (called *photons*) and electrons.

2.3.2 Compton Effect

From related experiments that measure the recoil of high-energy photons scattered on an electron Compton could demonstrate a simple relationship between the wavelength and momentum of a photon. X-rays are scattered from atoms by electrons that absorb the radiant energy, go into vibration and re-emit radiation at the same frequency. In practice a small component of the scattered X-rays emerges with a shift in wavelength that depends on the scattering angle θ only, independent of the target material,

$$\Delta\lambda = \lambda_C(1 - \cos\theta)$$

This effect was explained by Compton on the assumption that the outer electrons in an atom are so loosely bound as to appear virtually free. As an X-ray photon collides with such an electron, for example in a block of graphite, the conservation of momentum requires that

$$\vec{p}_X = \vec{p}_s + \vec{p}_e$$

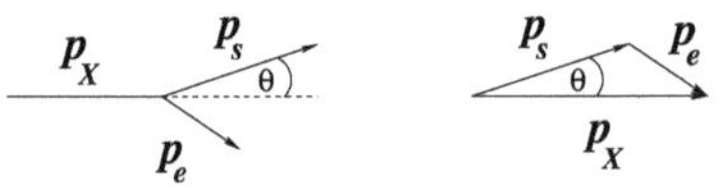

i.e.

$$p_e^2 = p_X^2 + p_s^2 - 2p_X p_s \cos\theta \tag{2.12}$$

The energy of the relativistic photon is $E_X = p_X c$, and the rest mass of the electron is m_e. Conservation of energy therefore requires

$$p_X c + m_e c^2 = p_s c + E_e$$

The energy of the scattered electron, E_e when written in the form

$$E_e = (p_X - p_s)c + m_e c^2$$

$$i.e. \qquad E_e^2 = (p_X - p_s)^2 + 2m_e c^2 (p_X - p_s)c + m_e^2 c^4$$

appears remarkably like the energy of a relativistic particle. In fact, to reproduce the observed experimental results it was necessary to assume that the scattered electron behaves relativistically, with energy E_e given by:

$$E_e^2 = p_e^2 c^2 + m_e^2 c^4$$

to give

$$p_e^2 = (p_X - p_s)^2 + 2m_e c(p_X - p_s)$$

On combining this result with (2.12) it follows that

$$m_e c\left(\frac{1}{p_s} - \frac{1}{p_X}\right) = 1 - \cos\theta$$

As photon momentum $p = E/c$, the quantum assumption $E = h\nu$ implies that $p = h\nu/c = h/\lambda$. This relationship between mechanical momentum and wavelength is an example of electromagnetic wave-particle duality. It reduces the Compton equation into:

$$\frac{m_e c}{h}(\lambda_s - \lambda_X) = 1 - \cos\theta$$

or

$$\Delta\lambda = \frac{h}{m_e c}(1 - \cos\theta)$$

This expression confirms that the experimental results can only be explained by treating X-rays as consisting of photons with energy $E = h\nu$ and momentum $p = h/\lambda$. More surprisingly, the fundamental constant $\lambda_C = h/m_e c$, being independent of X-ray wavelength and the scattering material, appears as an intrinsic wavelength associated with a free electron of momentum $p_C = m_e c$, in exact parallel with the X-ray photon.

The component which is scattered without a change in wavelength results from scattering by tightly bound electrons which recoil as part of the atomic core of mass $M >> m_e$, which means that the Compton shift becomes negligibly small. For the same reason there is no shift for light in the visible region where the photon energy is not sufficient even to overcome the binding energy of a valence electron. For highly energetic γ-rays, only Compton scattering is observed, as the photon energies are large compared to the binding energies of all electrons.

2.3.3 Electron Diffraction

The observation that the wavelength of light is linked to the particle-like momentum of a photon prompted de Broglie to postulate the likelihood of an inverse situation whereby particulate objects may exhibit wave-like properties. Hence, an electron with linear momentum p could under appropriate conditions exhibit a wavelength $\lambda = h/p$. The demonstration that an electron beam was diffracted by periodic crystals in exactly the same way as X-radiation confirmed de Broglie's postulate and provided an alternative description of the electronic stationary states on an atom. Instead of an accelerated particle the orbiting electron could be described as a standing wave. To avoid self-destruction by wave interference it is necessary to assume an integral number of wavelengths to span the orbit at a radial distance r, *i.e.* $n\lambda = 2\pi r$, such that $nh/p = 2\pi r$, which rearranges into the Bohr conjecture, $pr = nh/2\pi$.

De Broglie's thesis, while elucidating the problem of stationary states, introduced a new debate about the wave or particle nature of an electron, which still rages among scientists and philosophers alike. The logical way forward for the new quantum theory, surely was to interpret both *particle* and *wave* as classical concepts which can be used to approximate the features of quantum behaviour without implying that quantum objects are either particles or waves. Further research should have focused on finding the nature of whatever quantum entity exists instead. However, this approach is nowhere considered seriously. The unspoken consensus still regards the electron as a (classical) particle, mysteriously endowed with wave properties[4]. This contrived description came to replace separate wave and particle models only after general acceptance of Born's probability proposal.

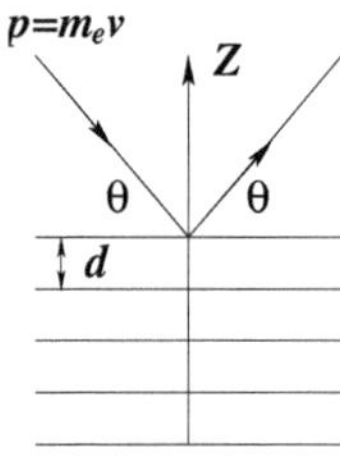

Figure 2.9: *Sommerfeld derivation of Bragg's law*

The particle school, not surprisingly, originated with Heisenberg who used Sommerfeld quantization to argue [9] that electron diffraction occurs when a particle impinges on a periodic grating constituted by crystal planes, spaced with regular separation of d, perpendicular to z, as in Figure 2.9. In terms of Sommerfeld's rules, motion of the crystal in the periodic direction z can only occur on absorption of momentum, quantized as a function of the periodicity d.

$$\int_0^d p_z \mathrm{d}z = nh \quad \text{or} \quad p_z = nh/d$$

An electron that impinges and reflects on the crystal at a glancing angle θ, transfers momentum to the crystal, on impact and on rebound, to the amount of $2mv\sin\theta$, only if it matches the quantized momentum states of the crystal, *i.e.*

$$2mv\sin\theta = nh/d \quad \text{or} \quad n\left(\frac{h}{p}\right) = 2d\sin\theta$$

[4]Advanced theories of physics describe the electron as either a zero-dimensional mass point or an equivalent plane wave, subject to mathematical manipulation, but not to visualization.

By interpreting the quantity $h/p = \lambda$ as a wavelength, the result agrees with Bragg's law that describes X-ray diffraction in terms of wave theory, $n\lambda = 2d\sin\theta$, and with experimental observation.

2.3.4 Wave Packets

Another popular approach to wave-particle duality, which originated with Schrödinger, was to view the quantum particle as a wave structure or *wave packet.* This model goes a long way towards the rationalization of particle-like and wave-like properties in a single construct. However, the simplified textbook discussion, which is unsuitable for the definition of quantum wave packets, relies on the superposition of many waves with a continuous spread of wavelengths, defines a dispersive wave packet, and therefore fails in modelling an electron as a stable particle.

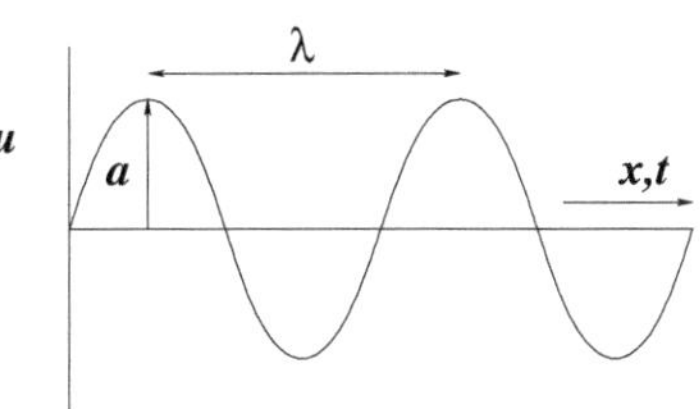

An ordinary plane wave of definite wavelength λ spreads over all space and can obviously not be used to describe the motion of a particle-like pulse, which is localized over a comparatively narrow region. One way to construct a pulse or wave packet that resembles an extended particle is to combine a number of waves with slightly different wavelengths and with amplitudes and phases chosen so that the waves interfere constructively over a limited region of space. The principle is readily demonstrated in one dimension, using the real part of the general wave equation

$$\begin{aligned} u_z &= a\cos k(x - ct) \\ &= a\cos(kx - \omega t) \end{aligned}$$

The constant k is known as the *wave vector,* $k = 2\pi/\lambda$. The wavelength $\lambda = c\tau$. The *period* $\tau = 1/\nu$ is the inverse of the *frequency* and the *angular frequency* $\omega = 2\pi\nu$. The *wave number* $\bar{\nu} = 1/\lambda$.

A wave packet can be formed by combining two harmonic waves of equal amplitude a, but slightly different frequencies, defining a total disturbance

$$z = a\cos(k_1x - \omega_1t) + a\cos(k_2x - \omega_2t)$$

Using the identity

$$\cos\alpha + \cos\beta = 2\cos\tfrac{1}{2}(\alpha + \beta)\cos\tfrac{1}{2}(\alpha - \beta)$$

it is readily shown that

$$z = 2a\cos\left(\frac{k_1 + k_2}{2}x - \frac{\omega_1 + \omega_2}{2}t\right) \times \cos\left(\frac{k_1 - k_2}{2}x - \frac{\omega_1 - \omega_2}{2}t\right)$$

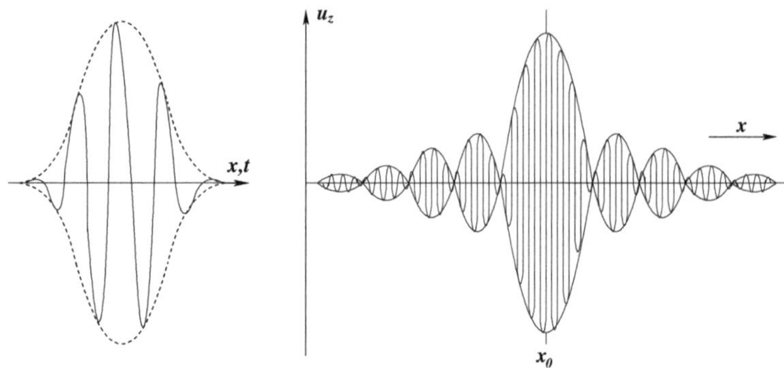

Figure 2.10: *Wave packets made up of two harmonic waves (left) and the superposition of many waves (right) over a continuous range of wavelengths.*

The first cosine term has the same form as the component waves with the average wavelength and frequency of the originals and moving with velocity $v = (\omega_1 + \omega_2)/(k_1 + k_2)$. In the case of electromagnetic waves $\omega_1 = ck_1$ and $\omega_2 = ck_2$, so that $v = c(k_1 + k_2)/(k_1 + k_2) = c$, equal to the velocity of the original waves.

The amplitude of the composite wave is no longer a periodic function because of the factor defined by the second cosine term. The total wave packet moves along without change in shape providing the component waves have the same velocity. The only instance where this is known to apply is for electromagnetic photons in vacuum. In all other cases, for instance electrons, the velocity depends on wavelength (and k).

Densely populated wave packets are constructed by the superposition of more and more cosine waves and integration over a range of wavelenghts:

$$\int_{k_0-\Delta k}^{k_0+\Delta k} \cos(kx - \omega t)(d)k = \frac{2\sin \Delta k(x - x_0)}{(x - x_0)} \cdot \cos k_0(x - x_0)$$

setting $(\omega t)_0 = x_0$. Plotted as a function of $(x - x_0)$ this wave packet has the form shown in Figure 2.10. The amplitude of oscillation reaches a maximum at $x = x_0$ and goes to zero as $x - x_0 = \pi/\Delta k$, after which it is a rapidly decreasing oscillatory function. The wave function u_z is concentrated in a packet and oscillates rapidly as a function of $(x - x_0)$. The intensity of the wave is proportional to the square of the maximum amplitude of oscillation. The same integration with respect to the time Δt required for a pulse to pass a given point, over the range of angular frequences ω_0, leads to the same result as before, now as a function of time.

Such a function will have a large pulse near $t = t_0$ and it *disperses* with time. In the two-component system the pair of dispersive waves have different velocities ($\omega_1 k_1 \neq \omega_2 k_2$) and the profile of the wave packet moves with a velocity $(\omega_1 - \omega_2)/(k_1 - k_2)$, which is different from the *phase velocity* $(\omega_1 + \omega_2)/(k_1 + k_2)$ of the rapidly oscillating part. Velocity of the wave packet is known as the *group* velocity. If the components are not too different $v_\phi = \omega/k$ and $v_g = (\omega_1 - \omega_2)/(k_1 - k_2) = \mathrm{d}\omega/\mathrm{d}k$. In terms of wavelength

$$v_g = \frac{\mathrm{d}\nu}{\mathrm{d}(1/\lambda)} = -\lambda^2 \frac{\mathrm{d}\omega}{\mathrm{d}k}$$

For matter waves with $\lambda = h/p$ and $\nu = E/h$ the group velocity is

$$v_g = -\left(\frac{h^2}{p^2}\right)\left(\frac{\mathrm{d}E}{h}\right)\left(-\frac{p^2}{h\mathrm{d}p}\right) = \frac{\mathrm{d}E}{\mathrm{d}p} = \frac{\mathrm{d}(p^2/2m)}{\mathrm{d}p} = p/m = v$$

equal to the particle velocity.

The group velocity of de Broglie matter waves are seen to be identical with particle velocity. In this instance it is the wave model that seems not to need the particle concept. However, this result has been considered of academic interest only because of the dispersion of wave packets. Still, it cannot be accidental that wave packets have so many properties in common with quantum-mechanical particles and maybe the concept was abandoned prematurely. What it lacks is a mechanism to account for the appearance of mass, charge and spin, but this may not be an insurmountable problem. It is tempting to associate the rapidly oscillating component with the Compton wavelength and relativistic motion within the electronic wave packet.

2.3.5 Matter Waves

It is of more than passing interest to note that de Broglie's relationship always leads to the Sommerfeld quantization rules. Consider a classical particle which is constrained in its motion by the energy barriers $a(E)$ and $b(E)$. If the particle is to be described in terms of de Broglie waves, a steady state can be reached only if the waves reflected at each barrier are in phase with those waves that approach the barrier, so as to create standing waves. The general effect will be the survival of a discrete set of waves at specific frequencies, and hence energies. The total number of waves going back and forth between the barriers is given by

$$2\int_a^b \frac{1}{\lambda(q)}\mathrm{d}q = 2\int_a^b \frac{p}{h}\mathrm{d}g = \oint \frac{p}{h}\mathrm{d}q$$

Equating this action potential to an integer leads directly to the Sommerfeld conditions for periodic systems.

In essence, the particle concept remains central in all versions of quantum theory. The wave part only serves to determine the probabilities for particles to take up certain states. In the de Broglie–Bohm interpretation of quantum phenomena the particle follows the equivalent of a classical trajectory guided by a pilot wave. The analogy of a ship that sails under its own power, but guided by a radio signal, is often used to explain the association between particle and wave. Still, it remains a mere analogy. In the final analysis all debates on wave-particle duality is informed by Young's two-slit experiment. The interference pattern generated by a beam of light or electrons, directed to pass through a two-slit screen, changes in a fundamental way if either of the slits is closed. Although the particle can only pass through any one of the slits, the mere presence of the second slit has a decisive influence on the final destination of the particle as observed on a recording screen. In the de Broglie–Bohm interpretation the pilot wave passes through both slits and sets up, by Huygens construction, an interference pattern that directs the particle, passing through one of the slits, to its final destination. Closure of one slit destroys the interference pattern of the pilot wave and all particles are simply directed through the remaining slit. The assumption works, but the guidance mechanism remains undefined. Whether the guiding wave is described as active information (Bohm), a propensity (Popper) or some other statistical operation (Born), visualization of the dual construct remains problematical.

A somewhat different conception, due to Einstein, of quantum particles as condensations of the electromagnetic field, points at an alternative possibility. Viewed more generally, elementary particles may be understood as persistent distortions in space-time, or the vacuum. The vacuum is made up of stuff, traditionally known as the aether, which carries an all-pervading unitary field that contains electromagnetic, gravitational and other, still to be discovered, components. In this sense an electron represents a semi-permanent, flexible knot of indefinite shape, but fixed topology, embedded in and made up of the same stuff as the aether. In response to environmental pressure it may be deformed to resemble a point particle or extend into a pulsating fluid, best characterized in terms of internal wave motion (*e.g.* with Compton frequency) that may extend into the neighbouring aether. Properties such as mass, charge and spin are fixed by a characteristic topology of the distortion. This distortion, like the aether, is a continuous whole without parts.

The response of a topological distortion to experimental study would be of the dual wave-particle kind. It generates a local wave field in its immediate environment and this field constitutes the wave that appears to guide a particle through two slits, into an interference pattern. The wave motion originates in the fluctuation of internal mass and charge densities. An observable particle-like trajectory coincides with centre-of-mass displacements. Mass and charge densities may also vary in response to external fields, like that of an atomic nucleus. The formation of a hydrogen atom happens when the electron wraps itself around a proton in an arrangement dictated by the interaction between proton and electron wave fields.

Because quantum theory is supposed only to deal with observables it may be, and is, argued as meaningless to enquire into the internal structure of an electron, until it has been observed directly. To treat an electron as a point particle is therefore considered mathematically sufficient. However, an electron has experimentally observed properties such as the Compton wavelength and spin, which can hardly be ascribed to a point particle. The only reasonable account of such properties has, to date, been provided by wave models of the electron.

2.3.6 Historical Note

A little-known paper of fundamental importance to modern atomic theory was published by Hantaro Nagaoka in 1904 [10]. Apart from oblique citation, it was soon buried and forgotten. With hindsight it deserved better than that. It contained the seminal ideas underlying the nuclear model of the atom, the standing-wave nature of orbital electrons and radiationless stationary states. It was so far ahead of contemporary thinking that later imitators either failed to appreciate its significance, or pretended to be unaware of it.

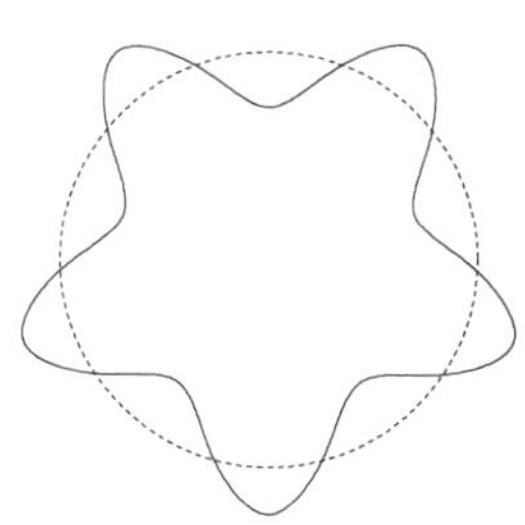

Hagaoka used the mechanical description of Saturn's rings as a model for orbital electrons to explain the radiation formulae for line and band spectra and to propose a speculative mechanism for radioactive decay.

An atom is pictured as a large number of electrons circling a positive nucleus at equal angular distances on a circle. Such a structure resembles the planet Saturn with its rings of particles, revolving with nearly the same velocity on stable orbits about an attracting centre. Because of interelectron repulsion the model differs from the Saturnian system with mutually attracting satellites, but the mathematical formalism is largely the same. To overcome the stability problem, caused by radiation from accelerated charges, a strategy to compensate for lost energy had to be devised. The proposed solution was found in a closed orbit with a radius vector that fluctuates around an equilibrium value.

The moment of inertia of a ring of particles, $\sum mr_k^2$, was used as the criterion for stability to define a closed orbit that combines circular motion with simple harmonic displacements. A more general discussion that substantiates the derivation is given by Goldstein [11]. The frequency of revolution is obtained in the form of a square root, defined by a set of integers,

$$\pm\nu = \omega_o + n^2 A + n^4 B \ldots (A < 0\,,\ B > 0\,,\ n = 1, 2, 3 \ldots)$$

interpreted by Nagaoka to show that "waves of equal frequency travel round the ring in opposite senses, so long as the particles are not acted upon by extraneous forces". This arrangement strictly defines a standing wave along the circumference of the orbit with wavelength $\lambda = 2\pi r/n$. The small oscillations may be radial or normal to the plane of the orbit. While this pattern remains stationary there are no radiation effects. When the equilibrium is disturbed the electronic energy increases to a higher level, defined by increased n. On relaxation the same amount of energy, $E_1 - E_2$, is emitted. The interesting implication is that an atom in a non-radiating state (*e.g.* the ground state) has zero orbital angular momentum – the assumption that will be shown to underlie Hund's empirical rule, section 4.7.4.

The calculated frequency of oscillation, expressed as a function of a row of integers, shows a close qualitative resemblance with observed line spectra and with the digital formulae of Balmer and others.

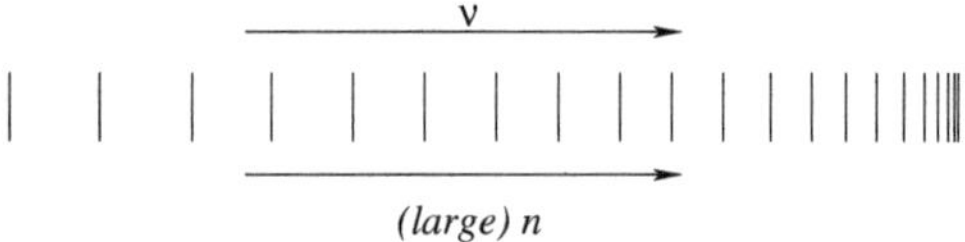

The nuclear concentration of mass anticipated Rutherford's model of the atom, and Bohr's planetary model by a decade. The spectral integers, linked to a standing-wave pattern, predates de Broglie's proposal by two decades.

All conclusions, drawn before the importance of Planck's quantum of action was appreciated, were strictly qualitative. Introduction of the quantum condition was Bohr's innovation, and it could have been more effectively combined with Nagaoka's stable orbits, rather than with electrodynamically unstable orbits. Whereas Bohr's was a one-electron theory, Nagaoka proposed a model for all atoms, with electrons spread across a set of concentric rings. Developing this into a quantum model remains an intriguing possibility.

A truly remarkable feature of Nagaoka's model is his interpretation of diatomic band spectra. The calculated frequency of transversal oscillation, of the form

$$\pm\nu' = \omega_o' - a_j n^2 + b_j n^4 + \ldots, \; a_j > 0$$

closely resembles the empirical formulae used to interpret the vibrational band structure of electronic transitions. It means that a diatomic molecule is also approximated by a heavy core, surrounded by an orbiting valence shell, like the arrangement considered in section 2.7.3, although an ellipsoidal Saturnian analogue could be more appropriate in this instance. A magnetic field, applied perpendicular to the orbital plane, exerts a radial force on orbiting electrons and hence line spectra, but not band spectra exhibit a Zeeman effect.

Like the periodic table of the elements (Chapter 4) and gaps in the asteroid belt, the spacing of Saturn's rings fits a numerical pattern based on the golden ratio [12].

2.4 Orbital Angular Momentum

The early quantum-mechanical atomic models of Bohr and Rutherford were based on the assumed quantization and conservation of the orbital angular momentum (o-a-m) of the electrons on electrodynamically unstable orbits around the nucleus. The de Broglie alternative that pictures the electron as a standing wave seems to overcome the stability problem, albeit at the expense of angular momentum conservation. The conceptually simpler idea of an electron as a pulsating continuous fluid, wrapped around the nucleus, eliminates the wave-particle dilemma and the requirement of an orbiting particle to generate the angular momentum. The argument is the same as that used by Laplace to study tidal motion as a hydrodynamic property of a liquid sphere (planet) stabilized under gravity. In the case of the hydrogen atom the electrostatic interaction is of the same inverse-square form as gravity and the dynamic effects of the electron fluid may be considered independent of electric charge.

2.4.1 Laplace's Equation

Disturbance of the fluid sphere[5] results in *harmonic* displacements, governed by Laplace's equation, which assumes a minimum average gradient of some

[5] For instance, by the moon.

potential function. In the notation of vector algebra gradient is defined as the first derivative of the potential; in one dimension,

$$\nabla V = \frac{\mathrm{d}}{\mathrm{d}x}(V)$$

and it has a minimum where the second derivative vanishes:

$$\frac{\mathrm{d}}{\mathrm{d}x}(\nabla V) = \nabla^2 V = \frac{\mathrm{d}^2 V}{\mathrm{d}x^2} = 0$$

It is shown by substitution that this equation is satisfied by the straight line, $y = mx + c$, with first and second derivatives $\nabla y = m$, $\nabla^2 y = 0$. A string which is stretched between two points is known to assume this linear shape. Such a string, when disturbed, goes into *harmonic* vibration with displacements described by sine functions,

$$u_n = A_n \sin(n\pi x/l)$$

in which l is the length of the string and n is an integer. These functions are special solutions of the Helmholtz equation

$$\frac{\mathrm{d}^2 u}{\mathrm{d}x^2} + k^2 u = 0$$

It has the general solution $u = A\cos kx + B\sin kx$. No displacement occurs at $x = 0$ and $x = l$ where the string is pinned down. Acceptable solutions of the differential equation must obey these boundary conditions, with the consequence that, $u(0) = 0$ only if $A = 0$, and $u(l) = 0$ only if $k = n\pi/l$, *i.e.* $\sin kx = \sin n\pi = 0$ for integer n.

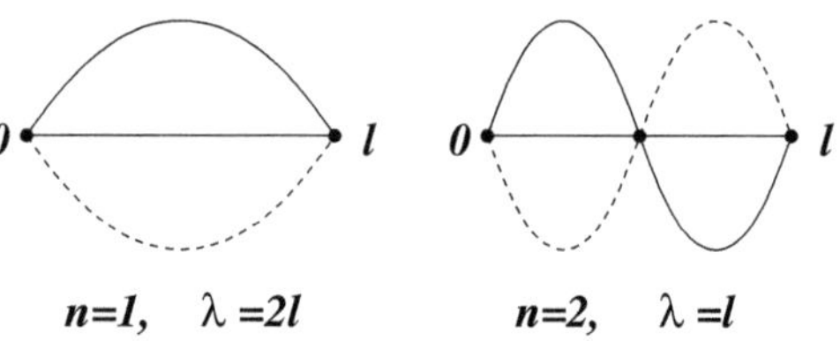

The vibration therefore sets up standing waves with wavelengths restricted by integers (n), such that $\lambda = 2l/n$ or $l = n\lambda/2$. As the string vibrates each point on the string goes up and down and passes periodically through zero. At special points, *e.g.* $l = 1/2$ at $n = 2$ there is never any displacement. Such a position is called a *node*. The points of maximum displacement are called *antinodes*. The vibration $n = 1$ is known as the *fundamental* and in the theory of music it provides the principal tone or pitch. Vibrations of higher n are called *harmonics* and in musical terms they provide the quality or *timbre* of the notes. These harmonic vibrations, or normal modes, were first identified by Pythagoras.

The two-dimensional analogue of a vibrating string is a vibrating membrane such as a drumhead. Two-dimensional harmonics are commonly demonstrated by spreading sand on a vibrating drumhead. The grains collect along the *nodal* curves where vibration disappears. When struck in the centre the same normal modes of vibration of the string are generated, but with circular symmetry. When struck off-centre, the drum vibrates differently and the displacements no longer have circular nodes, but display one or more nodal lines, known as angular nodes, instead. These normal modes appear

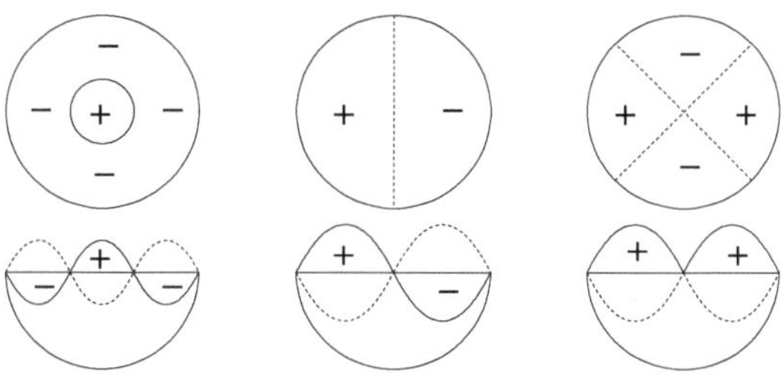

Figure 2.11: *Harmonic vibrations on a drumhead shown in projection, at the top, and perpendicular to the drumhead (below)*

as solutions of Lapace's equation in two dimensions

$$\frac{\partial^2 V}{\partial x^2} + \frac{\partial^2 V}{\partial y^2} = 0$$

In plane polar coordinates

$$\nabla^2 V = \frac{\partial^2 V}{\partial r^2} + \frac{1}{r}\frac{\partial V}{\partial r} + \frac{1}{r^2}\frac{\partial^2 V}{\partial \varphi^2} = 0$$

When multiplied by r^2 this equation separates into two parts, in one of which, V, is a function of r and in the other of φ:

$$r^2\frac{\partial^2 V}{\partial r^2} + r\frac{\partial V}{\partial r} = -\frac{\partial^2 V}{\partial \varphi^2}$$

If, in addition $V = R(r)\cdot\Phi(\varphi)$, the variables can be completely separated on dividing by V after differentiation. To satisfy the equation each side must now be separately equal to the same constant, k^2 say, to yield:

$$r^2\frac{\mathrm{d}^2 R}{\mathrm{d}r^2} + r\frac{\mathrm{d}R}{\mathrm{d}r} - k^2 R = 0$$

$$\frac{\mathrm{d}^2\Phi}{\mathrm{d}\varphi^2} + k^2\Phi = 0$$

with solutions $R = e^{\pm kr}$, $\Phi = e^{\pm ik\varphi}$. The cyclic boundary condition $\Phi = \Phi + 2\pi$ requires that $k = m$, an integer, giving rise to the observed angular nodes. Radial nodes, which occur for $m = 0$, are shown on the left in Figure 2.11. As for the vibrating string the radial wave depends on solution of a 2D Helmholtz (wave) equation $\nabla^2 V = k^2 V$. Separation of the variables leads to two equations, as before, an angular equation identical to that obtained from Laplace's equation, and a modified radial equation,

$$r^2\frac{\mathrm{d}^2R}{\mathrm{d}r^2} + r\frac{\mathrm{d}R}{\mathrm{d}r} + (k^2r^2 - m^2)R = 0$$

recognized as Bessel's equation. For $m = 0$ the solutions are the Bessel functions $J_m(kr)$, which correctly describe the radial modes.

As a generalization of these observations it follows that vibrations in a central field (*i.e.* around a special central point) are of two types, radial modes and angular modes. Laplace's equation separates into angular and radial components, of which the angular part accounts in full for the normal angular modes of vibration. Radial modes are better described by the related radial function that separates out from a Helmholtz equation. It is noted that the one-dimensional oscillator has no angular modes.

What is of further interest here, as a model of the hydrogen atom and its angular momentum, is the vibration of a three-dimensional fluid sphere in a central field. As in 2D the wave equation separates into radial and angular parts, the latter of which determines the angular momentum and is identical with the angular part of Laplace's equation.

Laplace's equation in three dimensions reads

$$\nabla^2 V = \frac{\partial^2 V}{\partial x^2} + \frac{\partial^2 V}{\partial y^2} + \frac{\partial^2 V}{\partial z^2} = 0$$

The V may be interpreted as a gravitational potential, velocity potential or, in the present context, as a circulation potential. The differential equation may be solved by separation of the variables under the assumption that the potential may be defined as the product of three one-dimensional potentials, or cartesian components of $V = X(x){\cdot}Y(y){\cdot}Z(z)$. This solution is substituted into the equation and after differentiation, divided by V, to give

$$\frac{1}{X}\frac{\mathrm{d}^2X}{\mathrm{d}x^2} + \frac{1}{Y}\frac{\mathrm{d}^2Y}{\mathrm{d}y^2} + \frac{1}{Z}\frac{\mathrm{d}^2Z}{\mathrm{d}z^2} = 0$$

Each term is a function of one variable only and hence independent of the two other variables, which implies that each term is independently equal to some constant, *e.g.* $\nabla^2 X = KX$. By defining the constant as a squared quantity each equation is a one-dimensional Helmholtz equation, solved by an exponential function. For $K = k_1^2$, $X(x) = c\exp(\pm k_1 x)$. Noting that $e^a.e^b = e^{(a+b)}$, the overall solution follows as

$$V(x,y,z) = e^{k_1x+k_2y+k_3z}$$

with

$$k_1^2 + k_2^2 + k_3^2 = 0 \tag{2.13}$$

This condition requires at least one of the constants to be a complex quantity, such that $(ik)^2 = -k^2$ and $i = \sqrt{-1}$, unless

$$k_1 = k_2 = k_3 = 0 \tag{2.14}$$

An interesting situation arises as one of the constants is equal to zero. The solution that involves the two remaining terms then requires $k_1^2 + k_2^2 = 0$, *e.g.* $k_1 = k$, $k_2 = ik$. Hence

$$X = c_1 e^{\pm kx} \quad , \quad Y = c_2 e^{\pm iky}$$

or

$$V_k = c_1 e^{\pm k(x+iy)} \tag{2.15}$$

$$V_z = c_2 z + c_3 \tag{2.16}$$

2.4.2 Angular Momentum

The various solutions of Laplace's equation describe the angular momentum that can arise in the fluid body if V is defined as a circulation potential. The condition (2.14) defines a spherically symmetrical state without angular momentum, recognized as an atomic s-state. Conditions (2.15) and (2.16) describe a state with a well-defined component of angular momentum in the z-direction and variable components in the x, y plane. This state occurs for an atom in an applied magnetic field. While the total angular momentum vector precesses about the Z-axis (Figure 2.12) only one component, in the field direction, is well defined. The L_x and L_y components fluctuate together, such that $L^2 = L_x^2 + L_y^2 + L_z^2$. The important conclusion to draw from this analysis and condition (2.13) is that the orbital angular momentum of a free atom cannot have well-defined components in more than one polar direction.

Systems of spherical symmetry are more amenable to analysis in a spherical polar, rather than cartesian, coordinate system. In this case the Laplacian operator is a function of (r, θ, φ) rather than (x, y, z). Separation of

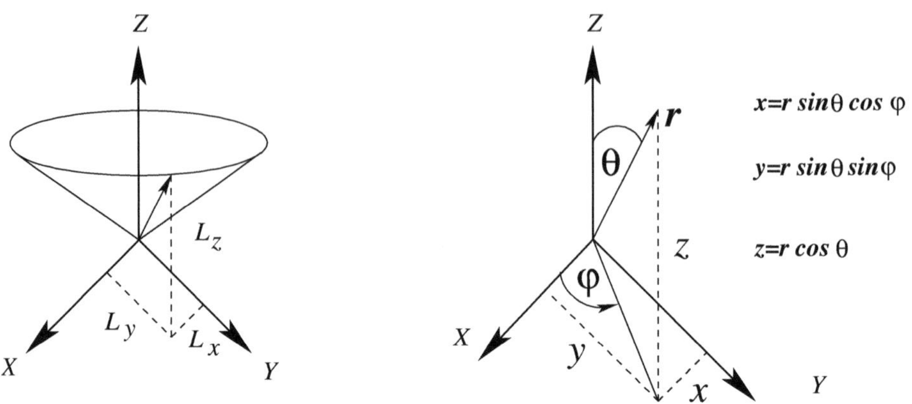

Figure 2.12: *Definition of the components of angular momentum in cartesian and in spherical polar coordinates.*

the variables is possible, as before, if the potential function is defined as the product $V(r,\theta,\varphi)=R(r).\Theta(\theta).\Phi(\varphi)$. Separation of radial and angular parts is achieved by equating both parts to the same constant c. The resulting angular equation becomes ill-conditioned at arbitrary polar points, unless the separation constant is defined as $c=l(l+1)$, for integer l. Separation of the angular part produces two equations in Θ and Φ. The latter reads

$$\frac{\mathrm{d}^2\Phi}{\mathrm{d}\varphi^2}=-m^2\Phi$$

It has the same form as the cartesian components and the solution, $\Phi = ke^{\pm im\varphi}$, describes rotation about the polar axis in terms of the orbital angular momentum vector L_z, specified by the eigenvalue equation

$$L_z\Phi=km\Phi$$

with an arbitrary scale factor k. In order for $\Phi(\varphi)$ to be single-valued it is required that $\Phi(\varphi)=\Phi(\varphi+2\pi)$, which is possible only if m is an integer. The $2l+1$ independent solutions Y_l^m for each l are distinguished by $|m|<l$. This result shows that Laplace's equation defines orbital angular momenta within classical theory as discrete quantities, correct to within an arbitrary constant. The physically meaningful value of this constant was correctly conjectured by Bohr, albeit for the wrong reason, as $k=\hbar$, such that

$$L_zY_l^m=m\hbar Y_l^m$$

and likewise for total angular momentum

$$L^2 Y_l^m = l(l+1)\hbar^2 Y_l^m$$

so that

$$L_x^2 + L_y^2 = [l(l+1) - m^2]\hbar^2$$

All of these predictions are supported by spectroscopic measurement. The maximum conceivable value of $L_z = L$, which corresponds to the Bohr model, implies that, in all cases, $L_z < L$ and $L_x + L_y > 0$. This result refutes the orbital assumption of the Bohr conjecture, but the quantization condition is upheld.

2.4.3 Surface Harmonics

The fact that these semi-classical results agree in all respects with those obtained from Schrödinger's amplitude equation

$$\nabla^2 \psi = -\frac{2m}{\hbar}(E-V)\psi$$

which, in this form is the eigenvalue equation for kinetic energy $T = E - V$, is hardly surprising. In the Bohm interpretation, which leads to the prediction of zero kinetic energy in stationary states, the equation reduces to that of Laplace, as considered before.

Any solution V_l of Laplace's equation of degree l is called a *solid* spherical harmonic, and it has the form $V_l = r^l Y_l^m(\theta, \varphi)$. For $r = 1$, $V_l = Y_l^m$ so that Y_l^m is the value of the solid harmonic at points on the surface of the unit sphere defined by the coordinates θ and φ, and hence Y_l^m is called a *surface harmonic* of degree l. Surface harmonics are orthogonal on the surface of the unit sphere and not at $r = 0$, as commonly assumed in the definition of atomic orbitals.

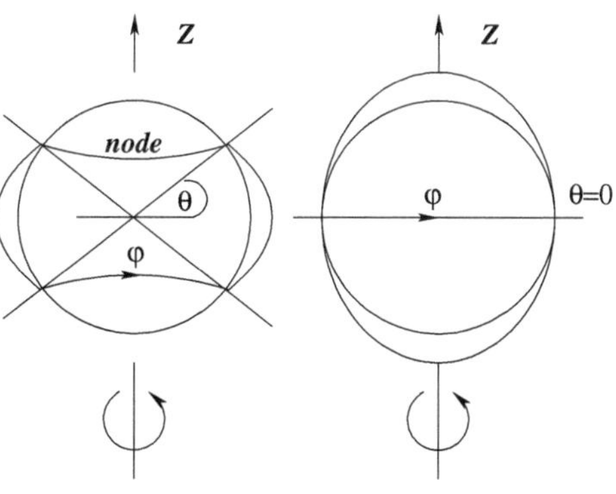

Figure 2.13: *Spherical surface harmonics for* $l = 1$, $m = \pm 1$ *(left) and* $m = 0$.

The surprising conclusion to be drawn from this analysis is that a quantitative description of orbital angular momentum, as detailed as the wave-mechanical model, can be obtained within the old quantum theory.

The only assumption, in addition to Bohr's conjecture, is that the electron appears as a continuous fluid that carries an indivisible charge. As already shown, Bohr's conjecture, in this case, amounts to the representation of angular momentum by an operator $L \to i\hbar\partial/\partial\varphi$, shown to be equivalent to the fundamental quantum operator of wave mechanics, $p \to -i\hbar\partial/\partial q$, or the difference equation $(pq - qp) = -i\hbar(I)$, the assumption by which the quantum condition enters into matrix mechanics. In view of this parallel, Heisenberg's claim [13](page 262), quoted below, appears rather extravagent:

> ... it seems sensible ... to concede that the partial agreement of the (old) quantum rules with experience is more or less fortuitous. Instead it seems more reasonable to try to establish a theoretical quantum mechanics, analogous to classical mechanics, but in which only relations between observable quantities appear.

Instead of this claim, more careful analysis would have shown that the basic premises of the old quantum theory remained intact and by approaching the problem from a somewhat different angle nothing fundamentally new was brought to the table. Schrödinger introduced yet another approach on exactly the same basis. The post-1926 (new) quantum theory, as a logical extension of earlier work, may not be as revolutionary as generally assumed. It managed to put certain aspects of quantum behaviour in clearer perspective, but at the same time created the false impression of an inevitable break with established philosophies of reality and causality.

The new insights will be examined next, leaving the complications aside, for the time being.

2.5 The Quantum Theory

As it became apparent that the mechanical behaviour of electrons is fundamentally different from that of macroscopic objects, several alternatives to account for these differences from first principles, were proposed. It could eventually be shown that all of these proposals are essentially equivalent, differing only in mathematical formalism. However, at the time, the situation must have appeared dauntingly complicated to non-specialists and many of the concepts, which were introduced to describe supposedly distinctive features of *quantum systems*, acquired unwarranted special significance in colloquial use.

2.5.1 The Uncertainty Principle

The famous *uncertainty principle* is a typical example of a concept, well known to classical physics, which entered into the vocabulary of laymen and philosophers and created an aura around quantum theory, now widely believed to describe a mysterious netherworld that defies comprehension. The philosophical fall-out has been enormous, to the point where quantum theory is dragged into debates on free will, cosmology and determinism; even into theological discourse.

In physics the concept is known as the property of all pairs of *conjugate* variables, such as position and momentum, mathematically related by Fourier transformation. The de Broglie formula that relates the momentum of matter waves to wavelength

$$p = h/\lambda$$

defines the same conjugate pair, using slightly different notation. The shorter the wavelength, the better is the definition of (particle) position, and the faster the motion through that position. Simultaneous estimates of particle position and momentum must inevitably reflect a degree of error for both quantities. In the de Broglie picture this uncertainty translates into the relationship which is often presented as a basic tenet of quantum mechanics:

$$\Delta p \Delta q \sim h$$

The only non-classical feature of this relationship is the appearance of Planck's constant. The uncertainty itself is well-known to occur in any classical system described in wave formalism. The reason why it never became an issue in classical physics is because h is too small to cause measurable effects in macroscopic systems.

2.5.2 The Measurement Problem

Volumes have been written about the red herring known as Schrödinger's cat. Without science writers looking for sensation, it is difficult to see how such nonsense could ever become a topic for serious scientific discussion. Any linear differential equation has an infinity of solutions and a linear combination of any two of these is another solution. To describe situations of physical interest such an equation is correctly prepared by the specification of appropriate *boundary conditions*, which eliminate the bulk of all possible solutions as irrelevant. Schrödinger's equation is a linear differential equation of the Sturm-Liouville type. It has solutions, known as *eigenfunctions*, the sum total of which constitutes a *state* function or *wave* function, which carries all information about a system of interest. The cat problem assumes the animal to be described by such a wave function. In particular, it has eigenfunctions corresponding to a dead and to an alive cat, respectively. When confined to a closed box, with some lethal device that could be triggered by a chance event, it is not known if the cat is dead or alive until the box is opened and examined by an observer. The correct way to describe the state of the cat in this situation is by superposition (linear combination) of the two relevant eigenfunctions. As far as the outside observer is concerned the cat is said to be half-dead and half-alive until the box is opened. It is only when the observer sees the cat, that the wave function collapses into either of the two eigenfunctions. The logical error here is that the wave function is interpreted, not only as describing the state of the cat, but actually deciding it. In fact, the wave function collapses only in the mind of the observer and certainly cannot kill a cat or keep it alive. The poison decides that. Any uncertainty is finally resolved when the cat dies and some clever observer, unbeknownst to the primary observer, secretly used an x-ray or nmr probe to monitor this event, rather than wait for the box to be opened. The wave function appears to collapse twice – once for each observer. Schrödinger's cat therefore demonstrates, not some magical quality of quantum mechanics, but rather that observers have no effect on the outcome of physical events.

All of the many supposed ghost-like properties associated with quantum theory are of the same ilk. The most bizarre is perhaps the many-worlds theory, which states that whenever an observer opens an eye on the world a wave function collapses and the universe splits into two. The observer persists in only one of these and keeps on creating an endless number of irrelevant universes. Any serious student of quantum theory should be advised to ignore this sensational stuff at the outset. There are enough poorly understood real issues to keep the keenest minds occupied.

2.5.3 The Quantum Limit

When first confronted with the oddities of quantum effects Bohr formulated a correspondence principle to elucidate the status of quantum mechanics relative to the conventional mechanics of macroscopic systems. To many minds this idea suggested the existence of some classical/quantum limit. Such a limit between classical and relativistic mechanics is generally defined as the point where the velocity of an object $v \to c$, approaches the velocity of light. By analogy, a popular definition of the quantum limit is formulated as $h \to 0$. However, this is nonsense. Planck's constant is not variable.

A simple demonstration of the correspondence principle is provided by the Bohr atomic model that allows for the transfer of an electron between neighbouring quantum orbits by discontinuous energy changes. For small quantum numbers the energy differences are substantial and give rise to dramatic measurable spectroscopic effects of the Balmer type. For larger quantum numbers the spacing between energy levels becomes smaller and approaches a continuum as $n \to \infty$, shown on an energy rather than a radial scale in the diagram.

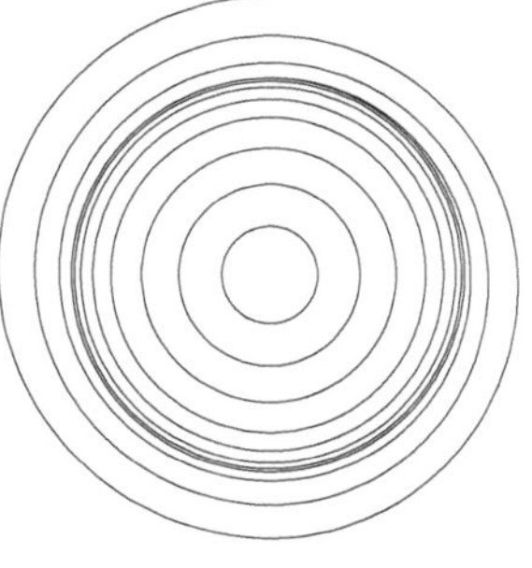

An electron in an orbit of such a high quantum number approaches the nucleus by transfer through a succession of closely spaced lower energy levels, radiating energy at each transfer. Energy is lost and radiated away continuously as the electron spirals in by continuous progress to smaller orbits, exactly as required by classical electrodynamics. The quantum effects have become so small as to be no longer noticeable. In this domain individual energy pulses, defined as $\Delta E = h\nu$, are immeasurably small. This argument demonstrates the crucial point that there is no quantum limit, apart from recognizing that quantum effects are important only at the microscopic level. Wave functions of the universe that feature in theoretical physics are presumably of a different kind. On the other hand, chemical interactions clearly happen in the quantum domain, and cannot be understood in any other terms.

The models of Bohr, Sommerfeld and de Broglie provide a firm basis for the further development of a quantum theory of chemistry by re-assessment of the more advanced theories of quantum physics. However, there is little support for such a pursuit, not if we find statements like the following [14], put out by one of the world's leading publishers of academic science:

> After 1926, however physicists went from ignorance to almost complete understanding of the equations governing simple atoms. The power of quantum mechanics was so enormous that all of chemistry could, in principle, be reduced to a series of equations.
>
> Although the Schrödinger wave equation is difficult to solve for increasingly complicated atoms and molecules, we could, if we had a large enough computer, deduce the properties of all known chemicals from this equation. Quantum mechanics, however, is even more powerful than an ordinary cookbook because it also allows us to calculate the properties of chemicals that we have yet to see in nature.

This dogma has clearly not been fulfilled. All advances in chemistry happen at the bench, as it should, but without the theoretical understanding, even of common events such as intramolecular rearrangement. No calculation can predict chemical reactions.

In a more humble approach to chemical theory it may be more fruitful to follow Einstein, who, when asked (*ca.* 1950) to explain the nature of an increasing number of elementary particles, is purported to have replied:

> I would be happy just to know what an electron is.

Only then can there be any hope of understanding atoms, molecules and the rest.

2.5.4 Wave Mechanics

The first objective of quantum theory is indeed aimed at the electron. The wave-mechanical version of quantum theory, which is the most amenable for chemical applications, starts with solution of Schrödinger's wave equation for an electron in orbit about a stationary proton. There is no rigorous derivation of Schrödinger's equation from first principles, but it can be obtained by combining the quantum conditions of Planck and de Broglie with the general equation[6] for a plane wave, in one dimension:

$$\begin{aligned} u &= A\cos(2\pi x/\lambda - 2\pi t/\tau) + B\sin(2\pi x/\lambda - 2\pi t/\tau) \\ &= a\exp 2\pi i(x/\lambda - \nu t) \end{aligned}$$

With the quantum conditions $E = h\nu$, $p = h/\lambda$, the wave-mechanical analogue becomes:

$$\Psi = a\exp 2\pi i(px/h - Et/h)$$

To establish how this wave function describes the position and motion of a particle it is differentiated with respect to x and t respectively:

$$\frac{\partial\Psi}{\partial x} = \left(\frac{2\pi ip}{h}\right)\Psi$$

$$\frac{\partial\Psi}{\partial t} = \left(\frac{2\pi iE}{h}\right)\Psi$$

The variation of ψ as a function of particle position resembles a standing wave, which is seen to move without change in form. The partial derivatives

[6]The complex exponential form derives from de Moivre's relation: $\cos\theta + i\sin\theta = \exp(i\theta)$.

can be rearranged into eigenvalue equations of Sturm-Liouville type:

$$\frac{h}{2\pi i}\frac{\partial}{\partial x}(\Psi) = p\Psi$$

$$\frac{h}{2\pi i}\frac{\partial}{\partial t}(\Psi) = E\Psi$$

The eigenvalues of linear momentum and energy, respectively, are generated by the differential operators acting on a wave function:

$$p \rightarrow -i\hbar\frac{\partial}{\partial x} \quad ; \quad -i\hbar\nabla \text{ in three dimensions}, \quad E \rightarrow -i\hbar\frac{\partial}{\partial t}$$

The total energy of a mechanical system is also given by the sum of its kinetic and potential energies, $E = p^2/2m + V$. The operator equivalent of this classical expression, known as the Hamiltonian operator, is obtained by substituting the momentum operator into this equation, *i.e.*

$$H = \left[\frac{1}{2m}\cdot\frac{\hbar}{i}\nabla\left(\frac{\hbar}{i}\nabla\right) + V\right] = -\frac{\hbar^2}{2m}\nabla^2 + V$$

Schrödinger's equation is obtained by equating the temporal and space expressions for the total energy,

$$\begin{aligned} i\hbar\frac{\partial\Psi}{\partial t} &= \left(\frac{\hbar^2}{2m}\nabla^2 - V\right)\Psi \\ H\Psi &= E\Psi \end{aligned}$$

A time-independent Schrödinger equation, or amplitude equation, is obtained by substituting $\Psi = \psi\exp(-iEt/\hbar)$.

$$\frac{\partial\Psi}{\partial t} = -(iE/\hbar)\psi e^{-iEt/\hbar}$$

Hence

$$i\hbar\frac{\partial\Psi}{\partial t} = E\psi e^{-iEt/\hbar} = \left(-\frac{\hbar^2}{2m}\nabla^2 + V\right)\psi e^{-iEt/\hbar}$$

which is traditionally rearranged into the form

$$\nabla^2\psi + \frac{2m}{\hbar^2}(E - V)\psi = 0$$

Only the potential energy remains unspecified.

2.5.5 Schrödinger's Equation

The equation is assumed valid for any system. It turns into an eigenvalue equation for the total energy once the potential energy has been correctly specified, *i.e.*

$$H\psi = \left(-\frac{\hbar^2}{2m}\nabla^2 + V\right)\psi = E\psi \tag{2.17}$$

This equation can be solved by separation of variables, provided the potential is either a constant or a pure radial function, which requires that the Laplacian operator be specified in spherical polar coordinates. This transformation and solution of Laplace's equation, $\nabla^2\psi = 0$, are well-known mathematical procedures, closely followed in solution of the wave equation. The details will not be repeated here, but serious students of quantum theory should familiarize themselves with the procedures [15].

In summary, the total solution reduces to a product function

$$\psi_{nlm} = R_{nl} \cdot Y_l^m(\theta, \varphi)$$

The principal quantum number is restricted to be a positive integer, $n = 1, 2, 3, \ldots$. The *azimuthal* quantum number l is restricted to the integer values $l < n$. The *magnetic* quantum number has $2l + 1$ allowed values for each l, such that $-l \leq m \leq l$. These numbers quantify the eigenvalues of the total energy E, the square of total angular momentum L^2 and the projection L_z of the angular momentum in the polar direction. Of these, only the energy depends on the potential V, which means that the angular momentum in any central field always satisfies the equations

$$L^2 Y_l^m = -l(l+1)\hbar^2 Y_l^m$$
$$L_z Y_l^m = m\hbar Y_l^m$$

Important examples of chemical interest include particles that move in the central field on a circular orbit (V constant); particles in a hollow sphere ($V = 0$); spherically oscillating particles ($V = \frac{1}{2}kr^2$), and an electron on a hydrogen atom ($V = 1/4\pi\epsilon_0 r$). The circular orbit is used to model molecular rotation, the hollow sphere to study electrons in an atomic valence state and the three-dimensional harmonic oscillator in the analysis of vibrational spectra. Constant potential in a non-central field defines the motion of a free particle in a rectangular potential box, used to simulate electronic motion in solids.

Eigenvalues of energy and angular momentum emerge naturally on solution of Schrödinger's equation. The angular momenta, in all instances, depend only on the eigenfunctions Y_l^m, which are identical in central fields. The energy depends on radial eigenfunctions $R(r)$, obtained as solutions of the radial equation, which for the hydrogen atom, in e.s.u., reads

$$\frac{\mathrm{d}^2R}{\mathrm{d}r^2}+\frac{2}{r}\frac{\mathrm{d}R}{\mathrm{d}r}+\left[\frac{2m}{\hbar^2}\left(E-\frac{e^2}{r}\right)-\frac{l(l+1)}{r^2}\right]R=0 \tag{2.18}$$

For those solutions without angular dependence ($l=0$) the equation reduces to

$$\frac{\mathrm{d}^2R}{\mathrm{d}r^2}+\frac{2}{r}\frac{\mathrm{d}R}{\mathrm{d}r}+\frac{2mER}{\hbar^2}-\frac{2me^2R}{\hbar^2 r}=0 \tag{2.19}$$

For large r this equation becomes

$$\frac{\mathrm{d}^2R}{\mathrm{d}r^2}+\frac{2mER}{\hbar^2}=0$$

which is the Helmholtz equation

$$\frac{\mathrm{d}^2R}{\mathrm{d}r^2}+k^2R=0$$

with $k^2=2mE/\hbar^2$, and solutions $R=a\exp(-ikr)+b\exp(ikr)$. To avoid divergence of R as $r\to\infty$ it is necessary to put $b=0$. Substitution of $R=a\exp(-ikr)$, $\mathrm{d}R/\mathrm{d}r=-ikR$, $\mathrm{d}^2R/\mathrm{d}^2r=-k^2R$ into (2.19) gives:

$$-k^2-\frac{2ik}{r}+k^2-\frac{2me^2}{\hbar^2 r}=0$$

i.e.

$$2ik=\frac{2me^2}{\hbar^2}\quad,\quad k^2=-\frac{m^2e^4}{\hbar^4}=\frac{2mE}{\hbar^2}$$

It follows that $(1/ik)\equiv(1/\kappa)=\hbar^2/(me^2)=a_0$, the radius of the first Bohr orbit, and $E_1=-me^4/(2\hbar^2)=-W_1$, the ionization energy of the hydrogen atom. This solution is known as the *ground state.*

Considerably more work is required to find the complete solution

$$R_{nl}(r)=Ce^{-r\kappa_n}(2\kappa r)^lL_{n+l}^{2l+1}(r)$$

such that $\hat{H}R(r)=E_nR(r)$, $\kappa_n=na_0$, $E_n=-me^4/(n\hbar)^2$ and the L are associated Laguerre polynomials. The constant C is chosen such that $R(r)$ is normalized to unity, over all space, *i.e.* $R^*(r)R(r)\mathrm{d}\tau=1$. The first few radial functions are:

$$\begin{aligned}
R_{10} &= 2\left(\frac{1}{a_0}\right)^{3/2} e^{-r/a_0} \\
R_{20} &= \frac{1}{\sqrt{2}}\left(\frac{1}{a_0}\right)^{3/2}\left(1-\frac{r}{2a_0}\right) e^{-r/2a_0} \\
R_{21} &= \frac{1}{2\sqrt{6}}\left(\frac{1}{a_0}\right)^{5/2} re^{-r/2a_0}
\end{aligned}$$

The solutions, $E_n < 0$, are called bound states and are the only solutions of chemical importance.

The total wave function has the property of predicting the eigenvalues of all dynamic variables when operated on by the appropriate operator, for instance the operator $-i\hbar\nabla$, associated with linear momentum, or $-\hbar^2\nabla^2/2m$ associated with kinetic energy.

2.5.6 Quantum Probability

The most serious limitation of wave mechanics is the complexity of any wave equation that describes interacting particles and prevents application to atoms other than hydrogen. To separate the equation and solve for any situation of interest it is necessary that it be reduced to a one dimensional one-particle problem. In the case of the hydrogen atom this is done by assuming the proton to be of infinite mass and therefore stationary. The only molecular system that can be treated in the same way is H_2^+, if the two protons are clamped to remain at a fixed distance apart.

The second complication is that the equation, as traditionally interpreted, only handles point particles, but produces eigenfunction solutions of more complex geometrical structure. By analogy with electromagnetic theory the square of the amplitude function could be interpreted as matter intensity, but this is at variance with the point-particle assumption. The standard way out is to assume that ψ^2 represents a probability density rather than intensity. Historical records show that this interpretation of particle density was introduced to serve as a compromise between the rival matrix and differential operator theories of quantum observables, although eigenvalue equations, formulated in either matrix or differential formalism are known to be mathematically equivalent.

In quantum-mechanical application the characteristic eigenvectors of the Schrödinger equation are called *wave functions*. The casual conclusion that matrix mechanics is a corpuscular theory of matter, and therefore distinct from wave mechanics, although of historical interest, persists even today. The rival argument may be forgotten, but the prejudice remains. The historical consensus accepts that cloud-chamber tracts confirm the particle nature of matter and that wave-like phenomena, such as diffraction, are the consequence of the uncertainty principle and the statistical outcome of quantum events. However, the precise meaning of this probability density has never been agreed.

A common interpretation is that it refers to the outcome of a large number of identical measurements, rather than a single event. This interpretation implies that there is no unique solution to one-particle problems. It leaves the possible shapes of atoms and molecules completely undefined and the nature of stationary states even more perplexing. Chemists have tried to chip away at the problem by defining regions of space, called "orbitals", in which the probability of finding an electron at a given point, is statistically predicted as $\psi^2(xyz)$. A necessary condition seems to be that each measurement must reveal a different position of the electron within the orbital. Such an electron can hardly be considered stationary. It is either accelerated along an erratic path as it moves at high speed from point to point or, like Schrödinger's cat, it hangs in limbo until it is assigned a definite position by an observer. The experts see no need to elaborate on this issue and manipulate their orbitals as if all interior points are occupied by the same electron. Unconcerned about measurements and observers they proceed to reduce their orbitals to pairs of electrons in inflatable bags that resemble Sommerfeld-type elliptic orbits, in order to rationalize preconceived ideas about chemical bonding. The fanciful stories, which have grown from the vague notion of orbitals, make up the bulk of undergraduate chemistry teaching and a large part of what is believed to be "quantum chemistry".

2.5.7 The Periodic Table

One of the benefits that quantum theory has for chemistry is an improved understanding of elemental periodicity, spectroscopy and statistical thermodynamics; topics which can be developed without reference to the nature of electrons, atoms or molecules. The success of these applications depend on approximations to model many-electron atoms on the hydrogen solution and the recognition of *spin* as a further component of electronic angular momentum, subject to the secondary condition known as (Pauli's) exclusion principle.

In its most general form the exclusion principle posits that all electronic wave functions are anti-symmetrical; which means that the wave function for systems comprising more than one electron must change sign on interchanging the positions of any two of these electrons. What started out as an *ad hoc* postulate, turned out to be one of the pillars on which quantum theory is based. It is an example of an *irreducible emergence*[7]. It does not feature in the mathematical formulation of quantum mechanics and only emerges when the theory is applied to describe the structure of matter. It cannot be inferred from the properties of a single electron or from the interaction properties of a pair of electrons. It is a holistic property which asserts that *fermions* [8] form world collectives with totally antisymmetric wave functions. The slightest disturbance of one electron affects, in principle, all other electrons in the universe. The exclusion principle, in this sense, is an irreducible concept that applies to the totality of electrons in the universe.

The way in which the exclusion principle determines the order of hydrogen-like energy-level occupation in many-electron atoms, is by dictating a unique set of quantum numbers, n, l, m_l and spin m_s, for each electron in the atom. Application of this rule shows that the sub-levels with $l = 0$, 1, 2 can accommodate no more than 2, 6, 10 electrons respectively. In particular, no more than two electrons with $m_s = \pm\frac{1}{2}$, can share the same value of m_l. Each principal level accommodates $2n^2$ electrons.

It is fair to assume that, in a series of elements, stepwise increase of atomic number would be accompanied by progressive occupation of electronic sub-levels of higher energy, in the order shown in Table 2.14. This assumption is known as the Aufbau procedure and is considered to be the theoretical basis of the periodic table of the elements. Those familiar with the layout of the periodic table will notice that the predicted sequence is obeyed up to atomic number 18, which is followed by the $4s$ rather than the $3d$ sublevel. Again, at $Z = 29$ and $Z = 37$ there is confusion over the identification of $4s$ and $5s$ levels. The situation gets worse with the appearance of f sublevels. The glib explanation, usually offered to belittle these discrepancies, is to invoke interelectronic effects that do not appear in the hydrogen problem. This explanation is certainly not sufficient for the correct identification of the predicted 10-member transition-metal series that appear to run from 21,

[7]Emergent properties are well-known in biological systems. In this case there are various levels of explanation, ranging from the atomic, through molecular, cellular, tissue, organ, body, herd, to the eco-system level. Succeeding levels are characterized by specific emergent properties which cannot be predicted from the properties of systems at the lower levels. One example of an emergent property is human consciousness, which is associated with the global nervous system, but cannot be predicted from the properties of individual neurons which compose the system, or from a knowledge of their interaction.

[8]Quantum-mechanical particles with half-integer spin, such as electrons, protons and neutrons. Particles with integer spin such as photons, helium atoms or hydrogen molecules, are *bosons* and they are not subject to the exclusion principle.

Figure 2.14: *Energy spectrum of the hydrogen electron.*

Level	n	Sub level	id	Electrons	
				Sub	Total
K	1	$l = 0,\ m_s = \pm(1/2)$	s	2	2
L	2	$l = 0,\ m_s = \pm(1/2)$	s	2	
		$l = 1,\ m_l = 0, \pm1,\ m_s = \pm(1/2)$	p	6	8
M	3	$l = 0,\ m_s = \pm(1/2)$	s	2	
		$l = 1,\ m_l = 0, \pm1,\ m_s = \pm(1/2)$	p	6	
		$l = 2,\ m_l = 0, \pm1, \pm2,\ m_s = \pm(1/2)$	d	10	18
N	4	$l = 0,\ m_s = \pm(1/2)$	s	2	
		$l = 1,\ m_l = 0, \pm1,\ m_s = \pm(1/2)$	p	6	
		$l = 2,\ m_l = 0, \pm1, \pm2,\ m_s = \pm(1/2)$	d	10	
		$l = 3,\ m_l = 0, \pm1, \pm2, \pm3,\ m_s = \pm(1/2)$	f	14	32

39 and 71, but come to a close at 28, 46 and 78, respectively. The honest assessment is that, like the Sommerfeld model, the Schrödinger model is in general agreement with the observed periodic structure, but inadequate to account for the details.

2.6 Atomic Shape

2.6.1 Chemical Affinity and Shape

The way in which atoms join up to form molecules has traditionally been interpreted in terms of the perceived properties and shape of atoms. An early theory of chemical affinity was summarized by Kekulé [16], freely translated here, in terms of the concepts atom, molecule, radical and basicity (also called atomicity):

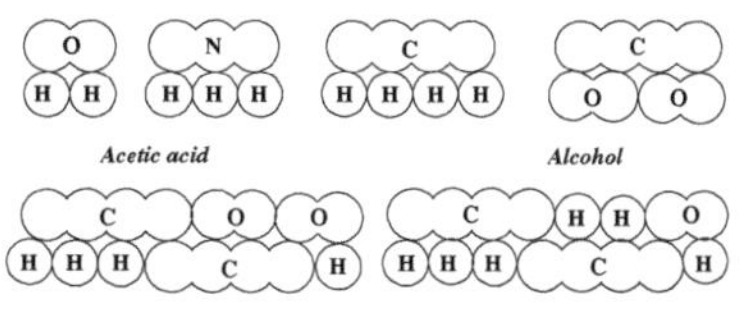

> The reason for combination of atoms into molecules must, at this time, be ascribed to an internal attractive force of atoms that we call affinity or chemical relationship.
>
> That part of a molecule that remains intact during decomposition we call a radical.
>
> The way in which atoms are stacked together in molecules and

> radicals depend on the basicity or atomicity of the elements concerned.
>
> This stacking can be symbolized by graphical representation of basicity according to size. In this sense, size is not related to actual atomic dimension but rather to the number of chemical units that it represents, such as the equivalent number of hydrogen atoms.

Some illustrative examples are shown above. Despite his disclaimer Kekulé's structural formulae are clearly the harbingers of their modern equivalents. The proposed tetrahedral structure of carbon, which followed, ignored the good advice and amounts to no more than a geometrical rearrangement of Kekulé's diagrams, as shown here to represent the actual size and shape of carbon in methane.

The vehement opposition generated by van't Hoff's proposals was evidently directed against the notion that atoms possessed non-spherical three-dimensional structures. The equivalent, more careful, formulation of Le Bel referred explicitly to the symmetry of molecules, such as methane, and is free of this criticism. It is unfortunate that it was the van't Hoff picture which became established, first in terms of Sommerfeld's elliptic orbits and later on in Pauling's hybrid orbitals.

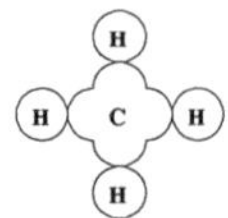

2.6.2 Orbiting Electrons

In both cases the predicted atomic shape is based on the assumed orbital motion of p-electrons. It is therefore of interest to examine the classical orbits of electrons with non-zero angular momentum. The classical orbits that correspond to rotating angular momentum vectors with $m = 0$ and 1, respectively, are shown in figure 2.15. An electron with $m = 0$ does not rotate around a fixed axis but performs three-dimensional rotation in spherical mode [7]. This rotation is independent of the direction of an applied magnetic field and is described equally well as L_x, L_y, L_z, or by any other choice of axes. For $m = 1$ the average rotation about the Z-axis occurs in a plane parallel to the XY plane at an uncertain z-coordinate. This rotation is strongly direction dependent. For $m = -1$ the vector diagram inverts and the rotation is in the opposite direction. This classical picture is totally different from the Sommerfeld model, but agrees well with the wave-mechanical result.

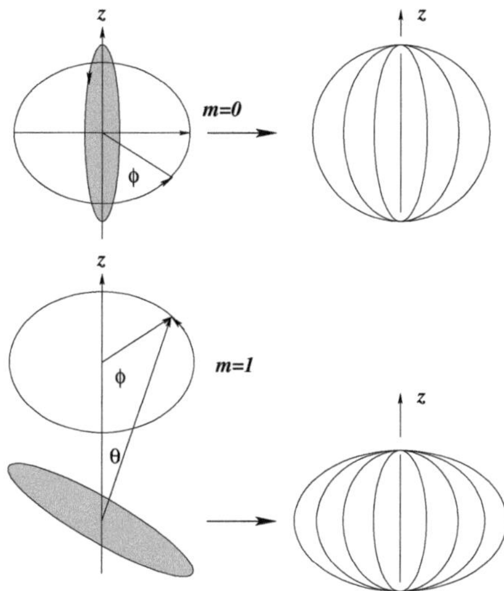

Figure 2.15: *Classical circular orbits of p-electrons with angular momentum $\hbar m$ in the direction of z. The orbit (shown shaded) rotates with the perpendicular angular momentum vector to sweep out the surfaces shown on the right.*

2.6.3 Hybrid Orbitals

The remarkable accord between the postulates of van't Hoff and Sommerfeld's elliptic orbits must, no doubt have convinced many sceptics of a more fundamental basis of both phenomena to be found in atomic shape. The new quantum theory that developed in the late 1920's seemed to define such a basis in terms of the magnetic quantum number m_l.

A polar plot of the squared function $\Phi^2 = N\cos^2\theta$, for $m_l = 0$ and θ the angle between the radius vector and the polar (Z) axis, is interpreted to show that a boundary surface which encloses most of the electron density described by Φ^2 is elongated in the Z-direction as shown schematically in Fig. 2.16. The enclosed region is called a $2p_z$ orbital.

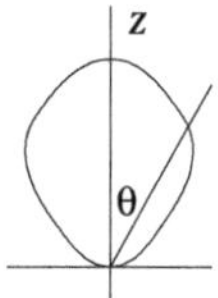

Figure 2.16: *Polar plot of Φ^2 for $m_l =$* 0

Furthermore, linear combinations of $\Phi_1 \pm \Phi_{-1}$ of the complex pair of eigenfunctions $m_l = \pm 1$ define new functions with the same shape as p_z, but directed along the X and Y axes, respectively, and dubbed p_x and p_y orbitals. This new set of p-functions are claimed to be equivalent to the original set.

In contrast to the four tetrahedrally oriented elliptic orbits of the Sommerfeld model, the new theory leads to only three, mutually orthogonal orbitals, at variance with the known structure of methane. A further new theory that developed to overcome this problem is known as the theory of orbital hybridization. In order to simulate the carbon atom's basicity of four an additional orbital is clearly required. The only possible candidate is the $2s$ orbital, but because it lies at a much lower energy and has no angular momentum to match, it cannot possibly mix with the p-eigenfunctions on an equal footing. The precise manoeuvre to overcome this dilemma is never fully disclosed and appears to rely on the process of chemical resonance, invented by Pauling to address this, and other, problems. With resonance, it is assumed that, linear combinations of an s and three p eigenfunctions produce a set of hybrid orbitals with the required tetrahedral properties.

This contrived solution of the problem was gratefully embraced by the chemical community. It has never been challenged successfully for almost eighty years, despite many glaring defects.

The first problem is that the so-called p_x and p_y orbitals are not equivalent to p_z – they are identical to it, in a rotated state. The procedure $Y_1^1 + Y_1^{-1} \rightarrow x/r$, is equivalent to a 90° rotation of the axes, about y:

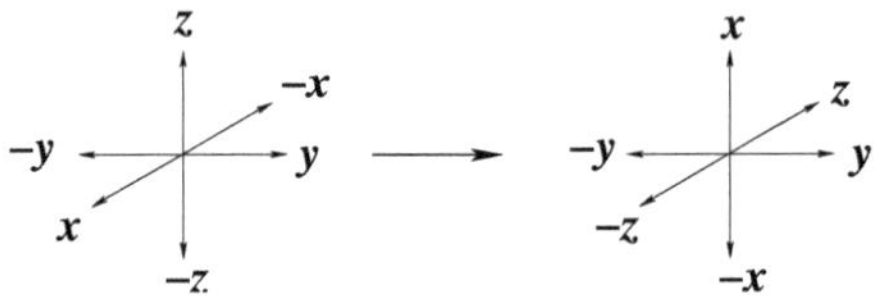

which further implies:

$$Y_1^1 \propto \frac{x+iy}{r} \rightarrow \frac{-z+iy}{r} \quad ; \quad Y_1^0 \propto \frac{z}{r} \rightarrow \frac{x}{r} \quad ; \quad Y_1^{-1} \propto \frac{x-iy}{r} \rightarrow \frac{-z-iy}{r}$$

In fact, it can be shown that any linear combination of p eigenfunctions is equivalent to rotation of the axes through some angle and that any rotation of the axes can be expressed as a linear combination of the eigenfunctions. More importantly, as p_z is rotated into p_x, its original meaning disappears, and a new complex pair of eigenfunctions appear, orthogonal to p_x.

The second problem is equally devastating. With the original choice of axes p_z is real and has $m_l = 0$. After rotation, p_x becomes real and assumes $m_l = 0$. The same with p_y. The upshot is three p functions with the same set of quantum numbers n, l and m_l, which is forbidden by the exclusion principle.

The mixing of s and p functions introduces another anomaly. In order to prepare the atom for hybridization the s electron is assumed to be promoted to the p level. This process requires extra angular momentum to raise l from 0 to 1 and is therefore forbidden by the law of angular momentum conservation.

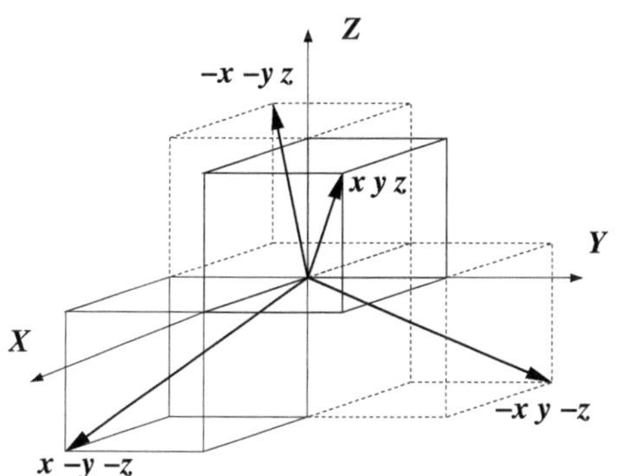

Figure 2.17: *Body diagonals of a cubic lattice in tetrahedral array.*

At the end of the day there remains the question of why the so-called sp^3 hybridization appears to work so well. The answer is immediately obvious on inspection of the four required combinations of p_x, p_y and p_z to shape the tetrahedral carbon atom. The vectorial addition $\vec{x}+\vec{y}+\vec{z}$ generates a vector from the origin, through the centre of a unit cube, along a body diagonal. To define a tetrahedron three more vector sums are required as shown in Figure 2.17. The proposed linear combination of orbitals has exactly this same form, except for the irrelevant addition of the s orbital. It is clear that hybridization is not a quantum-mechanical operation and amounts to nothing more than the geometrical definition of a tetrahedron. Final definition of the tetrahedral carbon atom is supposed to be achieved by linear combination of the four tetrahedral orbitals. Inspection of the diagram shows that this operation merely restores the original polar direction along z. The quantum-mechanical tetrahedral atom is a mirage.

Not only is hybridization an artificial simulation without scientific foundation, but even the assumed "orbital shapes" that it relies upon, are gross distortions of actual electron density distributions. The density plot shown above, like all textbook caricatures of atomic orbitals, is a misrepresentation of the spherical surface harmonics that describe normal excitation modes of atomic charge distributions. These functions are defined in the surface of the charge-density function, as in Fig. 2.13, and not at $r = 0$, as shown in Figure 2.16.

All free atoms are spherically symmetrical and deviations occur in polarizing fields, which may be of electromagnetic or ligand type. The spherical symmetry results from quenching of orbital angular momentum vectors, as first pointed out by Sommerfeld. To meet this requirement it is necessary that whenever an odd number of electrons occur on a degenerate sublevel one of these should have $m_l = 0$ and all others should occur as antisymmetric pairs with $\pm m_l \neq 0$. This requirement has been shown to generate Hund's rule through the exclusion principle – see section 4.7.4.

Closely associated with hybridization is the whole idea of covalent bonding, the agent which is assumed to link the atoms in a molecule. The exaggerated directional properties assumed for pure and hybrid orbitals alike, feature prominently in the theory of chemical bonding as a function of overlap. Like hybridization, this topic is more than a theory, having become the central dogma of chemistry.

Counter Arguments

In response to a previously published analysis of orbital hybridization [17] I was castigated by an international group of theoretical chemists for not understanding the subtleties of modern quantum chemistry. Their rebuttal of my views is summarized by three statements, from an extensive e-mail message forwarded to me by their spokesman [19]:

1. Orbital hybridization is not nowadays viewed as an intrinsic aspect of chemical bonding. That view, if it was ever held, is by now long out of date.

2. The Pauli exclusion principle is a simplified exposition, intended for chemists with no understanding of quantum mechanics, and applied to a particular system, the many-electron atom. Like all simplified explanations it should not be taken too literally.

3. p_x, p_y and p_z are linearly independent functions and the quantum numbers are not needed.

The savants clearly reject the arguments on which my conclusions were based and, in answer, I see no point in repeating this material. Instead I refer to an independent authority [20] to identify some concepts that are used in modern quantum chemistry:

Two examples of polyatomic calculations, on H_2O and NH_3, are outlined and explained in detail. In both cases the analysis starts from an assumed molecular structure of known symmetry. The transformation properties of the *atomic orbitals* on each atomic centre, under the symmetry operations of the group, are examined next. The atomic orbitals are defined as $1s$, $2s$, $2p_x$, $2p_y$ and $2p_z$. Nothing can be more explicit – these are the occupied atomic orbitals of a 'many-electron' atom. This configuration violates the exclusion principle[9]. Although the quantum numbers may not be needed,

[9]The extended formulation of the exclusion principle embraces Pauli's original statement, without invalidating it in any way

by what magic have they disappeared? They have disappeared because this configuration has no quantum-mechanical meaning and is based on a preconceived classical structure, which is enforced by the assumed symmetry group. If the *p*-functions are to be used merely as linearly independent, they should surely be part of some complete orthonormal set. However, only one of them can feature in a set of orthogonal spherical harmonics. Constructing a molecular-orbital scheme in terms of the symmetry species:

$$(1a_1)^2(2a_1)^2(1b_2)^2(3a_1)^2(1b_1)^2$$

does not change the situation in any way. Each of these simply represents one of the atomic orbitals and the final construct, not suprisingly is identical to the 'outdated' concept of sp^3 hybridization. See also section 6.4.3.

2.6.4 Atomic Structure

The internal electronic structure of atoms, as revealed by their atomic spectra and techniques such as X-ray and ultra-violet photoelectron spectroscopy, corresponds remarkably well with the shell structure foreshadowed by the hydrogen model. Simulation of these electronic shell structures was pioneered by Hartree [21], based on the principle of a self-consistent field (SCF). Many refinements and improvements of the method have been implemented over the years, but the central concept remains the same. The simplest application of the method, to the helium atom, illustrates the procedure well. One of the $1s^2$ electrons is assumed to move in the central field of the other electron and the nucleus. This field is approximated by the charge distribution of the hydrogen ground state and the eigenfunction of lowest energy for the electron moving in this field is then found by numerical integration.

The calculated eigenfunction, which differs from the $1s$ hydrogen wave function, must give a better description of an electron on the He atom and it is therefore used to recalculate the central potential field experienced by a second electron. The integration is repeated, using the updated field and an updated wave function is calculated. As this iteration is continued, the wave functions calculated in successive steps converge to a common form, which defines the SCF. The volume of calculation increases rapidly as this treatment is extended to larger and larger atoms and the independent-electron assumption becomes worse. Rather than start the calculation for each given atom from hydrogen wave functions, a lot of effort is saved by using the wave functions of the atom with one proton less, calculated before by the self-consistent procedure.

To take account of inter-electronic interactions the all-electron Hamiltonian (in atomic units) is formulated as

$$H = -\frac{1}{2}\sum_i \nabla_i^2 - \sum_i \frac{Z}{r_i} + \sum_{i>j} \frac{1}{r_{ij}}$$

The first term represents the sum of all electronic kinetic energies. In the second term, which represents the potential energy in the field of the nucleus, Z is the total nuclear charge and r_i is the distance between electron i and the nucleus. The third term, which represents the Coulombic repulsion between pairs of electrons, is the most difficult to calculate. A simplification that works well is to consider the charge cloud as a free-electron gas and to calculate the average interaction statistically. Software to computerize the necessary calculations is freely available. Numerical integration is carried out to some fixed distance from the nucleus and the calculation is extended to $r \to \infty$ by splicing in an analytic exponential function. It is of interest to note that no distinction is made between atoms with closed and open shells, on the assumption that all atoms are spherical. Based on quenching of angular momentum vectors this is a good assumption.

2.6.5 Compressed Atoms

The electronic configuration of free atoms is an important factor in the interpretation of atomic spectra, but less so for the understanding of chemical behaviour. Chemistry happens in crowded environments, which means that atomic electron densities fades to zero far from infinity. SCF wave functions are therefore not appropriate for atoms in a chemical environment. More suitable wave functions are obtained by terminating the SCF calculations at some fixed distance ρ from the nucleus, rather than infinity. The effect of such a new boundary condition is like applying hydrostatic pressure to the atom.

Actual calculations of compressed-atom densities, performed with suitably modified SCF software, show that the increased pressure raises all electronic energy levels, at different rates that depend on the shell structure. The effect is more pronounced on those levels of highest effective quantum number l and it is not uncommon for levels of different l to cross during compression. The interpretation of photoelectron spectra in terms of free-atom electron configurations may therefore be misleading in the study of surface chemistry and catalytic effects, for which they are routinely used.

Continued compression of any atom, simulated by decreasing ρ, eventually reaches a point where an electron crosses the ionization level and becomes effectively decoupled from interaction with the atomic core. The compression radius at which this happens is known as the ionization radius r_0. Individual values of r_0 are remarkably periodic and they faithfully reflect the trends predicted by the periodic table of the elements. These effects can be anticipated to be of decisive importance for the understanding of chemical reactivity.

2.7 Chemical Bonding

2.7.1 Classical Theory

The observation that chemical combination involves elements in rational proportions, supported the idea that molecules consisted of atoms, held together by intramolecular forces. Efforts to quantify such forces in terms of simple electrostatic interaction have been unproductive, giving rise to the consensus that intramolecular cohesion was a quantum effect. However, because of the complexity of molecular systems, the standard methods of quantum mechanics, restricted, as they are, to one-dimensional one-particle problems, cannot be applied directly to the problem of chemical bonding. The only visible progress beyond one-electron interactions was made possible by *ad hoc* combination of quantum arguments and the empirical classification schemes of 19th century organic chemistry. An important aspect of the schemes was to assign valencies of one, two, three and four to the four common elements of organic chemistry, hydrogen, oxygen, nitrogen and carbon, respectively. Together with the definition of monovalent radicals, or submolecular groups of atoms left unattacked during certain decompositions, it became feasible to classify all compounds by type, in terms of constitutional formulae, using connecting lines or dots between constituent radicals.

Constitutional formulae were designed ″on paper″, primarily to be ″in harmony″ with known chemical properties and without pretension to ″represent the symmetrical or spatial arrangement of the atom in a compound″ [22]. Not only was this stipulation gradually relaxed to represent three-dimensional structures, but the connecting lines were also soon after assumed to represent definite electronic links between atoms. This assumption opened the door for the introduction of semi-empirical quantum-mechanical characterization of chemical bonds. It is important to realize that chemical bonds have never been observed in any experiment and that they only exist as conjectures to interpret primitive molecular graphs. Their value as heuristic aids in the study of chemical change and composition is beyond dispute, but as a basis for the theoretical understanding of chemical cohesion they are of little value.

The observation [22] that several carbon atoms may be united by means of one, two or three valencies: C–C, C=C, C≡C, has also been incorporated, at face value, into electronic theories of chemical bonding and provided with a quantum gloss. In many cases, actual relationships within molecules are too complicated to be represented by simple graphs, and the supposed quantum effects assumed to be at work in these situations led to the invention of pseudo-scientific concepts such as hybridization and resonance. This patchwork is still featured as the quantum theory of chemical bonding.

2.7.2 Quantum Theory

The quantum content of current theories of chemical cohesion is, in reality, close to nil. The conceptual model of covalent bonding still amounts to one or more pairs of electrons, situated between two atomic nuclei, with paired spins, and confined to the region in which hybrid orbitals of the two atoms overlap. The bond strength depends on the degree of overlap. This model is simply a paraphrase of the 19th century concept of atomic valencies, with the incorporation of the electron-pair conjectures of Lewis and Langmuir. Hybrid orbitals came to be introduced to substitute for spatially oriented elliptic orbits, but in fact, these one-electron orbits are spin-free. The orbitals are next interpreted as if they were atomic wave functions with non-radial nodes at the nuclear position. Both assumptions are misleading.

There is no quantum-mechanical evidence for the localization of electron pairs between atomic nuclei, and atomic orbitals, in so far as they correspond to spherical surface harmonics, have their nodal curves in the surface of the density sphere. Sets of real hybrid orbitals are physically undefined. To understand intramolecular interactions as a quantum phenomenon it is necessary to approach the problem with the minimum of assumptions and to state all essential assumptions clearly and precisely at the outset.

The only model available for direct quantum-mechanical study of interatomic interaction is the hydrogen molecular ion H_2^+. If the two protons are considered clamped in position at a fixed distance apart, the single electron is represented by a Schrödinger equation, which can be separated in confocal elliptic coordinates. On varying the interproton distance for a series of calculations a complete mapping of the interaction for all possible configurations is presumably achieved. This is not the case. Despite its reasonable appearance the model is by no means unbiased.

The clamped-nuclei model is only valid under the assumption that the two protons behave classically and the electron quantum-mechanically. The rationale to justify this assumption is based on the difference in mass between proton and electron. For all practical purposes the protons should therefore move so sluggishly compared to the electron that their motion can simply be ignored when focussing on the electron. Under these circumstances the only relative proton motion amounts to classical vibration about an equilibrium internuclear separation. However, the inverse argument is equally valid. Because of the mass difference the tiny electron presents only a minor perturbation to the relative motion of the more massive pair of protons. Protons, like neutrons and electrons, also exhibit wave-like behaviour in diffraction experiments and carry the same quantum-mechanical component of angular momentum, called spin. To consider the relative motion of two protons as a classical problem therefore is a gross and unwarranted simplification. The quantum-mechanical two-proton system must, to some extent, exhibit quantized behaviour like an electron-proton pair. This mutual quantum interference implies quantized energy and angular momentum, as well as spin-orbit coupling.

2.7.3 Critique of the Model

The mere fact that these effects cannot be simulated by one-particle models does not discount their existence. These remarks should be read as a caveat not to interpret the well-known H_2^+ model and calculation as final, despite the excellent agreement with spectroscopic observation which, incidentally, relies on the same assumption of a vibrating diatomic molecule. It is not suggested that the calculated results be rejected, but to be used with the necessary caution. As a free molecule, H_2^+ is almost certainly of spherical symmetry, and the same probably applies to all diatomic molecules.

The only way in which such molecules can be demonstrated to occur as linear vibrating pairs of atoms, is by confinement as guest species in crystals. Even this situation is contingent on directed interaction with the host lattice, in the absence of which the guest appears structureless, or disordered. The general conclusion must be that protons, like electrons, appear as point particles only in close confinement. Protons and neutrons must, like electrons, logically be considered as distortions of the aether; as compressible and flexible fluids. Despite differences in mass and topological structure these different particles must therefore all have quantum-mechanical properties. In observation they would display the type of behaviour that seems to imply a dual wave-particle structure.

On returning to the H_2^+ one-electron calculation, the assumed linear vibrating structure, shown in Figure 2.18, has to be abandoned in favour of a quantized nuclear framework, and the calculated cylindrically symmetrical structure as suggested by contour maps of electron density should be rotated about all Eulerian angles to reveal the full spherical symmetry of the molecule in terms of both electron and proton charge densities[10].

Given such a structure there is no justification for assuming that the H_2^+ calculation may serve as a prototype for modelling covalent bonds by orbital overlap for electron pairs localized between point nuclei. To rationalize the assumption it is argued that the exact results with respect to equilibrium internuclear distance ($r_e = 2a_0$) and binding energy ($D_e = 2.79$eV) are adequately reproduced on simulating the interaction by the overlap of scaled ground-state hydrogen wave functions. The best result obtained by this technique yields $r_e = 2.0a_0$, $D_e = 2.36$eV. Using unscaled $1s$ functions, as used for H_2 and other diatomics, these values are $r_e = 2.49a_0$, $D_e = 1.76$eV. The predicted bond length is too long by a factor 1.25 and only 60% of the dissociation energy is accounted for. These results are not complicated by interelectronic effects that occur in all other molecules, which means that the vague qualitative resemblence between calculated and observed potential-energy curves constitutes anything but an adequate benchmark for the analysis of more complex bonding problems.

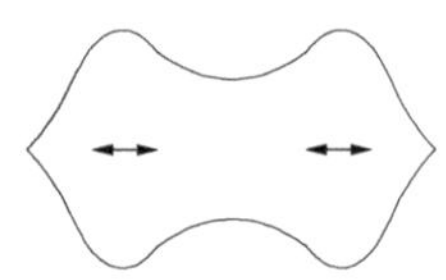

Figure 2.18: *Simplified one-dimensional working model of the H_2^+ molecule, assumed to eliminate three-dimensional computational complexities.*

It is surprising that a reasonable molecular wave function synthesized from two well-defined atomic wave functions should fail so comprehensively to account for the molecular properties. It shows that the atoms involved in formation of the molecule are not in their ground states. One way of improving the situation is in fact by using a linear combination of $1s$ and $2p_z$ functions to synthesize the molecular wave function. However, this procedure has no valid basis and cannot produce the final answer. The real reason for the failure is that atomic functions refer specifically to free atoms only. At

[10]Provocative experimental evidence, at variance with conventional theory, is provided by the estimates of molecular diameters for diatomic molecules. Bonding theory requires the concentration of valence densities between the nuclei to increase as a function of bond order, in agreement with observed bond lengths (1.097, 1.208, 0.741 Å) and force constants (22.95, 11.77, 5.75 $\mathrm{Ncm^{-1}}$) of the species N≡N, O=O and H–H respectively. Molecular diameters can be measured by a variety of techniques based on gas viscosity, heat conductivity, diffusion and van der Waals equation of state. The results are in excellent agreement at values of 3.75, 3.61 and 2.72 Å, for N_2, O_2 and H_2, respectively. Conventional bonding theory cannot account for these results.

the stage where the two atoms combine to form a molecule, they are far removed from their free state; in an excited state, conditioned by a potential field of vastly different symmetry.

Not even the SCF procedure can overcome this problem. In the case of atoms, the central field remains a valid and good approximation. Assuming a rigid linear structure in the molecular case is clearly not good enough, although it contains an element of truth. This inherent problem plagues all LCAO-SCF calculations to an even more serious extent.

The theory of chemical bonding is overwhelmed by a host of insurmountable obstacles: the real orbitals and hybrids of LCAO have no physical, chemical or mathematically useful attributes – certainly not in the quantum-mechanical sense; the distribution of electron density between atoms, in the form of spin pairs, is an overinterpretation of the empirical rules devised to catalogue chemical species; the structures, assumed in order to generate free-molecule potential fields, are only known from solid-state diffraction experiments; the assumption of directed bonds is a leap of faith, not even supported by crystal-structure analysis. The list is not complete.

Intramolecular cohesion evidently is a holistic quantum-mechanical property, which cannot be reduced to pairwise interactions. Molecules are formed during the interaction of atoms and/or radicals in their valence states, to be understood in terms of valence-state wave functions of appropriate symmetry.

Intramolecular rearrangements provide dramatic evidence to the inadequacy of the theories of directed valence bonds. The Beckmann rearrangement of ketoximes into acid amides, represented by the scheme:

$$\mathrm{R'(R)C{=}N{-}OH} \longrightarrow \mathrm{R'(HO)C{=}N{-}R} \longrightarrow \mathrm{R'C({=}O){-}NHR}$$

consists of the trans interchange of a hydroxyl group with an aryl or alkyl group, also with retention of chirality [23]. No conceivable mechanism, based on the conventional view of covalent bonding, can possibly account for these observations, without breaking or making bonds. The accepted mechanism [24] simply states that

> The migrating group does not become free but always remains connected in some way to the substrate.

This non-explanation simply refutes the model that it seeks to defend.

Chapter 3

The Quantum Quandary

3.1 The Classical Background

Apart from a few 'minor' unsolved problems the science of physics had reached such maturity towards the end of the 19th century that leading physicists could claim with confidence that no major new developments could be foreseen. Even Max Planck, who inaugurated one such development, is purported to have dissuaded young scientists from following careers in physics. This assessment, although misguided in retrospect, speaks for total accord among those scientists on the understanding of their science, without any need for, or arguments about alternative interpretations. Such agreement can only arise from absolute certainty on the basic premises of the subject. The cornerstone of 19th century physics, Newtonian mechanics as developed in the hands of Lagrange, Hamilton, Jacobi and others, was universally accepted and understood.

The situation, one hundred years on, could hardly be more different. The interpretation of quantum mechanics, which came to replace the Newtonian system, is as hotly disputed as ever and the common ground with the theory of relativity remains elusive and vague. The reason for the discord must lie somewhere in the transition from the classical to the new non-classical paradigm. What is proposed here, is to retrace the steps that led to the emergence of the new theoretical models, in an attempt to identify the point of conceptual bifurcation.

Theoretical quantum mechanics had its origin in two seminal papers, starting from apparently different points of view, and published independently and almost simultaneously by Heisenberg and Schrödinger, respectively. The major unsolved problem of physics, addressed by both, was to find a fundamental basis for the *ad hoc* quantum rules, formulated by Sommerfeld and others, to account for the frequencies, observed in the optical spectra of atomic hydrogen and other elements.

An obvious approach to the problem was to re-examine the theories of mechanics that failed to predict the observed frequencies and bring them into line with observation. It is therefore not surprising to find that both new theories were modified versions of classical Hamilton-Jacobi theory, adapted so as to incorporate the newly discovered quantum features of atomic spectra. To understand the type of reasoning that produced the quantum theory, it is necessary to make a short excursion into classical HJ theory. Some readers may prefer to skip this section on first reading. For an in-depth discussion more enthusiastic readers are referred to the authoritative text of Goldstein [11].

3.1.1 Hamilton-Jacobi Theory

Of fundamental significance in the development of this theory is Hamilton's principle of least action. It states that the action integral

$$S = \int_{t_0}^{t_1} L(q, \dot{q})\mathrm{d}t$$

has an extremum (*e.g.* minimum) value for any mechanical process.

Generalized Coordinates

The variables q and $\dot{q}$ represent generalized coordinates of particle positions and velocities. The advantage of generalized dynamic variables is their independence of coordinate system. This is of special importance when dealing with phenomena in which the motion of material particles is not observed directly, for instance in the study of electricity. Parameters, other than particle positions (*e.g.* currents) are observed here, although the behaviour of the system is really controlled by the motion of electrons which remains intrinsically concealed. The values assumed by the descriptive parameters, called generalized coordinates, are those connected with the position coordinates of the electrons. Another important application is in statistical mechanics.

It is assumed that the rectangular coordinates of a set of n particles may be defined in terms of $3n$ generalized position coordinates:

$$x_j = x_j(q_1, \ldots, q_n) \; ; \; y_j = y_j(q_{n+1}, \ldots, q_{2n}) \; ; \; z_j = z_j(q_{2n+1}, \ldots, q_{3n})$$

Similarly the velocity components of the jth particle will be of the form

$$\dot{x}_j = \sum_{k=1}^{n} \frac{\partial x_j}{\partial q_k}\dot{q}_k \quad , \quad etc. \tag{3.1}$$

The kinetic energy of the system will, by assumption, be the total energy of all particles, *viz.*

$$T = \frac{1}{2}\sum m_j(\dot{x}^2 + \dot{y}^2 + \dot{z}^2)$$

where the summation is over all particles in the system. Substituting from (3.1) this expression may be reduced to

$$T = \sum a_{jk}\dot{q}_j\dot{g}_k \tag{3.2}$$

with double summation over $(j, k = 1, n)$ and where the a_{jk} are functions of q.

Hamilton's Canonical Equations

The variable L is known as the Lagrangian function or *kinetic potential*, defined as

$$L = L(q_1, q_2, \ldots, q_{3n}\,;\, \dot{q}_1, \dot{q}_2, \ldots, \dot{q}_{3n}) = T - V\,,$$

the difference between the kinetic and potential energies for a mechanical system of n particles. Noting that T is a function of velocities only and V is a function of coordinates only, it follows that

$$\frac{\partial L}{\partial \dot{q}_j} = \frac{\partial T}{\partial \dot{q}_j} \quad \text{and} \quad \frac{\partial L}{\partial q_j} = -\frac{\partial V}{\partial q_j}$$

Newton's equations of motion

$$\frac{\mathrm{d}}{\mathrm{d}t}\left(\frac{\partial T}{\partial \dot{x}_j}\right) + \frac{\partial V}{\partial x_j} = 0 \quad (j = 1, \ldots, n)$$

with similar expressions for y and z, therefore take the form:

$$\frac{\mathrm{d}}{\mathrm{d}t}\left(\frac{\partial L}{\partial \dot{q}_j}\right) - \frac{\partial L}{\partial q_j} = 0\,, \quad (j = 1, \ldots, 3n) \tag{3.3}$$

The Lagrangian equations can be further simplified to contain only first derivatives by defining the *Hamiltonian* function,

$$H = \sum_{k=1}^{3n} p_k\dot{q}_k - L(q_k, \dot{q}_k)$$

Whereas V is a function only of coordinates, the momentum $p_k = m_k\dot{q}_k$, becomes

$$p_k = \frac{\partial T}{\partial \dot{q}_k} = \frac{\partial L}{\partial \dot{q}_k} \quad (k = 1, 2, \ldots, 3n) \tag{3.4}$$

From (3.3) follows that

$$\dot{p}_k = \frac{\mathrm{d}}{\mathrm{d}t}\left(\frac{\partial L}{\partial \dot{q}_k}\right) = \frac{\partial L}{\partial q_k} \quad (k = 1, 2, \ldots, 3n)$$

The derivatives of H are now obtained directly as

$$\begin{aligned} \frac{\partial H}{\partial p_k} &= \dot{q}_k \\ \frac{\partial H}{\partial q_k} &= -\frac{\partial L}{\partial q_k} = -\dot{p}_k \end{aligned}$$

giving the Hamiltonian or *canonical* form of the equations of motion. The pair (q_k, p_k) is said to constitute a *conjugate* pair.

From (3.4) and (3.3) it follows that:

$$p_k = \frac{\partial T}{\partial \dot{q}_k} = \sum_j a_{kj}\dot{q}_j + \sum_i a_{ik}\dot{q}_i$$

and hence that

$$\sum_k p_k\dot{q}_k = \sum_{k,j} a_{kj}\dot{q}_j\dot{q}_k + \sum_{i,k} a_{ik}\dot{q}_i\dot{q}_k = 2T$$

The Hamiltonian reduces to

$$H = \sum_k p_k\dot{q}_k - L = 2T - (T - V) = T + V = E\,,$$

the total energy.

Cyclic Coordinates

One situation in which solution of Hamilton's equations becomes trivial is when H is a constant of the motion and where the coordinates q_k do not appear in the Lagrangian. Such coordinates are said to be cyclic or ignorable. In this special case the Lagrangian equation of motion reduces to

$$\frac{\mathrm{d}}{\mathrm{d}t}\frac{\partial L}{\partial \dot{q}_k} = 0 \quad \text{or} \quad \frac{\mathrm{d}p_k}{\mathrm{d}t} = 0\,,$$

which means that the p_k are equal to the integration constants α_k. As shown before $\dot{p}_k = \partial L/\partial q_k = -\partial H/\partial q_k$. This means that cyclic coordinates are also absent from the Hamiltonian, which becomes $H = H(\alpha_j)$. The Hamiltonian equation for $\dot{q}_j$ reduces to

$$\dot{q}_j = \frac{\partial H}{\partial \alpha_j} = \omega_j \tag{3.5}$$

where the ω_j are functions of α_j only and therefore constant in time, so that (3.5) integrates to $q_j = \omega_j t + \beta_j$. The variables ω_j are recognized as frequencies, explaining the term cyclic coordinate.

Despite its simplicity this demonstration may be considered of little importance as very few mechanical systems have only cyclic coordinates. What is more important is the fact that any system can be described by more than one set of generalized coordinates, with the possibility of transforming non-cyclic into cyclic coordinates.

A simple example is provided by the coordinates that may be used to describe the motion of a particle in a plane. As generalized coordinates one can use either Cartesian coordinates: $q_1 = x$, $q_2 = y$, or plane polar coordinates: $q_1 = r$, $q_2 = \varphi$, as in Figure 2.6. The two sets are equally valid and the coordinates of choice becomes a matter of convenience. In the presence of a central force the coordinate φ becomes cyclic, while both x and y are non-cyclic. The number of cyclic coordinates thus depends on the choice of generalized coordinates, and for each problem there may be one particular choice for which all coordinates are cyclic.

Conversion from the set of generalized coordinates q_j to another, presumably more suitable, set Q_j is described by the transformation equations

$$Q_j = Q_j(q, t) \tag{3.6}$$

In the Hamiltonian formulation generalized momenta are also independent variables at the same level as the generalized coordinates and should also feature in the transformation, which, more appropriately, should be formulated as

$$\begin{aligned} Q_j &= Q_j(q, p, t) \\ P_j &= P_j(q, p, t) \end{aligned} \tag{3.7}$$

Canonical Transformation

Equations (3.6) and (3.7) are said to define a transformation in *configuration* and in *phase space*, respectively.

The only acceptable transformations are those that produce a new canonical pair Q_j and P_j of generalized coordinates and momenta. This requirement is satisfied if there exists some function, *e.g.* $K(Q, P, t)$ such that the transformed equations of motion are of the Hamiltonian form:

$$\dot{Q}_j = \frac{\partial K}{\partial P_j} \quad , \quad \dot{P}_j = -\frac{\partial K}{\partial Q_j}$$

and K represents the transformed Hamiltonian.

Transformation to all-cyclic coordinates is not always feasible and another alternative is to consider the transformation from initial conditions (q_0, p_0) to the coordinates (q, p) at time t. The transformation equations:

$$q = q(q_0, p_0, t) \quad , \quad p = p(q_0, p_0, t)$$

solve the mechanical problem directly for they give the coordinates and momenta as a function of the initial values and the time. The same equations are used for the forward and inverse transformations and to ensure that the constant starting parameters are described correctly, *i.e.* $\dot{Q}_j = \dot{P}_j = 0$, for the inverse transformation, it is necessary that $K \equiv 0$.

The transformed Hamiltonian, defined by the generating function F,

$$K = H + \frac{\partial F}{\partial t},$$

for zero K satisfies the equation

$$H(p, g, t) + \frac{\partial F}{\partial t} = 0 \tag{3.8}$$

By defining $F = F(q, P, t)$, with total derivative

$$\frac{\mathrm{d}F}{\mathrm{d}t} = \frac{\partial F}{\partial q}\dot{q} + \frac{\partial F}{\partial P}\dot{P} + \frac{\partial F}{\partial t}$$

equation (3.8) rearranges into

$$p\dot{q} - L + \frac{\mathrm{d}F}{\mathrm{d}t} - \frac{\partial F}{\partial q}\dot{q} - \frac{\partial F}{\partial P}\dot{P} = 0$$

In this equation the coefficients of unrelated variables must vanish independently, and given that $\dot{P} = 0$, this condition leads to:

$$p = \frac{\partial F}{\partial q} \text{ and } \frac{\mathrm{d}F}{\mathrm{d}t} = L \tag{3.9}$$

The final form of (3.8):

$$H\left(q, \frac{\partial F}{\partial q}, t\right) + \frac{\partial F}{\partial t} = 0 \tag{3.10}$$

is known as the *Hamilton-Jacobi equation*, with solutions S, known as *Hamilton's principal function.* Integration of the Lagrangian

$$\int_{t_0}^{t_1} L\mathrm{d}t = F$$

defines the solutions of (3.10) as the action function, S, defined before. Writing $H = p^2/2m + V$, leads to the common form of the HJ equation:

$$\frac{1}{2m}\left(\frac{\partial S}{\partial q}\right)^2 + V + \frac{\partial S}{\partial t} = 0 \tag{3.11}$$

From the complete time derivative of S:

$$\frac{\mathrm{d}S}{\mathrm{d}t} = \frac{\partial S}{\partial t} + \frac{\partial S}{\partial q}\frac{\partial q}{\partial t} = L$$

it follows that

$$\frac{\partial S}{\partial t} = L - p\dot{q} = -H\,,$$

in integrated form: $S = -Et + W$, where the integration constant $W(p,q)$ is known as Hamilton's *characteristic* function.

Hamilton's Characteristic Function

Substituting the form

$$S(q_j, \alpha_j, t) = W(q_j, \alpha_j) - \alpha_1 t$$

into the HJ equation for S, when H is not a function of time:

$$\frac{\partial S}{\partial t} + H\left(q_j, \frac{\partial S}{\partial \dot{q}}\right) = 0$$

reduces the equation to

$$H\left(q_j, \frac{\partial W}{\partial \dot{q}}\right) = \alpha_1 \tag{3.12}$$

which no longer involves the time. The integration constant α_1 is equal to the constant value (E) of H.

The function W can also be shown to be a generating function $W(q, P)$ of a canonical transformation in its own right, in which the new momenta are all constants of the motion α_j, and $\alpha_1 = H$. The equations of transformation are $p_j = \partial W/\partial q_j$, $Q_j = \partial W/\partial \dot{P}_j = \partial W/\partial \alpha_j$, hence $H(q_j, \partial W/\partial q_j) = \alpha_1$, which is identical to (3.12). The characteristic function is seen to generate a canonical transformation in which all momenta are conserved and hence all new coordinates are cyclic.

Substitution of $H = -\partial S/\partial t$, from (3.10) into (3.11) gives another form of the HJ equation:

$$\left(\frac{\partial S}{\partial q}\right)^2 = (\nabla W)^2 = 2m(E - V) \tag{3.13}$$

in terms of the characteristic function. At any instant the equation $S(q,t)=$ constant, defines a surface of constant W at a fixed position in configuration space.

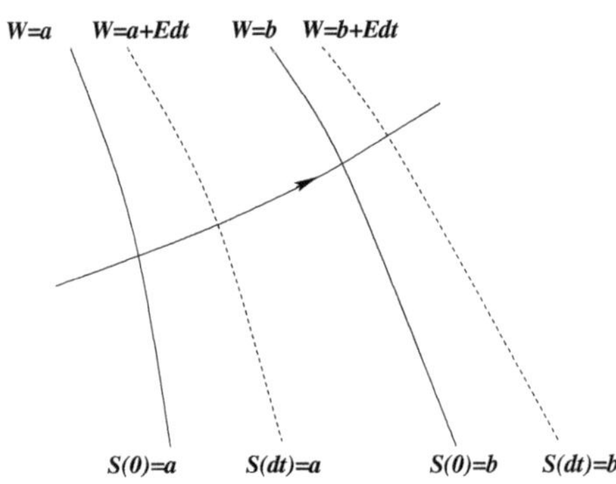

At $t = 0$ the surfaces $S = a, b$ coincide with $W = a, b$ respectively. However, at time dt the surfaces $S = a, b$ now coincide with surfaces for which $W = (a, b) + E\mathrm{d}t$. The surface $S = a$ has therefore moved from $W = a$ to $W = a + E\mathrm{d}t$, *i.e.* $\mathrm{d}W = E\mathrm{d}t$. To emphasize the parallel between Sommerfeld's quantization rules and the HJ equation, the latter is reformulated [25] as a differential field equation of the action potential, W, in the same way that fluid motion is described by a velocity potential, or the propagation of a wave front. The surfaces of constant S may thus be considered as wave fronts propagating in configuration space. Let s measure the distance normal to the moving surface. Writing $\mathrm{d}W = |\nabla W|\mathrm{d}s$, gives the velocity of the wave front

$$u = \frac{\mathrm{d}s}{\mathrm{d}t} = \frac{1}{|\nabla W|}\frac{\mathrm{d}W}{\mathrm{d}t} = \frac{E}{|\nabla W|}$$

From (3.13) follows $u = E/\sqrt{2(E-V)}$, the result that was used in Schrödinger's first formulation of wave mechanics.

The unmistakable similarity between the eikonal equation of geometrical optics:

$$(\nabla\phi)^2 = n^2$$

for index of refraction n, and the HJ equation (3.13) is interpreted to indicate that geometrical optics is the classical simplication of wave optics in the small wavelength limit, in the same way that classical mechanics relates to wave mechanics. The rays which in geometrical optics are orthogonal to the wave fronts, correspond to particle trajectories orthogonal to the surfaces of constant S. This argument is central to the Bohmian interpretation of quantum phenomena.

The characteristic function W plays the same role as the eikonal ϕ and $\sqrt{2m(E-V)}$ serves as the index of refraction. It becomes clear why Huygens' wave theory and Newton's formulation were able to account equally well for the phenomena of reflection and refraction.

Phenomena that depend on wavelength, such as interference and diffraction, cannot be explained in geometrical optics, just as quantum phenomena cannot be explained by classical mechanics. There is a duality of particle and wave even in classical mechanics, but the particle aspect is so dominant that the wave aspect can easily be ignored. The wave aspect is obviously not ignored in physical (wave) optics. By the same argument it would seem appropriate that wave aspects should supersede the particle concept in quantum theory. This has not been generally accepted and it will be argued here to have produced a distorted view of the nature of the electron, with a lasting negative effect on the development of theoretical chemistry. It still remains to show how Heisenberg mechanics relates to Hamilton-Jacobi theory, leading into the mechanics of periodic systems.

3.1.2 Periodic Systems

In the theory of atomic structure and spectra systems with periodic motion are of obvious importance. In most cases it is not the details of the orbit, but the frequencies of the motion that are important. This aspect is conveniently handled by a variation of the HJ procedure that considers suitably defined constants J_j, which form a set of n independent functions of the integration constants, α_j.

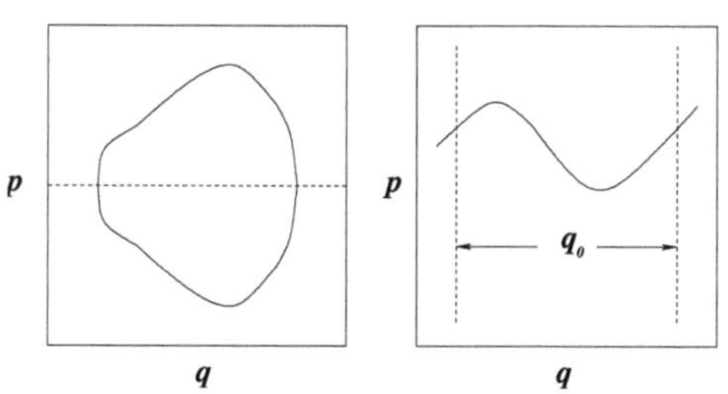

The procedure is demonstrated by a system with one degree of freedom, assumed to be conservative, with Hamiltonian $H(q,p) = \alpha_1$. Solving for the momentum, the relation $p = p(q, \alpha_1)$ defines the equation of the orbit traced out by the system point in two-dimensional phase space for constant $H = \alpha_1$. The graphical details for two types of periodic motion are shown in the diagram.

The closed orbit on the left represents a system in vibration (also called libration) and the open orbit on the right corresponds to rotation. Vibrational motion occurs when the initial position lies between two zeros of the kinetic energy. Both p and q are periodic functions of time and the system point retraces its steps periodically. In the second type of periodic motion p is a periodic function of q, with period q_0. Alternatively, when q increases by q_0, the configuration of the system remains unchanged. The most familiar example is a rigid body that rotates about an axis, with q as angle of rotation. Increasing q by 2π produces no essential change in the state of the system. The values of q are not bounded and can increase indefinitely.

Either type of periodic motion can be described by the variable J, designed to replace α_1 as the constant (transformed) momentum. This action variable is defined as

$$J = \oint p\mathrm{d}q \tag{3.14}$$

where the integration is to be carried out over a complete period of libration or rotation. By definition, J, which always has the dimensions of angular momentum, is some function of α_1: $\alpha_1 \equiv H = H(J)$. Hence Hamilton's characteristic function can be written as $W = W(q, J)$. The generalized coordinate conjugate to J, known as the angle variable w, is defined by the transformation equation

$$w = \frac{\partial W}{\partial J} \tag{3.15}$$

Correspondingly, the equation of motion for w is

$$\dot{w} = \frac{\partial H(J)}{\partial J} = \nu(J) \tag{3.16}$$

where ν is a constant function of J alone.

Equation (3.16) has the solution, linear with time:

$$w = \nu t + \beta \tag{3.17}$$

As q goes through a complete cycle of libration or rotation

$$\Delta w = \oint \frac{\partial w}{\partial q}\mathrm{d}q$$

which by (3.15) can be written as

$$\Delta w = \oint \frac{\partial^2 W}{\partial q \partial J}\mathrm{d}q \quad .$$

Because J is a constant, the derivative with respect to J can be taken outside the integral sign:

$$\Delta w = \frac{\mathrm{d}}{\mathrm{d}J}\oint \frac{\partial W}{\partial q}\mathrm{d}q = \frac{\mathrm{d}}{\mathrm{d}J}\oint p\mathrm{d}q = 1 \quad ,$$

using (3.14). The equation states that w changes by unity as q goes through a complete period, but according to (3.17), if τ is the period of a complete cycle of q, then $\Delta w = 1 = \nu\tau$ and hence the constant ν is the reciprocal of the period, $\nu = 1/\tau$ and thus is the frequency associated with the periodic motion of q. It is therefore possible to obtain the frequency of periodic motion without solving the complete equations of motion. As noted, J has the dimensions of an angular momentum and the coordinate conjugate to angular momentum is angle, w, as assumed.

The most important problem that has been solved in detail by using action-angle variables is the Kepler problem of planetary motion. The details of the analysis are not important in the present context, but the form of the Hamiltonian for rotation in a central potential $V(r) = -k/r$, obtained as

$$H = -\frac{2\pi^2 mk^2}{J_3^2} \tag{3.18}$$

played a significant part in the development of the Sommerfeld atomic model.

The Kepler model was ceased upon by Sommerfeld to account for the quantized orbits and energies of the Bohr atomic model. By replacing the continuous range of classical action variables, restricting them to discrete values of

$$J - \oint \mathrm{pd}q = nh \quad ,$$

according to Bohr's conjecture, Sommerfeld could show that the allowed electronic levels in the hydrogen atom are correctly predicted by (3.18) on setting $k = Ze^2$ and $J_3 = nh$:

$$E = -\frac{2\pi^2 mZ^2 e^4}{n^2 h^2}$$

This success set the scene for Sommerfeld's "royal road to quantization", that only required solution of the classical problem using action-angle variables and replacing the J-variables with integral multiples of h.

Multiply-periodic Systems

Action-angle variables can also be introduced for certain types of motion in systems with many degrees of freedom, providing there exists one or more sets of coordinates in which the HJ equation is completely separable. If only conservative systems are considered Hamilton's characteristic function should be used. Complete separability means that the equations of canonical transformation have the form

$$p_j = \frac{\partial W(q_j; \alpha_1, \alpha_2, \ldots, \alpha_n)}{\partial q_j}$$

such that each p_j is a function of the q_j and the n integration constants α_k: $p_j = p_j(q_j, \alpha_{1\to n})$. As the counterpart of the one-dimensional case, $p = p(q, \alpha_1)$, this equation represents the orbit equation of the projection of the system point on the (q_j, p_j) plane in phase space. This characterization of the motion does not mean that each q_j and p_j will necessarily be periodic functions of the time, which repeat their values at fixed time intervals. The motion is described as *multiply-periodic*. Even for such systems action-angle variables can be used to evaluate all frequencies of the motion without finding a complete solution.

In analogy with the one-dimensional analysis, the J_j are defined over complete periods of the orbit in the (q_j, p_j) plane, $J_j = \oint p_j \mathrm{d}q_j$. If one of the separation coordinates is cyclic, its conjugate momentum is constant. The corresponding orbit in the (q_j, p_j) plane of phase space is a horizontal straight line, which may be considered as the limiting case of rotational periodicity, for which the cyclic q_j always has a natural period of 2π, and $J_j = 2\pi p_j$ for all cyclic variables.

By the same procedure followed in the one-dimensional case the total change in angle variable is obtained as

$$\Delta w = \sum_k \frac{\partial}{\partial J_j} \oint_{m_k} p_k(q_k, J)\mathrm{d}q_k$$

where the index m_k indicates the integration is over m_k cycles of q_k. In the most general case any pair (q_k, p_k) must be a periodic function of each w_j with period unity, *i.e.* the q's and p's are multiply-periodic functions of the w's with unit periods. Such a multiply-periodic function can always be represented by a multiple Fourier expansion, which for q_k, say, would appear as

$$q_k = \sum_{j_1=-\infty}^{\infty}, \ldots, \sum_{j_n=-\infty}^{\infty} a(k)_{j_1,\ldots,j_n} e^{2\pi i(j_1 w_1 + j_2 w_2 + \cdots + j_n w_n)}$$

where the j's are n integer indices running from $-\infty$ to ∞. As before, each $w_k = \nu t + \beta_k$. Each angle variable represents an intrinsic orbital frequency and in the case of a radiating atom it appears that there must exist some link between the orbital frequencies and the frequencies of radiation. Should multiple Fourier series fail to account for the frequencies that occur in atomic spectra, a feasible alternative may be quantization of the procedure along the lines successfully used by Sommerfeld. The Heisenberg quantum mechanics was based on such an assumption.

3.1.3 Conclusion

The Hamilton-Jacobi form of the classical equations of motion has been shown to have provided the basis for the quantum-mechanical formulations according to Sommerfeld, Heisenberg, Schrödinger and Bohm. Each of these formulations inspired its own peculiar interpretation of quantum effects, despite their common basis. Each of the different points of view still has its adherents and the debates about their relative merits continue. Closer scrutiny shows that the Sommerfeld and Heisenberg systems assume quanta to be particles in the classical sense, although Heisenberg considered electronic positions to be fundamentally unobservable.

Schrödinger and Bohm both accepted that quantum motion follows a wave pattern. To account for wave-particle dualism the interpretation of matrix mechanics, developed by Heisenberg and others, was extended on the assumption of probability densities. Schrödinger developed the notion of wave structures to simulate particle behaviour, but this model has been rejected almost universally and apparently irretrievably, in favour of probabities, arguably prematurely and for questionable reasons. Bohm's attempt to revive the wave interpretation advocated a literary interpretation of wave-particle dualism in the form of a classical particle accompanied and piloted by a quantum wave.

The major source of confusion in understanding quantum theory stems from several demonstrations of the equivalence of matrix mechanics and wave mechanics. Despite the mathematical equivalence of the two systems, disparate interpretations have persisted, although rarely recognized by uncritical readers and the authors of introductory texts. The 'orthodox' interpretation of the Copenhagen school, based on the particle assumption and matrix mechanics, is now generally accepted as the only valid consensus reading of quantum theory. Efforts to look beyond the standard clichés are considered a waste of time. Widespread ignorance about the nature of electrons is considered an inevitable consequence of the uncertainty principle – calculation is all that counts, even when the calculated numbers have no recognizable physical meaning.

Most chemists are not even aware that the interpretation of quantum theory may be in dispute. If quantum theory stipulates that atoms grow orbital hair, then so be it. Such things are needed to rationalize chemical bonds after all. But surely, there must be another, logical way out. The known physical world is not unreasonable and if chemistry is to be understood quantum-mechanically, quantum theory better be reasonable. The purpose of what follows is to find that reasonable model, if it exists.

3.2 The Copenhagen Orthodoxy

The orthodox or Copenhagen interpretation of quantum theory originated with three seminal papers published in 1925-26 by Heisenberg, Born and Jordan and an independent paper by Dirac (1926); all of these are available in English (translation) in a single volume [13]. A detailed summary was published by Heisenberg [9]. The primary aim of these studies was to formulate a mathematical system for the mechanics of atomic and electronic motion, based entirely on relations between experimentally observable quantities. An immediate consequence of this stipulation was that the motion of electrons could no longer be described in terms of the familiar concepts of space and time, but rather in terms of state functions constructed from matrix elements that relate to the Fourier sums over observed spectroscopic frequencies. The procedure became known as matrix mechanics.

3.2.1 Matrix Mechanics

It was known from experiment that all the spectral lines of an element could be represented as the differences of a relatively small number or *terms*. If these terms are arranged in a one-dimensional array $\{T_j\} = T_1, T_2, \ldots,$ the atomic frequencies form a two-dimensinal array of elements $\nu(nm) = T_n - T_m$, *e.g.*:

$n \backslash m$	1	2	3
1	0	$\nu_{12} = T_1 - T_2$	ν_{13}
2	ν_{21}	0	ν_{23}
3	ν_{31}	ν_{32}	0

This formulation is consistent with the combination principle observed by Ritz:

$$\nu_{nk} + \nu_{km} = (T_n - T_k) + (T_k - T_m) = T_n - T_m = \nu_{nm}$$

The Bohr frequency condition relates the characteristic frequencies of an atom to a set of characteristic energies

$$\nu_{nm} = (1/h)(W_n - W_m)$$

linking spectroscopic terms to discrete energy levels.

Recall that the canonical equations of classical mechanics can be used to derive the Hamiltonian expression for the total energy of a system from the momenta p_k and positional coordinates q_k:

$$\dot{p}_k = -\frac{\partial H}{\partial q_k} \quad , \quad \dot{q}_k = \frac{\partial H}{\partial p_k} \quad (k = 1, 2, \ldots)$$

On the basis of Bohr's correspondence principle, Heisenberg postulated that a comparable set of *quantum-mechanical* canonical relations must exist, and that these could be derived for multiply-periodic systems by specifying the quantum equivalents of coordinate and momentum variables as square arrays of elements which depend on two sets of integers.

In classical theory particle coordinates in a multiply-periodic system may be written as a Fourier sum of harmonic terms, of the form

$$q_k = \sum_{n_j=-\infty}^{\infty} a(n_j)e^{2\pi i\left(\sum n_j\nu_j\right)t}$$

over constant amplitudes and the fundamental frequencies of the motion. For quantum systems observed frequencies are not functions of individual states or orbits, but arise as a difference between two stationary states. Rather than use the Fourier terms of classical theory, a new kind of term written as a function of two quantum numbers, $q(nm)\exp[2\pi i\nu(nm)t]$ was used as elements to form two-dimensional arrays to represent the generalized coordinates and momenta:

$$q_k = \left(q(nm)e^{2\pi i\nu(nm)t}\right) \quad , \quad p_k = \left(p(nm)e^{2\pi i\nu(nm)t}\right)$$

When forming the product pq another square array

$$(pq)_k = \left(pq(nm)e^{2\pi i\nu(nm)t}\right) \tag{3.19}$$

should be obtained, provided that all elements of the same frequency are added together, *i.e.* those specified by the combination principle, $\nu(nk) + \nu(km) = \nu(nm)$. The new amplitudes are obtained by summation of the terms

$$pq(nm) = \sum_k p(nk)q(km)\,,$$

to generate the elements of the new array (3.19).

This empirical multiplication rule was soon identified as equivalent to the mathematical rule for the multiplication of matrices, with two important implications:

(i) To ensure that the matrix elements have real values it is required that each element $q_k(mn) = q_k^*(nm)$ be equal to its complex conjugate. Matrices with this symmetry are known as *Hermitian* matrices. It means that $q(nm)q(mn) = |q(nm)|^2$ and $\nu(nm) = -\nu(mn)$.

(ii)Matrix multiplication is non-commutative, $pq(nm) \neq qp(nm)$. The quantum condition enters the theory through this property of matrices. From the combination principle

$$\nu(nm) = \nu(nk) - \nu(mk)$$

and the frequency condition

$$\nu(nm) = \frac{1}{h}(H_n - H_m)$$

it follows that (H) is a diagonal matrix, $H_{nm} = 0$ for $n \neq m$, and $\dot{H} = 0$. Taking time derivatives:

$$\frac{\mathrm{d}}{\mathrm{d}t}\left(q(nm)e^{2\pi i\nu_{nm}t}\right) = 2\pi i\nu(nm)q(nm)e^{2\pi i\nu_{nm}t}$$

or simply

$$\dot{q} = 2\pi i\nu q \quad \text{and} \quad \dot{p} = 2\pi i\nu p \tag{3.20}$$

Consider the diagonal array of energy values H_n multiplied with a non-diagonal square array $\big(q(nm)\big)$. Form the *commutator* of the matrices (H) and (q), $[H, q] = \big(Hq - qH\big)$. By the rules of matrix multiplication it has as (nm)th element the sum

$$\sum_k H_n\delta_{nk}q(km) - \sum_k q(nk)H_k\delta_{km}$$

$$\delta_{km} = \begin{cases} 1 & \text{for} \quad k = m \\ 0 & \text{for} \quad k \neq m \end{cases}$$

The only surviving term in the first sum is for $k = n$ and in the second $k = m$, to give

$$\begin{aligned} (H_n - H_m)q(nm) &= \big(Hq - qH\big)(nm) \\ \text{or} \quad h\nu(nm)q(nm) &= \big(Hq - qH\big)(nm) \\ \text{Substituting } \dot{q}: \quad \frac{h}{2\pi i}\dot{q}(nm) &= \big(Eq - qE\big)(nm) \end{aligned} \tag{3.21}$$

Equation (3.21) is the matrix equation of motion and it can be generalized into the coupled equations:

$$\Big(FQ - QF\Big) = \frac{h}{2\pi i}\frac{\partial F}{\partial P}$$

$$\Big(PF - FP\Big) = \frac{h}{2\pi i}\frac{\partial F}{\partial Q}$$

where (F) is a matrix function of two other matrices (P) and (Q), such that P and Q satisfy the expression:

$$\Big(PQ - QP\Big) = \frac{h}{2\pi 1}\delta_{pq} = \frac{h}{2\pi i}(I) \tag{3.22}$$

(I) is the unit matrix. To validate this procedure, set $F = P$ and again $F = Q$, thereby reproducing equation (3.20). The momentum-position commutator is the best known example of (3.22) in the form $(pq - qp) = h/(2\pi i)(I)$, *i.e.* $[p, x] = i\hbar \quad , \quad \hbar = h/2\pi$.

Having found the quantum-mechanical equation of motion the behaviour of all periodic systems such as vibrators, rotators and atomic electron distributions can in principle be solved. The prescription for handling any such problem is to find a set of matrices $Q_1, Q_2, \ldots, Q_n; P_1, P_2, \ldots, P_n$, which satisfy the commutation rules $Q_mQ_n - Q_nQ_m = 0$, $P_mP_n - P_nP_m = 0$, $P_mQ_n - P_nQ_m = i\hbar(I)$, and to render the matrix $H(Q_1, \ldots, Q_n; P_1, \ldots, P_n)$ diagonal. When such a set of matrices has been found, the diagonal elements of H will be the measurable values in question. General acceptance of matrix quantum mechanics followed immediately on the demonstration, achieved independently by Pauli and Dirac [13], that this procedure leads, with few assumptions, to the Balmer formula for the hydrogen spectrum.

3.2.2 The Interpretational Problem

Once the mathematical formalism of theoretical matrix mechanics had been established, all players who contributed to its development, continued their collaboration, under the leadership of Niels Bohr in Copenhagen, to unravel the physical implications of the mathematical theory. This endeavour gained urgent impetus when an independent solution to the mechanics of quantum systems, based on a wave model, was published soon after by Erwin Schrödinger. A real dilemma was created when Schrödinger demonstrated the equivalence of the two approaches when defined as eigenvalue problems, despite the different philosophies which guided the development of the respective theories. The treasured assumption of matrix mechanics that only experimentally measurable observables should feature as variables of the theory clearly disqualified the unobservable wave function, which appears at the heart of wave mechanics.

A worse dilemma was created by the user-friendly nature of wave mechanics, arising from the relative ease of manipulating differential equations, compared to the diagonalization of matrices. Most physicists who had eagerly anticipated the appearance of a generally applicable quantum theory immediately turned to wave mechanics. The Copenhagen school must have perceived this as a dangerous development that could potentially pollute the purity of quantum mechanics and they started to develop an interpretational structure that would eliminate deviant perceptions created by wave mechanics.

The most serious immediate problem was the appearance of (patently 'unobservable') position coordinates in the wave-mechanical Hamiltonian, which allowed an estimate of electronic charge density $\rho(\boldsymbol{x}) = \psi^2$, while complicating the notion of discontinuous quantum jumps. To override the obvious, but unacceptable, interpretation of wave mechanical variables it was therefore considered necessary to introduce an alternative interpretation commensurate with the postulates of matrix mechanics. This programme was successfully pursued and resulted in the muddled interpretation of wave mechanics which still continues to plague students and scholars alike. It distorts a beautiful comprehensible picture by insisting on incorporating into the theory ill-defined foreign concepts such as probability density, uncertainty relations, a measurement problem, role of the observer, collapse of the wave function – all of these part of an apology for matrix mechanics – to effectively suppress the real meaning of wave mechanics. Small wonder that Schrödinger, on invitation to Copenhagen (1926), reacted ([1], page 261) to the insistence that he adopt Bohr's contrived interpretation by saying that he would rather not have published his papers on wave mechanics, had he been able to foresee the consequences.

To ensure that the heretical ideas of wave mechanics remain permanently suppressed it was necessary to show that matrix mechanics provided a complete and infallible description of the atomic world. The famous debates against Einstein were obviously conducted by Bohr to defend this position. It is generally agreed that Bohr prevailed in this confrontation, and his stance, no longer considered in dispute, was accepted as the orthodox interpretation of all quantum theory.

Because of the obvious validity of matrix mechanics it was natural to assume that the initial premises on which Heisenberg first approached the quantum problem also had to be valid beyond dispute. The same courtesy is not extended to Schrödinger, who after all, produced the most useful version of the theory. The damage can clearly not be undone, but the relevance of the Copenhagen orthodoxy to wave mechanics can certainly be re-examined. Whereas the contribution of orthodox quantum theory to the understanding of chemistry has been minimal, there is the realistic hope that unfettered use of wave mechanics can only be an improvement on the current situation.

3.2.3 The Copenhagen Model

The Copenhagen model is universally acknowledged as the ruling interpretation of quantum theory, although an authorized complete statement of the underlying principle does not appear to exist. In fact, such a statement is probably no longer needed as the Copenhagen interpretation, or orthodoxy, is so widely accepted as synonymous with quantum theory itself that, in effect, it now is the quantum theory. Many of the seminal beliefs about the nature of quantum measurements, disturbance of the system by experimental probes, the decisive effect of the observer on the outcome of experimental measurement, disappearance of physical reality, and the conviction that the ultimate theory of quantum mechanics can never be superseded or even supplemented, are largely forgotten, but the certainty, in the words of Niels Bohr ([26], page 36), that "all competent physicists know that the Copenhagen interpretation is correct, since it has been proved by experiment", lingers on.

The prudent way forward is clearly not by launching yet another attempt to challenge the orthodoxy, but rather to stop chasing windmills and to examine what is left of wave mechanics without the dictates from Copenhagen. Maybe there emerges a theory of relevance to chemistry.

In order to understand the antagonism against wave mechanics it is important to know the basic idea that inspired Max Born and his collaborators (Heisenberg, Jordan and Pauli) to develop a quantum mechanics. It was the firm conviction that in atomic physics classical things like particles and waves have no meaning and the only legitimate variables are observable quantities such as spectral frequencies. In the words of Heisenberg [27], " quantum mechanics does not represent particles, but rather our knowledge, our observations, or our consciousness, of particles". Along came Schrödinger with the suggestion that non-classical particles could be viewed as wave structures and he demonstrated how the dynamic properties of such particles are to be simulated by wave packets. A serious, but not sufficient, objection against this claim was that *only certain* wave packets are non-dispersive[1] and more seriously, that Schrödinger tried to turn wave mechanics into a branch of classical physics. This last objection is obviously absurd, and also not sufficient to counter Schrödinger's argument. More drastic measures were needed and Born came to the rescue with his probability interpretation of the wave function, stating ([1], page 256):

> It is necessary to drop completely the physical pictures of Schrödinger which aim at a revitalization of the classical continuum theory, to retain only the formalism and to fill that with new physical content.

This single item was originally referred to as the Copenhagen interpretation, which developed from there with each new argument against Schrödinger.

[1]Virtually every textbook states that (all) wave packets disperse and therefore cannot serve as a serious particle model.

Interpreting the square of the wave amplitude as a probability for finding the particle, certainly describes a particle, and not ′ our knowledge, our observations, or our consciousness′ of the particle. In the words of Popper ([26], page 10):

> From that moment chaos ruled in the Copenhagen camp.

The generally accepted notion of wave-particle duality, which predates Heisenberg and Bohr, could be reconciled with the probability interpretation, but the fuzziness associated with waves remained unexplained in the orthodox tradition. The proclamation of the quantum-mechanical uncertainty principle was intended to take care of the oversight. A more serious indictment of the orthodox tradition is hard to imagine, short of the blunt statement by Nobel physicist, Murray Gell-Mann [28]:

> Niels Bohr brainwashed a whole generation of theorists into thinking that the job was done 50 years ago.

Without rehashing the arguments it can be stated here with certainty that there is nothing non-classical about the notion of uncertainty, which is considered with awe among orthodox theorists, but in fact, is a characteristic of any wave theory or Fourier transform. To explain how the uncertainty principle complicates quantum measurement it was argued that it reflects the way in which any measurement interferes with the position and motion of an elementary particle, meaning that it had a sharp position and momentum before the interference. To avoid self-contradiction this complication was blamed on the influence of the observer. The new argument could next be used to explain to Schrödinger how wave packets (considered as irrelevant before) can only collapse when observed by an intelligent observer, and eventually led on to the many-worlds interpretation of quantum theory, moving out of the ambit of serious science.

The final *coup de grâce* against any alternative to the orthodox formulation was supposed to be delivered by John von Neumann (1932) with ′mathematical proof′ that dispersion-free states[2] and hidden variables are impossible in quantum mechanics. He concluded [29] that:

> It is therefore not, as often assumed, a question of reinterpretation of quantum mechanics – the present system of quantum

[2] *E.g.* sharp momentum and position states. At least one of such a conjugate pair is dictated by Heisenberg's principle to be statistically uncertain.

> mechanics would have to be objectively false in order that another description of the elementary process than the statistical one be possible.

On analysis by Bell [30] the proof was shown to rely on the assumption that dispersion-free states have additive eigenvalues in the same way as quantum-mechanical eigenstates. Using the example of Stern-Gerlach measurements of spin states, the assumption is readily falsified. It is shown instead that the important effect, peculiar to quantum systems, is that eigenvalues of conjugate variables cannot be measured simultaneously and therefore are not additive. The uniqueness proof of the orthodox interpretation therefore falls away.

The conclusion reached here is that Bohr, Born and their followers convinced themselves of an unbridgeable gap between classical mechanics and quantum mechanics, despite accepting a correspondence principle on occasion. Without wave mechanics it was easy to make their point on the back of discontinuities and quantum jumps. They refused to accept that a logical transition between classical and non-classical behaviour could possibly be contemplated and any suggestion with such a smell had to be stamped out. At the end of the day the scientific world ended up with a theory they could use, but were forbidden to understand. Opponents of the system have been left with precious little room to maneuvre, not even an alternative terminology. *Wave mechanics* has been ordained as part of *quantum mechanics*, the name coined for Born's original programme, *quantum theory* is the synonym of the Copenhagen orthodoxy, and by *quantum chemistry* is understood the computational programme based on probabilities and three-dimensional constructs, called orbitals – by definition at variance, not only with Schrödinger's model, but also the Copenhagen interpretation. To avoid meaningless debate none of these descriptors can be used to describe the envisaged model for chemistry, based on the original conceptions of Schrödinger. Taking a cue from Karl Popper's statement ([26], page 126):

> ... the wave theory offers the possibility of explaining matter by something which is not matter (and more general than matter);

it may be safe to continue this enquiry into *the wave model of chemical matter*, without further callenging the quirks of quantum physics. No attempt will therefore be made to explore the notion of *quantum logic* [31], which is a bold attempt to cast the illogical structure of the Copenhagen orthodoxy into some non-Boolean mathematical structure, apart from noting that the operations of quantum logic are largely necessitated by the probability interpretation.

3.3 The Schrödinger Interpretation

Widespread disagreement on the understanding of chemical matter can, to a large extent, be traced back to the confusion over the nature of the electron. The intuitive concept of elementary *particles* completely dominates theoretical thinking, although the ultimate quantum entity, the photon, is a particle in name only. The carrier of the electromagnetic field is defined unequivocally by Maxwell's field equations, of the form

$$\nabla^2\Phi = \mu\epsilon\frac{\partial^2\Phi}{\partial t^2}$$

which is immediately recognized as a wave equation. All of optics is based on solutions to this equation without agonizing over photon statistics. No other differential equation has been studied in more detail than the wave equation. Plane-wave solutions that describe linear propagation is the most common, but not the most informative on the common processes of emission and absorption of radiative energy. These phenomena are better appreciated in terms of spherical waves that allow propagation in all directions with equal probability. It is a curious fact that, despite the dramatic demonstration by the theory of special relativity (1905) of the equivalence of matter and energy ($E = mc^2$), it took another twenty years to realize that there is a physically real wave aspect associated with material motion.

The accepted theory of matter, and of electrons in particular, is a fusion of two incongruous notions, formerly known as quantum mechanics and wave mechanics respectively. In its purest form pioneering quantum mechanics rejected the reality of both particles and waves, but it developed into a theory of potential particles that only manifest themselves in a statistical pattern, under observation. Schrödinger based his description of an electron entirely on his wave-mechanical results for the hydrogen atom and various oscillators. His antagonists framed their criticism of his proposals as if these referred to free electrons, something which Schrödinger only attempted much later. One commentator [32], quoting out of context, even suggested that Schrödinger admitted failure and lost interest in the issue. For rejecting the idea of a particle that performs quantum jumps, the Copenhagen protagonists argued against Schrödinger by insinuaton rather than fact. The fact remains that Schrödinger's early interpretation of wave mechanics has never been refuted on scientific grounds.

On the other hand, the nature of an electron is considered meaningless in the Copenhagen circle. Until the internal structure of an electron has been measured by some device, they consider it to have no place in scientific discourse. If the only interest in an atom is spectroscopic observation, there is no need to calculate or know anything more about that atom than the spacing between energy levels. It is a mystery why this philosophy should be considered appropriate for chemistry.

Theoretical chemistry is informed by the observation of charge density distributions in crystals and molecules, and these do not appear as sets of discrete points. Conventional wisdom implicates the time scale of diffraction experiments for building up, what appears to be diffuse charge densities, by the statistical accumulation of data points. However, not only does quantum theory deny the existence of electronic positions and paths, but until it has been shown experimentally that the data points appear sequentially, the statistical argument is no more persuasive than the wave-mechanical.

3.3.1 The Negative Reaction

The attack on Schrödinger has been analyzed in great detail by Popper [26]. The details will not be repeated here, apart from a few pertinent statements:

1. p.42: ... what we are seeking in science is not so much usefulness as *truth: approximation to truth; explanatory power, and the power of solving problems;* and *thus understanding.*

2. p.42: The view that theories are *nothing but* instruments or calculating devices, has become fashionable among quantum theorists, owing to the Copenhagen doctrine that quantum theory is *intrinsically ununderstandable* because we can understand only *classical* 'pictures', such as 'particle pictures' or 'wave pictures'. I think this is a mistaken and even vicious doctrine.

3. p.125: In these discussions, a certain element of irritation or exasperation is sometimes discernible, directed against Schrödinger who, it is indicated, has 'a personal attachment to the waves', and 'is in love with this idea' – quoting Born.

4. p.125(footnote): Both criticize Schrödinger: Born, for relying on the intuitively understandable 3-dimensional waves which result from 'the method of second quantization' ... and by Heisenberg for his 'misunderstanding of the usual' (orthodox) 'interpretation': a misunderstanding which consists in 'overlooking the fact' that in the orthodox interpretation, the 3-dimensional waves of Klein, Jordan and Wigner 'have a continuous density of energy and momentum, like a Maxwell field'. Thus the two defenders of the orthodox view are here seen to agree only in one point: that Schrödinger must be wrong. But their arguments contradict each other.

5. p.137: Born, so far as I can see does not answer Schrödinger's question (Are there quantum jumps?) at all. Instead, he discusses a completely different question which he introduces in the title of his section 2: '*Are there atoms?* ' He claims that Schrödinger's real intention is to deny that atoms exist. This seems to be an inexplicable misinterpretation of Schrödinger's views, and more specifically of what he says in the place criticized by Born. Schrödinger's theory may explain atoms (not as waves, but as wave *structures*); but it does not explain them away.

   While no answer to Schrödinger's question is given by Born, Heisenberg does give an answer. ... Thus his answer to Schrödinger amounts to the assertion that there *are* quantum jumps – exactly as there are quantum jumps in the classical theory.

   These answers strongly suggest that there really appears to be no evidence for quantum jumps. But if this is so, why not say so? And why attack Schrödinger when he says so?

6. p.140: The answer to the second question – whether the existence of particles is implied in the interpretation of the present quantum theory – of course depends upon the interpretation chosen. Born, Pauli and Heisenberg clearly imply that the existence of particles is part of the interpretation which they accept, which is the orthodox or Copenhagen interpretation, and that they differ in this point from Schrödinger. I find this a little surprising.

   As against this view, I should have thought that Schrödinger asserted with great force the real existence of atoms or particles, even though he tried to explain them as wave structures. I did not know that, by saying that a table or chair is made of timber, I should expose myself to the criticism that I have denied the existence of discrete tables and chairs, and that I have done so because I am 'in love' with timber (as Born says of Schrödinger's attitude towards waves).

7. p.158: As to Schrödinger, many of the physicists of the new generation know him mainly from the text-books, as the famous author of the wave equation. This is a pity. For Schrödinger's *Collected Papers on Wave Mechanics* are a classic.

   Schrödinger is not only the real father of the *formalism* of the quantum theory, he is first of all a physicist who tries to understand the physical world in which we live. ... he is the first to have shown us that matter may one day be explained as a disturbance of something that is not in turn material.

The suspicion that Schrödinger's interpretation of wave mechanics was suppressed and rejected by quantum physicists for non-scientific reasons, is inescapable. Because of this inherent bias the form of wave mechanics which became established as the basis of theoretical chemistry has, understandably, never been assessed independently for this purpose. The point electron that jumps between quantum states with statistical probability fails to explain chemical behaviour with the same authority that it enjoys in physics. Nevertheless, the Schrödinger alternative is dismissed out of hand by chemists. A typical expert on quantum chemistry declares [33]:

> Schrödinger's original interpretation of $|\psi|^2$ was that the electron is "smeared out" into a charge cloud. If we consider an electron passing from one medium to another, we find $|\psi|^2$ is nonzero in both mediums. According to the charge-cloud interpretation, this would mean that part of the electron was reflected and part transmitted. However, experimentally one never detects a fraction of an electron; electrons behave as indivisible entities. This difficulty is removed by the Born interpretation, according to which the values of $|\psi|^2$ in the two mediums give the *probabilities* for reflection and transmission.

This argument fails to distinguish between an electron and its wave function. The wave function describes the behaviour of an electron and covers all eventualities allowed by the initial and boundary conditions. Solutions of the wave equation for given momentum and potential energy are therefore valid for any initial position of the electron. In this sense ψ describes an ensemble of electrons with individual trajectories that depend on their initial positions. Stated differently, the wave function defines a potential field that governs the wavelike motion, not of a specific electron, but of any electron with given momentum, *i.e.* an ensemble[3] with variable phase relations. The description may be restricted to a single electron by specifying its initial position and hence its phase when reaching the boundary between the two media. This phase relationship determines whether it penetrates into the next medium or whether it is reflected. The transmission probability, defined as $|\psi^2|$ [33], does not refer to a single electron, but to the ensemble average. Despite their uniform momenta only some electrons are transmitted and others are not.

[3]A statistical ensemble is not to be confused with a collection of separate entities. It refers to all possible outcomes on replicating the same trial under different starting conditions.

3.3.2 The Positive Aspects

Trying to get away from the historical disputes, the best starting point is the collected English translations of Schrödinger's early papers [34], containing parts I–IV of his famous series in *Annalen der Physik*, together with five other papers.

Wave Structures

The most attractive feature of Schrödinger's original interpretation of atomic structure, because it flatly contradicted the orthodox view of quantum jumps, became so controversial that, for all practical purposes, it has ceased to be considered a viable alternative. In the case of the hydrogen atom, electronic stationary states are viewed as vibrational eigenstates, the only solutions to the eigenvalue problem. These eigenstates resemble three-dimensional wave structures, spherically surrounding the atomic nucleus. On excitation (or relaxation of excited states) interference between the affected states leads to the generation of *beats* (wave packets), which are either expelled (on relaxation) or absorbed (on excitation), with transition to a state of either lower or higher energy, appearing as photons. According to Schrödinger [34](I,III) the form of vibration can change continuously in space and time. A beat arises from the simultaneous existence of two proper vibrations, and lasts just as long as both are excited together. The sympathetic change of frequency that follows the applied field in the Stark effect provides experimental validation of this proposed model.

One of the most telling examples of Schrödinger's wave model of a quantum particle is his discussion of the Planck oscillator [34], p.41–44. A nearly-classical particle is simulated by summing the orthogonal Hermite-polynomial eigenfunctions over a narrow range of high quantum numbers. The result is a well-defined wave packet that moves between the oscillation limits with group velocity corresponding to the classical particle velocity and energy.

The wave group remains compact and of constant extension, but the breadth and number of the wavelets vary with time, becoming narrower and more numerous as the centre ($q = 0$) is approached, and smoothed out at the turning points; hence a function of velocity.

The wave structure at small quantum numbers is well known as the textbook example of an harmonic oscillator. The inference is almost self evident: particle-like behaviour only becomes pronounced on approaching the classical world, and changing into wave-like behaviour in the quantum domain.

In three dimensions a spatial wave group moves around an harmonic ellipsoid and remains compact, in contrast to the dispersive wave packets of classical optics. The distinction is ascribed to the fact that the quantum wave packet is built up from *discrete* harmonic components, rather than a continuum of waves. The wave mechanics of a hydrogen electron is conjectured to produce wave packets of the same kind. At small quantum numbers the wave spreads around the nucleus and becomes more particle-like, at high quantum numbers, as it approaches the ionization limit where the electron is ejected from the atom.

In order to avoid any hearsay or second-hand interpretation of Schrödinger's 'original' proposal, the following direct quote from [34],II,p.25 should put the record straight:

> What I now categorically conjecture is the following:
>
> The true mechanical process is realised or represented in a fitting way by the *wave processes* in q-space[4], and not by the motion of *image points* in this space. The study of the motion of image points, which is the object of classical mechanics, is only an approximate treatment, and has, as such, just as much justification as geometrical or "ray" optics has, compared with the true optical process. A macroscopic mechanical process will be portrayed as a wave signal of the kind described above, which can approximately enough be regarded as confined to a point compared with the geometrical structure of the path. We have seen that the same laws of motion hold exactly for such a signal or group of waves as are advanced by classical mechanics for the motion of the image point. This manner of treatment, however, loses all meaning where the structure of the path is no longer very large compared with the wave length or indeed is comparable with it. Then we *must* treat the matter strictly on the wave theory, *i.e.* we must proceed from the *wave equation* and not from the fundamental equations of mechanics, in order to form a picture of the manifold of the possible processes. These latter equations are just as useless for the elucidation of the micro-structure of mechanical processes as geometrical optics is for explaining the *phenomena of diffraction.*

[4]Schrödinger consistently uses generalized coordinates, which reduce to simple 3D Euclidean coordinates in the case of one electron.

To date, no direct evidence that Schrödinger ever deviated from this point of view could be found in the literature. On the contrary, he revisited the issue of electron structure on several occasions, by sharpening the wave picture through the discovery of *Zitterbewegung* and a critical evaluation of *second quantization.*

Second Quantization

The procedure, known as second quantization, developed as an essential first step in the formulation of quantum statistical mechanics, which, as in the Boltzmann version, is based on the interaction between particles. In the Schrödinger picture the only particle-like structures are associated with waves in 3N-dimensional configuration space. In the Heisenberg picture particles appear by assumption. Recall, that in order to substantiate the reality of photons, it was necessary to quantize the electromagnetic field as an infinite number of harmonic oscillators. By the same device, quantization of the scalar ψ-field, defined in configuration space, produces an equivalent description of an infinite number of particles in 3-dimensional space [35, 36]. The assumed symmetry of the sub-space in three dimensions decides whether these particles are bosons or fermions. The crucial point is that, with their number indeterminate, the particles cannot be considered individuals [37], but rather as 'intuitively understandable 3-dimensional waves' – (Born) – with 'a continuous density of energy and momentum' – (Heisenberg).

Zitterbewegung

Another of Schrödinger's major achievements was the identification [38] of a superluminal trembling motion (Zitterbewegung) that occurs in the interior of an electron, by substitution of the relativistic Dirac Hamiltonian

$$H = c\alpha_1 p_1 + c\alpha_2 p_2 + c\alpha_3 p_3 + \alpha_4 mc^2$$

into the Heisenberg equation of motion,

$$-i\hbar\frac{\mathrm{d}A}{\mathrm{d}t} = HA - AH$$

leading to the solution along any coordinate

$$\left(\frac{\mathrm{d}x_k}{\mathrm{d}t}\right)^2 = c^2 \cdot I$$

in which I is a unit matrix. The problem is to explain the velocity expectation value $\pm c$, knowing the electron to move at more moderate velocities.

The explanation is that the calculated velocity is not linear and the α_k 4×4 matrices commute only with the momenta but not the Hamiltonian. Integration under these conditions yields the velocity as made up of two sums

$$x_k = \tilde{x}_k + \xi_k$$

The first sum is a linear function of time and defines the velocity that corresponds to the momentum p_k and reduces to the group velocity of a de Broglie wave packet. The second sum describes a rapidly changing periodic function of high frequency and small amplitude, superimposed on the linear motion. The amplitude of this Zitterbewegung is approximated as $h/4\pi mc = \lambda_C/4\pi$, of the same order as the Compton wavelength; the critical length dimension beyond which a wave packet cannot be compacted without changing the momentum by a monstrous amount mc. It means that each point in the feigned charge cloud is to a certain extent also dispersed, in the same way, within the small charge cloud. This small charge cloud has the linear dimension $h/4\pi mc$ and is closely connected to total angular momentum, including spin, $j = l + s$. The product of group and phase velocities of the de Broglie and Compton components of the electronic wave packet, $v_g v_\phi = c^2$, accounts for the superluminal $v_\phi > c$ Zitterbewegung.

3.3.3 The Wave Formalism

In order to avoid any possible confusion about terminology and the formulation of various *wave equations*, used by Schrödinger, the following summary may be helpful. In standard wave notation the wave equation in q-space reads [34]II:

$$\nabla^2\Psi \equiv \text{div grad } \Psi = \frac{1}{u^2}\frac{\partial^2\Psi}{\partial t^2}$$

and is valid for processes which only depend on time through a factor $\exp(2\pi i\nu t)$,

$$i.e. \quad \Psi = \psi e^{2\pi i\nu t} \quad , \quad \frac{\mathrm{d}^2\Psi}{\mathrm{d}t^2} = -4\pi^2\nu^2\psi$$

Using

$$u = \frac{\mathrm{d}s}{\mathrm{d}t} = \frac{E}{\sqrt{2(E-V)}}$$

as in HJ theory, giving

$$u = \frac{h\nu}{\sqrt{2(h\nu - V)}}$$

with $\nu = E/h$, there follows:

$$\nabla^2\psi + \frac{8\pi^2}{h^2}(h\nu - V)\psi = 0$$

and in cartesian space[5]

$$\frac{1}{m}\nabla^2\psi + \frac{8\pi^2}{h^2}(E - V)\psi = 0$$

commonly known as Schrödinger's amplitude equation. To obtain a proper wave equation the parameter E may be eliminated [34]IV by substituting from

$$\frac{\partial\Psi}{\partial t} = \pm\frac{2\pi i}{h}E\Psi$$

to give

$$\frac{1}{m}\nabla^2\Psi - \frac{8\pi^2}{h^2}V\Psi \mp \frac{4\pi i}{h}\frac{\partial\Psi}{\partial t} = 0$$

The more familiar form of this equation

$$\left(\frac{\hbar^2}{2m}\nabla^2 - V\right)\Psi = \pm\hbar i\frac{\partial\Psi}{\partial t}$$

is commonly known as the *time-dependent* Schrödinger equation. The dynamical equation for a free particle is the equivalent of Newton's equations of motion in classical microphysics, and is equally valid in either of its complex conjugate forms:

$$\frac{\hbar^2}{2m}\nabla^2\Psi = \frac{\hbar}{i}\frac{\partial\Psi}{\partial t}$$

$$-\frac{\hbar^2}{2m}\nabla^2\Psi^* = \frac{\hbar}{i}\frac{\partial\Psi^*}{\partial t}$$

By taking the difference of these equations, left multiplied by Ψ^* and Ψ respectively, and rearranging:

$$\frac{\partial}{\partial;t}(\Psi\Psi^*) = -\frac{i\hbar}{2m}\left[\frac{\partial}{\partial x}\left(\Psi\frac{\partial\Psi^*}{\partial x} - \Psi^*\frac{\partial\Psi}{\partial x}\right) + \frac{\partial}{\partial y}\left(\Psi\frac{\partial\Psi^*}{\partial y} - \Psi^*\frac{\partial\Psi}{\partial y}\right) + \frac{\partial}{\partial z}\left(\Psi\frac{\partial\Psi^*}{\partial z} - \Psi^*\frac{\partial\Psi}{\partial z}\right)\right] \quad (3.23)$$

in which the density function $\rho = \Psi\Psi^*$. The rhs of (3.23) is simplified by introducing the vector $\boldsymbol{j}$ with components

$$j_x = \frac{i\hbar}{2m}\left(\Psi\frac{\partial\Psi^*}{\partial x} - \Psi^*\frac{\partial\Psi}{\partial x}\right)$$

[5]In generalized coordinates the kinetic energy is a function of velocities and not of momenta, hence the factor $1/m$.

and similar expressions for j_y and j_z. Equation (3.23) is thereby transformed into

$$-\frac{\partial \rho}{\partial t} = \frac{\partial j_x}{\partial x} + \frac{\partial j_y}{\partial y} + \frac{\partial j_z}{\partial z} = \boldsymbol{\nabla} \cdot \boldsymbol{j}$$

the *divergence* of the vector $\boldsymbol{j}$. Hence

$$\frac{\partial \rho}{\partial t} + \text{div } \boldsymbol{j} = 0$$

This equation is an exact analogue of the continuity equation of hydrodynamics, and this allows definition of a current density

$$\boldsymbol{j} = \frac{\hbar}{2mi}(\Psi^* \nabla \Psi - \Psi \nabla \Psi^*)$$

A general expression for a one-electron wave function over all available states

$$\Psi = \sum_k c_k \psi_k e^{2\pi i \nu_k t}$$

may be used to calculate the current density over two states k and l:

$$j - \frac{\hbar e}{mi} \sum_{k,l} c_k c_l \left(\psi_l \nabla \psi_k - \psi_k \nabla \psi_l\right) e^{2\pi i(\nu_k - \nu_l)t}$$

If only a single eigenvibration is excited, the current components disappear and the distribution of electricity is constant in time. At the same time $\rho = \Psi\Psi^*$ becomes constant with respect to time. This is still the case when several vibrations with the same eigenvalue are excited, but the current density no longer vanishes, although there may be a *stationary* current distribution. Schrödinger [34]IV refers to this situation as "a return to electrostatic and magnetostatic atomic models", and a simple explanation of the lack of radiation in the normal state.

3.3.4 Summary

From a chemical perspective the great virtue of Schrödinger's interpretation is the one aspect that raised the most violent criticism from the Copenhagen school – the ease of reconciling familiar classical concepts with the new wave formalism. In this respect the optics analogy bears repetition. On the familiar scale of interaction of visible light with macroscopic objects, the recognition of phenomena such as reflection, refraction, imaging, aberration and even spectral resolution are readily accounted for in terms of rays of light that obey simple geometrical laws. The demonstration that these phenomena are explained equally well by considering the light beams as propagating wave fronts, could well be seen as an unnecessary complication without visible benefit. Applied to optical phenomena involving objects of a size comparable to the wavelength of the light however, the geometrical approach, based on the eikonal equation: $(\nabla\phi)^2 = n^2$, fails in the same way that the mechanical HJ equation $(\nabla W)^2 = 2m(E - V)$ fails when applied to objects of atomic size. In wave optics the eikonal equation is upgraded to a wave equation that accounts for phenomena such as interference and diffraction. In atomic physics the HJ equation is upgraded to the Schrödinger equation, which is a modified wave equation for matter waves.

Each mechanical variable is replaced by a wave variable:

$$\begin{aligned}
\text{the eikonal, } \phi &\rightarrow U, \text{ the electromagnetic wave amplitude} \\
\text{characteristic, } W(q,p) &\rightarrow \psi, \text{ amplitude function} \\
\text{principal, } S(q,p,t) &\rightarrow \Psi, \text{ time-dependent function}
\end{aligned}$$

In the optical case the transition amounts to taking wavelength into account – in the mechanical case Planck's constant becomes a factor, also in the short-wavelength region. There is no implication that the classical equations describe fundamentally different situations. They are simply less detailed than their non-classical analogues and more convenient to use in the macroscopic world. The two sets of equations deal with the same concepts at different levels of refinement. Apart from Planck's constant quantum theory does not introduce any additional concepts, unknown to classical theory, but it has the ability to explain some experimental results that baffled classical science.

3.4 The Hydrodynamic Alternative

3.4.1 Madelung's Model

Soon after first publication Schrödinger's wave-mechanical model was extended by Madelung [39] on the basis of the obvious correspondence with the classical theory of hydrodynamics, already pointed out in Schrödinger's original papers [34] (II,p.17). Writing the time dependence of Ψ in terms of an action function

$$\Psi = \psi e^{2\pi i\nu t} \rightarrow Re^{iS/\hbar}$$

and substituting this into

$$\frac{\partial\Psi}{\partial t} = \left(-\frac{i\hbar}{2m}\nabla^2 + V\right)\Psi \tag{3.24}$$

the latter separates into two real equations, which represent the wave function in the form of waves with R and S determining the amplitude and phase respectively:

$$\frac{\partial S}{\partial t}+\frac{(\nabla S)^2}{2m}-\frac{\hbar^2}{2m}\frac{\nabla^2 R}{R}+V=0 \tag{3.25}$$

$$\frac{\partial R^2}{\partial t}+\nabla\cdot\left(\frac{R^2\nabla S}{m}\right)=0 \tag{3.26}$$

These two equations resemble the pair of coupled differential field equations of hydrodynamics, which describe the irrotational flow of a compressible fluid by [40]:
Euler's equation of fluidal motion for the velocity field $\boldsymbol{v}$,

$$m\frac{\mathrm{d}\boldsymbol{v}}{\mathrm{d}t}=m\left(\frac{\partial \boldsymbol{v}}{\partial t}+(\boldsymbol{v}\nabla)\boldsymbol{v}\right)=-\nabla\left(V+\int\frac{\mathrm{d}p}{\rho}\right) \tag{3.27}$$

and the equation of continuity

$$\frac{\partial \rho}{\partial t}+\mathrm{div}\ (\rho\boldsymbol{v})=0 \tag{3.28}$$

Madelung assumed that R^2 represented the density $\rho(\boldsymbol{x})$ of a continuous fluid with stream velocity $\boldsymbol{v}=\nabla S/m$. Putting

$$\int\frac{\mathrm{d}p}{\rho}=-\frac{\hbar^2}{2m}\frac{\nabla^2 R}{R}, \tag{3.29}$$

$$m\left\{\frac{\partial(\nabla S/m)}{\partial t}+\left(\frac{\nabla S}{m}\cdot\nabla\right)\frac{\nabla S}{m}\right\}=-\nabla\left(V-\frac{\hbar^2}{2m}\frac{\nabla^2 R}{R}\right)$$

which, integrated once, leads to (3.25) and (3.26).

Madelung made the important observation that, irrespective of the overall validity of the hydrodynamic analogy, all linear combinations of the stationary solutions of the amplitude equation

$$\nabla^2\psi+\frac{8\pi^2 m}{h^2}(E-V)\psi=0 \tag{3.30}$$

are also solutions to the time-dependent equation (3.24), in which both R and S vary with time. It is shown that, in contrast to the stationary flow states of equation (3.30), both density and flux vary periodically for the non-stationary states of (3.24), with the same periodicity as $\nu_{ik}=(E_i-E_k)/h$, that results from the superposition of states i and k. This equation shows that radiation is not due to quantum jumps, but rather happens by slow transition in a non-stationary state.

From equations (3.27) and (3.29) Newton's second law is retrieved in the form

$$m\frac{\mathrm{d}\boldsymbol{v}}{\mathrm{d}t} = -\nabla V - \nabla V_q$$

in which the quantum potential

$$V_q = \frac{\hbar^2\nabla^2 R}{2mR}$$

Any stationary state is seen to be brought about by the force balance $\nabla V = -\nabla V_q$. The quantum force is a function of the pressure potential, or stress tensor, $\int \mathrm{d}p/\rho$ that produces inner forces in the continuum. At equilibrium, in stationary states, the potential energy remains constant, *i.e.*

$$\pm\frac{\hbar^2\nabla^2 R}{2mR} + V = k$$

which reduces to (3.30) for $R = \psi$ and $k = E$. To ensure that the wave function $\Psi = \exp(iS/\hbar)$ is single valued it is necessary that $S/\hbar = S/\hbar + 2\pi n$, integer n, which resembles the Sommerfeld quantization condition $\int p_\phi \mathrm{d}\phi = mh$.

3.4.2 Refinements of the Model

Although the quantum problem seems to be solved by the hydrodynamics of a continuous distribution of electricity with charge density proportional to mass density, this approach has never been accepted as a serious alternative, largely because of doubts raised by Madelung himself. The most important of these, concerns the self-interaction between the charge elements of an extended electron.

It is rather surprising that Madelung should interpret the continuous unit of electricity as made up of even smaller discrete elements, forgetting that such a cascading argument has no end. In the event, Madelung's model remained dormant until Takabayashi [41] re-interpreted the hydrodynamic fluid as an ensemble of trajectories, amenable to a statistical interpretation in the Born sense, and subject to fluctuation under the influence of a quantum potential, through a mechanism related to Brownian motion. Fluctuation is seen as arising from the random action of an outside medium, which in the case of a free electron implicates a virtual medium or an aether. In the final analysis, this is a return to the particle model.

In a further extension [42] of the model the Madelung fluid is assumed to be some kind of physically real fluid with an embedded particle, which takes the form of a highly localized inhomogeneity that moves with the local fluid velocity $\boldsymbol{v}(\boldsymbol{x},t)$. It could be a stable dynamic structure such as a small vortex which is carried along by the velocity field. To simulate the erratic motion of this 'particle', fluctuations of the supporting fluid, also interpreted as the ψ-field, but without disclosing its physical nature, are considered in elaborate detail. The electron of this model has four, rather than two complexions: that of particle, fluid, fluid element and wave field. In terms of the hydrodynamic model the particle vortex and the granular fluid elements are artificial additions that serve no other purpose, than to salvage the classical particle character of the electron. Without these spurious attributes the model is no less general, but perhaps more convincing.

The ultimate aim of Bohm and Vigier is clearly to elucidate the concept of a classical particle, which is guided along a causal trajectory by a quantum-mechanical pilot wave, without accepting that the hydrodynamic model makes no provision for a particle. The only alternative is to accept that in the case of a one-unit problem the electron corresponds to the entire fluid under consideration. This fluid is no more than a region of space-time, or a piece of the aether, isolated and confined by the immediate environment. From the wave-mechanical point of view [43], the proper vibrations allowed in the enclosure that holds the electron, and pictured as three-dimensional spherical harmonics, become discrete or quantized. At the highest level of confinement the harmonic nodal surfaces subdivide space into compartments of decreasing size, becoming point-like in the limit. This limiting point resembles the idea of a particle-like inhomogeneous vortex, but not swimming in the Madelung fluid – it is the Madelung fluid in close confinement. The particle trajectory that the pilot-wave model seeks to associate with the fluid may be interpreted as the path of its centre of gravity and the guiding wave is the proper vibrations in the fluid, correctly described by the ψ-function.

3.4.3 Implications of the Model

Despite his reservations about the self-energy of an electron Madelung did offer some guidance on the hydrodynamic model for many-electron systems. Of the three possibilities that

1. electrons flow together into a larger construct;
2. exclude themselves and transform with specific boundary conditions in to each other;
3. interpenetrate without fusing together;

he firmly opted for the third alternative. He envisaged a swarm of individuals with characteristic velocity potentials, moving without impediment in the same space. He therefore proposed, by implication, the Planck radiation model for electrons, which is the basic assumption for second quantization of the ψ-field.

This electronic model meshes well with Schrödinger's views [43]. For each type of fermion, an infinite number of energy levels, or vibrational states, are potentially possible, but only a finite number of these, starting from the lowest level, are activated at any instant. Each member of the finite set may be considered to be one fermion. To put this conclusion into perspective it is appropriate to assume that there is an electron associated with each proton and embedded in each neutron. Each of these three entities consists of a distinctive standing-wave packet, which is stabilized in the aether as a characteristic topological structure. The neutron only exists in close confinement, which causes the fusion of an electron and a proton by the creation of a new wave structure and appropriate distortion of the aether. As a free neutron decays the rearrangement, which now happens in a different environment, produces the original proton and electron together with a neutrino, in a process akin to the production of a photon on the return of an electron from an excited state to its ground state. A symmetry-related situation may be assumed to define antimatter.

Schrödinger, Einstein, Bohm and others who may have happened to support aspects of the model outlined here, were invariably accused of trying to revive a classical interpretation of non-classical events. The implied sin is that these individuals dared to recognize a causal structure where it is expressly forbidden by the Copenhagen doctrine. The exact opposite is probably closer to the truth: What is more non-classical than a 'particle' that consists of vibrations in the aether; a 'particle' with the ability to adapt its shape as dictated by the environment; a 'particle' that disappears into, or appears from, the wave structure of another 'particle'; a 'particle' with continuous non-local mass and charge densities; or a 'particle' which is different from the mass points of classical mechanics? However, this is an irrelevant argument. Whether a model is 'classical' or 'quantal' is of no consequence – what is important is that it leads to a reasonable interpretation of chemical phenomena.

The most attractive feature of the hydrodynamic model is that it obviates the statistical interpretation of quantum theory, by eliminating the need of a point particle. However, even Einstein, despite his famous insistence that "the old one does not play dice", and despite the convincing physical picture of the Schrödinger interpretation, remained convinced that an electron had to be a point particle. His first allegiance was, after all, with his own special theory of relativity that imposes an upper limit of c on the speed at which any signal can be transmitted.

All experimental evidence identifies the electron as an elementary unit. As it receives an electromagnetic signal and responds as a unit, it means that the signal must be transmitted instantaneously (non-locally) through the interior of the electron [44]. To be relativistically self-consistent the electron must by this argument therefore be viewed as a point particle. This is not the only non-local effect that Einstein found suspicious. He actually discovered that quantum mechanics apparently is a non-local theory and hence concluded that it had to be incomplete. Many years later it became possible to demonstrate experimentally through Bell's inequality that he was right, that quantum theory is essentially non-local, and that the theory of relativity may not have to hold within an electron. The Lorentz model [45], which pictures the electron as a flexible sphere, thereby regains its feasibility, especially when interacting with light of a wavelength many orders of magnitude larger than the diameter of the electron.

It is worth noting that the assumption of point-like particles leads to other insurmountable problems on assuming an electric potential energy $V \propto 1/r$ in the calculation of an electron's "self-energy". As $r \to 0$ this energy goes to infinity. The problem is resolved in quantum electrodynamics by subtracting an infinite vacuum contribution through renormalization, which assumes zero interaction within the classical radius of the electron. The rationalization of this assumption is that it works. The procedure correctly predicts several observed phenomena, but obviously remains unsatisfactory as a fundamental physical theory.

3.5 Bohmian Mechanics

David Bohm gave new direction to Madelung's proposal by using the decomposition of the wave equation for a radically new interpretation of quantum theory. He emphasized the similarity between the Madelung and Hamilton-Jacobi equations of motion, the only difference between them being the *quantum potential energy* term,

$$V_q = -\frac{\hbar^2 \nabla^2 R}{2mR}$$

Without this term the quantum equation becomes identical with the classical expression. The only factor which can cause V_q to vanish is the mass. Not surprisingly, massive macroscopic objects have $V_q \to 0$ and are predicted to behave classically whereas sub-atomic entities with appreciable quantum potential energy $V_q > 0$, are known to exhibit quantum behaviour. The clear implication that there is no sharply defined classical/quantum limit, but rather a transition region, was interpreted by Bohm to indicate that classical and quantum objects, close to this transition region, should have dynamical properties in common. Two such properties, orbital angular momentum and a mechanical trajectory are well defined classically, and argued to depend on hidden variables in the quantum case.

The idea of hidden variables is fairly common in chemical models such as the kinetic gas model. This theory is formulated in terms of molecular momenta that remain hidden, and evaluated against measurements of macroscopic properties such as pressure, temperature and volume. Electronic motion is the hidden variable in the analysis of electrical conduction. The firm belief that hidden variables were mathematically forbidden in quantum systems was used for a long time to discredit Bohm's ideas. Without joining the debate it can be stated that this 'proof' has finally been falsified.

An immediate success of Bohm's proposal concerns the trajectories traced out by quantum objects in a cloud-chamber experiment, problematic to understand in terms of the conventional views of quantum measurement theory. However, Bohm's conclusion that the quantum object must be a particle like its classical counterpart is not a necessary consequence of the hidden-variable argument. His proposal of how to associate quantum characteristics with this 'particle' is even less convincing. The idea, borrowed from de Broglie, is to associate a pilot wave with the particle. The function of the pilot wave is to transmit active information from the quantum-potential field to guide the particle along its trajectory. This chain of events may well be consistent with quantum behaviour but there is no conceivable mechanism of how such interactions could be effected. How does a structureless particle interact with a wave and process information? In this respect the hydrodynamic model is more convincing on assuming that the particle-like trajectory refers to the centre of gravity of the quantum object considered to be a compressible fluid with internal wave structure.

In the case of angular momentum there is no conflict between the classical and quantum descriptions for an electron fluid and continuity across the classical/quantum limit presents no problem.

3.5.1 Quantum Potential

Bohm's failure to give an adequate explanation to support the pilot-wave proposal does not diminish the importance of the quantum-potential concept. In all forms of quantum theory it is the appearance of Planck's constant that signals non-classical behaviour, hence the common, but physically meaningless, proposition that the classical/quantum limit appears as $h \to 0$. The actual limiting condition is $V_q \to 0$, which turns the quantum-mechanical

Hamilton-Jacobi equation

$$\frac{\partial S}{\partial t} + \frac{(\nabla S)^2}{2m} + V = \frac{\hbar^2 \nabla^2 R}{2mR} = -V_q$$

into its classical form.

In a region of constant quantum potential energy, the expression

$$\nabla^2 R = -\frac{2mRk}{\hbar^2} \tag{3.31}$$

implies the absence of external forces and constant kinetic energy, $k = E - V$. Write the amplitude in its more familiar form, $R \to \psi$, to retrieve Schrödinger's amplitude equation:

$$\nabla^2 \psi + \frac{2m}{\hbar^2}(E - V)\psi = 0$$

It will now be shown that the existence of quantum potential energy eliminates the need to allow for repulsion between sub-electronic charge elements in an extended electron fluid. An electron, whatever its size or shape is described by a single wave function that fixes the electron density at any point as

$$\rho(\boldsymbol{x}) = |R(\boldsymbol{x})|^2 \text{ and the unit charge at } e\int_{-\infty}^{\infty} R^2(\boldsymbol{x})\mathrm{d}x$$

The quantum potential energy however, depends on the wave function over the entire space occupied by the electron, *i.e.*

$$V_q = -\frac{\hbar^2}{2m}\int_{-\infty}^{\infty} \frac{\nabla^2 R(\boldsymbol{x})}{R(\boldsymbol{x})}\mathrm{d}x$$

which is a continuous function. The quantum potential energy associated with a pair of charge elements, $V_q(\boldsymbol{x}_1, \boldsymbol{x}_2)$, is therefore independent of the coordinates of these elements and depends holistically on the quantum state of the whole system. There is no distance-related interaction between charge elements and hence nothing that corresponds to the self-energy of quantum electrodynamics. Should the wave function therefore change in any way, through local distortion, the entire system responds instantaneously. A *non-local* connection is said to pervade the system.

The possibility of non-local interaction within quantum systems, so vividly illustrated here for a holistic electron, was first recognized by Einstein and others [46]. To avoid conflict with the theory of special relativity the effect was interpreted to mean that quantum theory was incomplete. More recently however, the non-local nature of quantum systems has been observed experimentally. The initial reservations against Madelung's hydrodynamic model therefore falls away.

There is no reason why the previous result should not be valid for all holistic systems, which accordingly are predicted to exclude the electromagnetic field[6]. It may well be the case for protons, and perhaps for atomic nuclei, but clearly not for atoms, in which case the electronic energy is determined by coulombic interaction. Another intriguing possibility is that the gravitational field may disappear at short range for the same reason.

Some insight into holistic systems is gained by considering a collection of electrons that appear to flow together in a conductor without fusing into a continuous condensate, as in superconduction, each described by an individual wave function. The total wave function for this current is defined as a product function

$$\psi = \prod_i^n \phi_i(\boldsymbol{x}_i, t)$$

and the quantum potential energy as the sum over n terms

$$V_q = \sum_i^n V_q^i(\boldsymbol{x}_i, t) \text{ in which } V_q^i = -\frac{\hbar^2}{2m}\frac{\nabla_i^2 R_i(\boldsymbol{x}_i, t)}{R_i(\boldsymbol{x}_i, t)}$$

A system of this type is not holistic, but partially holistic, which means that pairwise interaction occurs between the holistic units. The distinction drawn here between holistic and partially holistic systems is not in line with the terminology used in general philosophic discourse and in order to avoid any confusion it is preferable to distinguish between systems that interact either continuously, or discontinuously, with the quantum potential field. Quantum potential, like the gravitational potential, occurs in the vacuum, presumably with constant intensity. The quantum potential energy of a quantum object therefore only depends on the wave function of the object.

The archetype of quantum objects is the photon. It is massless, has unit spin, carries no charge, and responds to the quantum potential field. By comparison, an electron is a massive fermion with half-spin and unit negative charge. It responds to both classical and quantum potentials. The only property that these two entities have in common is their wave nature,

[6]The Meissner effect in superconductivity will be argued to be an example of such exclusion.

which hence defines the only possible mode of coupling with the quantum-potential field.

In terms of Wheeler-Feynman absorber theory [47] a standing wave, consisting of synchronized retarded and advanced components, originating at emitter and absorber respectively, constitutes the wave function of a photon and it extends over the entire region between emitter and absorber.

The advanced wave is emitted backward in time as the retarded wave arrives at A and it arrives back at E at the instant that the retarded wave is being emitted. The two sites E and A may be astronomical distances apart and in the same way that simultaneity is not defined relativistically, the concept of instantaneous response also looses its meaning. The interaction between E and A is therefore non-local, irrespective of time differences. On absorption of the photon energy the wave function collapses everywhere. The photon represents the handshake between emitter and absorber, and therefore has no velocity. The constant c refers to the transmission of radiant energy between E and A, and the photon exists for the duration of the transmission.

The electron which responds to both quantum and classical potential fields exhibits this dual nature in its behaviour. Like a photon, an electron spreads over the entire region of space-time permitted by the boundary conditions, in this case stipulated by the classical potential. At the same time it also responds to the quantum field and reaches a steady, so-called stationary, state when the quantum and classical forces acting on the electron, are in balance. The best known example occurs in the hydrogen atom, which is traditionally described to be in the product state $\psi_H = \psi_p \cdot \psi_e$, hence with broken holistic symmetry. In many-electron atoms the atomic wave function is further fragmented into individual quantum states for pairs of electrons with paired spins.

3.5.2 The Phase Factor

The appearance of a quantum-potential field is related to the *gauge*, or phase, transformation of a quantum wave. It is assumed that the wave field of a quantum object that moves through space suffers a change of phase, such that

$$\psi(x) \rightarrow \psi'(x) = e^{i\alpha}\psi(x) \tag{3.32}$$

In Euclidean (Minkowski) space the phase factor α is a constant and eqn. (3.32) is said to describe a *global* gauge transformation. Following Schrödinger [48] the complex field which describes an electron has a charge (q) associated with the field and the charge density, given by $\rho(x,t) = q|\psi(x,t)|^2$ is clearly invariant under phase transformation, and defines the global conservation of electric charge.

The situation is entirely different within the theory of general relativity, which is based on a curved Riemannian manifold rather than flat space with a globally fixed coordinate system. Each point now has its own coordinate system and hence its own gauge factor. By doing away with the rigid coordinate system the gauge factor necessarily becomes an arbitrary function of position, $\alpha(x)$. Because the phase has no real physical significance, it may be redefined locally by an arbitrary rotation, at every space-time point of the manifold, without changing the physical situation. This stipulation may seem to rule out local charge conservation, unless there is some compensating field that restores the invariance under local phase transformation. Such a compensating field may be represented by a set of functions $A(x)$ that depend on all space-time coordinates. The phases of a quantum wave at positions x and $x+\mathrm{d}x$ are restored into parallel alignment in the presence of a compensating field if the local values of these phases differ by an amount $A\mathrm{d}x$. The field that couples to the object is called the *gauge field*. Interaction between object and gauge field is known as the *gauge principle*.

Gauge fields (A) that restore local phase invariance are evidently closely related also to the quantum-potential field. The wave function of a free electron, with temporal and spatial aspects of the phase factor separated, may be written as

$$\Psi = \psi e^{-i\left(Et/\hbar + \xi(x)\right)}$$

Spatial gauge invariance results from the compensating field identified as the vector potential of the electromagnetic field, which couples with the charge. Temporal invariance depends on a dynamic phase factor and an interaction, equivalent to the total energy of the free electron, which by eqn. (3.31) is precisely the quantum potential energy. The compensating field under local Lorentz transformation is the gravitational field. Geometrical phase factors and other examples of anholonomy, rediscovered under various names, such as Berry's phase, simply restate the consequences of known gauge theory for curved space.

The conclusion that it is the geometry of the space-time manifold that generates all known fields that operate in the physical world seems to be unassailable, without further proof. One circumstance that spontaneously produces a constant potential in space is an interface. It has been speculated

[49] that the vacuum, as an interface between different states of matter, would generate a field like the quantum potential.

3.5.3 Stationary States

As in the hydrodynamic model the Bohmian interpretation assumes a wave function in polar form,

$$\psi = Re^{-iS/\hbar}$$

with velocity potential $\nabla S/m$. Any real wave function therefore defines an electron with zero kinetic energy. This is the case for a hydrogenic s-function, which consequently appears in a truly stationary state.

No other aspect of the Bohmian interpretation is treated with more scepticism. When pushed for an answer most chemists try to rationalize the accepted probability density model, of a single s-electron, by describing the electron as being in rapid motion on a random track[7]. The situation in a p-state with $m_l = 0$ appears even more mysterious. The quantum number $l = 1$ certainly defines non-zero angular momentum, which is hard to reconcile with a stationary electron. When rejecting the Bohmian interpretation for this reason it is well to remember that the conventional interpretation also fails to explain this situation. If an electron has angular momentum how can it fail to respond to an applied magnetic field?

According to standard quantum theory electrons are point particles, but because they do not follow classical trajectories nothing is known about their motion until they appear at statistically random destinations. The trick of the Copenhagen interpretation is that quantum behaviour cannot be modelled in terms of classical dynamics, it can only be measured. It is therefore not necessary to understand by what mechanism an electron acquires angular momentum, but if it can be measured (in a magnetic field) it must be there, without implying rotation or motion of the particle. Whoever understands this argument belongs to a very exclusive group of individuals, most of whom are deceased by now. What Bohm said was that the dynamic properties of electrons are hidden variables and that quantum particles follow trajectories exactly like their classical counterparts and their behaviour only appears to be dictated statistically[8]. It is therefore perfectly correct to deduce that an electron with quantum numbers $n = 2$, $l = 1$, $m_l = 1$ (for instance) has non-zero angular momentum and must be accelerated to maintain its orbital

[7] Author's survey amongst learned individuals of the profession.

[8] Remember: the old one does not play dice. – Einstein

motion. The point is that this electron state is never realized unless there is a second electron at $n = 2$, $l = 1$, $m_l = -1$, with exactly the opposite angular momentum. Because electrons are not distinguishable the resultant angular momentum is zero, which requires no acceleration[9]. The reason why this conclusion can be drawn is because degenerate hydrogenic levels are always $2l + 1$ in number, *i.e.* an odd number. The odd one has $m_l = 0$, which means that it has no angular momentum component in the field direction and therefore cannot be rotating around an axis. Its angular momentum, like spin, must then be due to a different cause, such as rotation in spherical mode, which does not lead to radiation.

3.6 Atomic Theory

The Schrödinger solution for an electron in the field of a stationary proton has, admittedly, provided the sole basis for a quantum-mechanical understanding of atoms and their chemical behaviour, but, at the same time, many misconceptions have been introduced and perpetuated by extrapolating in good faith from hydrogen to more complicated atoms and even molecules.

3.6.1 The Virial Theorem

A prime example is the so-called *quantum-mechanical virial theorem* that appears in countless chemistry textbooks. The theorem is purported to state that the relationship between the expectation values of kinetic and potential energies

$$2\langle T\rangle = -\langle V\rangle \qquad (3.33)$$

is of general validity. Where simple calculations show this not to be the case, a variety of *ad hoc* adjustments to *scaling* or *effective nuclear charge* are made to ensure the expected result, which are then used to compose spurious arguments about the role of kinetic and potential energies in chemical bonding [50]. The simple fact is [51] that only if the interaction potential is spherically symmetrical and proportional to r^s, and if the expectation values

[9]This degeneracy is lifted in a magnetic field, for instance in a Stern-Gerlach experiment.

exist, can it be shown that

$$\begin{aligned} 2\langle T\rangle &= \left\langle r\frac{\partial V}{\partial r}\right\rangle \\ &= s\langle V\rangle \end{aligned}$$

The case $s = 2$ corresponds to the harmonic oscillator, for which $\langle T\rangle = \langle V\rangle$ and $s = -1$ corresponds to the hydrogen atom, to yield (3.33). Even the general form of the virial theorem

$$2\langle T\rangle = \langle \boldsymbol{r}\cdot\boldsymbol{\nabla}V(\boldsymbol{r})\rangle$$

only applies under central forces, as in an ideal gas [11].

Whereas the quantum-mechanical molecular Hamiltonian is indeed spherically symmetrical, a simplified virial theorem should apply at the molecular level. However, when applied under the Born-Oppenheimer approximation, which assumes a rigid non-spherical nuclear framework, the virial theorem has no validity at all. No amount of correction factors can overcome this problem. All efforts to analyze the stability of classically structured molecules in terms of cleverly modified virial schemes are a waste of time. This stipulation embraces the bulk of modern bonding theories.

3.6.2 Electronic Structure

Another red herring is the use of the hydrogenic principal quantum number to describe the electronic structure of atoms in molecules and crystals. In the case of a free hydrogen atom the radial wave function diverges as $r \to \infty$, unless n is an integer greater than l, serving to truncate the infinite series solution that defines ψ. In most cases of chemical interest (*e.g.* in a molecule) r never approaches the infinite limit, because of environmental constraints, and the need of integer n falls away.

Having examined the leading interpretations of the quantum formalism, a more general theory of atomic structure, consistent with all points of view, could conceivably now be recognized. The first aspect, never emphasized in chemical theory, but fundamental to matrix mechanics, is that the observed frequencies that determine the stationary energy states of an atom, always depend on two states and not on individual electronic orbits. The same conclusion is reached in wave mechanics, without assumption. It means that an electronic transition within atoms requires the interaction between emitter and receptor states and the frequency condition: $\Delta E(\Delta n) = h\nu$, for all pairs in n. This condition by itself offers no rationale for the occurrence of the ground-state level in hydrogen. The simplest explanation is provided by the participation of a photon in the electronic transition.

The Bohr conjecture, as interpreted here demands a change of angular momentum: $\Delta L = (\Delta l)\hbar$, for any transition. Noting that photons carry angular momentum $\hbar$ immediately implies that $\Delta l = \pm 1$ for any electronic transition in an atom. This condition, by itself, is still not sufficient to account for the ground-state energy. However, acceptable solutions of the Schrödinger equation for the H electron only occur for $l < n$. An electronic s-state ($l = 0$) can therefore not relax any further and transitions with negative Δn become impossible. For $n = 1$ the $1s$ state defines the absolute minimum energy of an electron in an atom and ns defines the state of minimum energy for an electron with energy $E = -Ry/n^2$ in an atom.

The s-states are responsible for the appearance of atomic quantum levels because an s-electron lacks the angular momentum to generate a photon, which has to be emitted in order to lower the energy of the electron. Bohr was therefore correct in concluding that stationary states are caused by the quantization of orbital angular momentum, but mistaken in assuming that the H ground state has one unit of angular momentum, because there is nothing to prevent the emission of a photon from such a state. To my mind, this argument also refutes the particle model of an electron. Not even the probability interpretation can save it here. The s-state is spherically symmetrical and for the electron to have that symmetry it must perforce consist of a vibrating fluid at a fixed average radial distance from the nucleus. It is a standing wave packet with dimensions corresponding to those of de Broglie waves and internal wave structure of Compton wavelength.

Electronic transition between stationary states consists in the transfer of a photon by the Wheeler-Feynmann handshake mechanism which implies the photon to exist between the radial surfaces of the two vibrating states before emission or absorption, exactly as envisaged in Schrödinger's beat model for electron transition.

3.6.3 Compressive Activation

A convenient method to investigate the behaviour of atoms in realistic physical environments is by boxing in hydrogen [52], and other atoms [53], within impenetrable spheres, by computer simulation. A free electron in a crowded chemical environment is considered to be surrounded on all sides by the negatively charged outer electronic shells of neighbouring atoms and/or molecules. This environment approximates a uniform electrostatic field that prevents electronic charge density from extending indefinitely. The boundary condition of $\psi \to 0$ becomes modified to $\lim_{r \to r_0} \psi = 0$, in which $r_0 < \infty$. Increased pressure is simulated by decreasing the radius r_0 of the impenetrable sphere.

Simulated compression of the hydrogen atom raises all electronic energy levels relative to their free-atom values until the ionization limit is reached on compression to a well-defined critical radius. This response is not unexpected. The pressure is increased by supply of energy in the form of mechanical work. Under the modified boundary condition the electron no longer moves in a pure Coulombic field, with the result that n is no longer constrained to be an integer and the sub-level degeneracy on l is lifted. Sub-levels of constant n are observed to split in such a way that the largest l corresponds to the lowest energy. On compression to $r_0 = 0.1a_0$ the sub-level sequence becomes $1s < 2p < 3d < 2s < 4f < 3p < 3s < 4d < 5f < 4p$. Extreme compression is predicted to produce complete level inversion as in $4f < 3d < 2p < 1s$, *etc.* The leading coefficients do not specify principal quantum numbers, but rather the number of nodes of the relevant radial function. Because compression is isotropic the magnetic sub-level degeneracy is not split.

On compression of non-hydrogen atoms the energy levels, which in this case are occupied by electrons, respond in the same way. Apart from level crossings, interelectronic interactions now also lead to an internal transfer of energy and splitting of the magnetic sub-levels, such that a single electron eventually reaches the ionization limit on critical compression. The calculated ionization radii obey the same periodic law as the elements and determine the effective size of atoms in chemical interaction.

On compression to the ionization radius of an atom the equivalent of one electron becomes decoupled from the atomic core and finds itself in an impenetrable hollow sphere at constant potential, conveniently defined as $V = 0$. This problem, which is closely related to the problem of an electron confined to a one-dimensional finite line segment, has been studied in great detail. The Hamiltonian

$$-\frac{\hbar^2}{2m}\nabla^2\psi = E\psi$$

defines the Helmholtz equation

$$\nabla^2\psi + k^2\psi = 0\,, \quad \text{where } k = \sqrt{2mE/\hbar^2}$$

The radial part has solutions made up of spherical Bessel functions, $R = \sqrt{2kr/\pi}kj_l(kr)$. Since k^2 can take any positive real value the energy spectrum $0 < E < \infty$ is continuous. To ensure that the wave function vanishes at $r = a$, it is necessary that the product ka should coincide with some zero value of j_l. At the first zero of $j_0 = \sin(kr)/kr$, this condition implies $ka = \pi$, and hence

$$E_0 = \frac{\hbar^2\pi^2}{2ma^2} = \frac{h^2}{8ma^2} \tag{3.34}$$

$$\text{In general} \quad E_n = \frac{(nh)^2}{8ma^2}$$

The Bessel functions $\psi(r, \theta, \varphi)$ that vanish at $r = a$ for $l = 0$ define a Fourier series [54]. The ground-state series is the Fourier transform of $\sin(ka)/ka$, which is the box function

$$f(r) = \begin{cases} \sqrt{2\pi}/2a & \text{if} \quad |r| < a \\ 0 & \text{if} \quad |r| \geq 0 \end{cases}$$

For an atom, compressed to its ionization radius, the decoupled (valence) electron hence is spread across the sphere of radius r_0 at uniform density, and its wave function is the step function

$$\psi(r) = (\phi/V_0)^{\frac{1}{2}} \exp[-(r/r_0)^p] \,, \quad p >> 1$$

assuming that the scale factor, $\phi \leq 1$, compensates for an inaccessible core. Writing $V_0 = \frac{4}{3}\pi r_0^3$ and noting that ϕ increases with increasing r_0 and decreases with the number of radial nodes, both factors proportional to atomic size, *i.e.*

$$\phi = \frac{cr_0}{n} \tag{3.35}$$

the wave function reduces to

$$\psi(r) = \sqrt{3c/4\pi n}(1/r_0) \exp\left[-(r/r_0)^p\right] \tag{3.36}$$

It has been demonstrated empirically [55] that equations (3.35) and (3.36) are generally valid for s and p block elements.

The total energy of an electron in the potential field V of an atomic core tends to zero as $V \to 0$ on compression to r_0. The calculated energy of the decoupled valence electron can therefore only arise from the quantum potential and it will be argued that this energy represents the concept, intuitively defined before as the electronegativity of an atom. The wave function (3.36) of the valence electron will be argued to determine the chemical interaction of an atom with its environment.

3.7 Quantum Chemistry

What started out as an effort to understand chemical phenomena through quantum physics has resulted in a computational scheme, almost universally accepted as the ultimate theory of chemistry. Admittedly, computations are performed, depending on computer size, at different "levels of theory", but the thinking is left to the software. The main purpose of the exercise is to predict molecular properties computationally, starting from the quantum rules, interpreted to mean that the molecule constitutes a differential Sturm-Liouville system. For any molecular property of interest there must be a differential operator that projects eigenvalues from the state function of the system. Some of the better known operators are those for total energy: $-i\hbar\partial/\partial t$, kinetic energy: $-(\hbar\nabla)^2$, orbital angular momentum: $i\hbar\partial/\partial\varphi$ and linear momentum: $-i\hbar\partial/\partial q$. The almost insurmountable problem is to solve one of these differential equations for the state function Ψ, which has not been achieved for any chemical system apart from the electron on the hydrogen atom.

A common alternative is to synthesize approximate state functions by linear combination of algebraic forms that resemble hydrogenic wave functions. Another strategy is to solve one-particle problems on assuming model potentials parametrically related to molecular size. This approach, known as free-electron simulation, is widely used in solid-state and semiconductor physics. It is the quantum-mechanical extension of the classic (1900) Drude model that pictures a metal as a regular array of cations, immersed in a sea of electrons. Another way to deal with problems of chemical interaction is to describe them as *quantum effects*, presumably too subtle for the ininitiated to ponder. Two prime examples are , the so-called *dispersion interaction* that explains van der Waals attraction, and *Born repulsion*, assumed to occur in ionic crystals. Most chemists are in fact sufficiently intimidated by such claims to consider the problem solved, although not understood.

The head-on quantum-mechanical analysis of molecules as Sturm-Liouville problems is complicated, not only by mathematical complexity, but more seriously, by the non-existence of a differential operator associated with the fundamental concept of molecular structure. Without molecular structure no computational progress is possible. It is therefore surprising to find that virtually no effort has been made to simulate molecular structure quantum mechanically. The only progress has occurred through the use of classical mechanics and the definition of molecular force fields to simulate the interaction within pairs, triplets and larger groups of neighbouring atoms in a molecule. This approach is tantamount to breaking down the intramolecular holistic interaction into localized quantum-mechanical electronic interactions, empirically simulated in terms of classical force-field parameters. The method, known as *molecular mechanics*, rivals experimental studies in the accuracy of predicted structural parameters. Together with the fact that the potential-energy term in Schrödinger's equation is defined classically, it could mean that the concept of molecular structure is also classical and to be retained in the same form in quantum systems. This idea is in line with the seminal

Born-Oppenheimer theorem of quantum chemistry. In essence, it states that because of the large mass discrepancy between electrons and atomic nuclei, the latter may be considered to remain effectively stationary on the scale of electronic motion in a molecule. The major on-going effort to deny the classical nature of Born-Oppenheimer molecular structures has been singularly unproductive and the quantum-chemical optimization of classical structures is rather fatuous.

The tortuous process of *ab initio* LCAO-MO-SCF calculation, the flagship of computational chemistry, has been the subject of interminable reviews, *e.g.* [15], and will be described here in the briefest of outlines.

3.7.1 The *Ab-initio* Model

First-principle quantum-mechanical simulation of molecular properties calls for solution of the wave equation

$$H\Psi = E\Psi \tag{3.37}$$

in order to obtain the molecular wave function Ψ, which under appropriate operators produces the eigenvalues of all observable molecular properties, such as intramolecular fluctuation, orbital angular momentum and spectroscopic properties. In problems of molecular structure stationary state functions

$$\psi = \Psi e^{i(E/\hbar)t}$$

are easier to handle although, by definition, they remain complex functions in the coordinates of all nuclei and electrons. The stationary-state problem, even for the simplest of molecules, are still too complicated for *ab-initio* mathematical or digital solution.

An alternative strategy is to synthesize a molecular wave function, on chemical intuition, and progressively modify this function until it solves the molecular wave equation. However, chemical intuition fails to generate molecular wave functions of the required spherical symmetry, as molecules are assumed to have non-spherical three-dimensional structures. The impasse is broken by invoking the Born-Oppenheimer assumption that separates the motion of electrons and nuclei. At this point the strategy ceases to be *ab initio* and reduces to semi-empirical quantum-mechanical simulation. The assumed three-dimensional nuclear framework is no longer quantum-mechanically defined. The advantage of this model over molecular mechanics is that the electron distribution is defined quantum-mechanically. It has been used to simulate the H_2 molecule.

For larger molecules it is assumed that a molecular wave function, Φ, is an anti-symmetric product of atomic wave functions, made up by linear combination of single-electron functions, called orbitals. The Hamiltonian operator, H, which depends on the known molecular geometry, is readily derived and although eqn. (3.37) is too complicated, even for numerical solution, it is in principle possible to simulate the operation of H on Φ. After variational minimization the calculated eigenvalues should correspond to one-electron orbital energies. However, in practice there are simply too many electrons, even in moderately-sized molecules, for this to be a viable procedure.

To reduce the operation to a manageable size requires a number of drastic approximations. The most serious approximation is to trim the number of electrons, often to as little as one electron per atom. The total number of relevant electrons is then assumed to limit the size of the *basis set* [10]. The choice of *basis functions* in the set decides, not only the computational complexity, but also the final outcome of the simulation. While a lucky choice gives reasonable answers, an equally logical, alternative choice, could produce patent nonsense – see footnote 10.

Even the most eminent computational chemists often despair of the feasibilty of their model, as reflected by the following remarks[11] [56]:

> ".... let us ask what chemical properties we should be trying to calculate .... Molecular structure is required .... More important are energetics, dissociation energies or barrier heights .... One of the reasons that I became despondent was when IBM asked me to help calculate frequency dependent hyper-polarisabilities of medium size molecules .... Even at MP2 the code was so horrendous that I never wish to see another like it, and I believe it is impossible to consider meeting IBM's request with the standard methods of quantum chemistry .... It is also important that the quantum chemist (typified by the author) gets away from being a

[10]Any basis set, consisting of a *complete orthonormal set* of functions, should produce the correct eigenvalues after variational minimization, *e.g.*

$$\Phi(r) = \sum_{n=0}^{\infty} c_n f_n(r)$$

This assumption however, is strictly only valid for infinite basis sets.

[11]MP2 = Møller-Plesset second-order perturbation theory.

> small-molecule person, talking only to his friends and a few spectroscopists. We must stop validating Schrödinger's equation!"

In a similar vein, Clementi, the foremost quantum chemist of the world stated [58]:

> ".... there is the intriguing possibility that, whereas the papers from [Hartree, Fock, Hylleraas, Thomas, Fermi, Dirac] provide the base of today's quantum chemistry, perhaps a modified Heitler-London approach might become the base of tomorrow's theoretical chemistry. Alternatively stated, theoretical chemistry can not develop in a balanced fashion without greater attention to a formalism, where bonds, bond energies, and chemical structures are fully recognized as main operands."

To develop such a formalism an open-minded re-assessment of the situation is called for and again, reconsideration of astute insight, largely forgotten in the face of dogmatic certainty.

3.7.2 The Hellmann-Feynman Theorem

This famous, but underutilized, theorem provides an unerring guide to the examination of chemical interactions in Born-Oppenheimer molecules where concepts such as the chemical bond, bond energy and chemical structure retain their classical meaning. Given that kinetic energy of the electrons does not depend on the fixed nuclear coordinates it is readily demonstrated [59] that the force of attraction between the nuclei and the electronic charge

$$F(r) = -\int \psi^* \frac{\partial V}{\partial r} \psi \mathrm{d}\tau$$

only depends on the classical potential and the electronic wave function which defines the charge distribution.

The theorem has the important implication that intramolecular interactions can be calculated by the methods of classical electrostatics if the electronic wave function (or charge distribution) is correctly known. The one instance where it can be applied immediately is in the calculation of cohesive energies in ionic crystals. Taking NaCl as an example, the assumed complete ionization that defines a $(Na^+Cl^-)_n$ crystal, also defines the charge distribution and the correct cohesive energy is calculated directly by the Madelung procedure.

Two further, equally important, situations in line with the Hellmann-Feynman requirement can be envisaged. If, by some means, the wave function of a valence electron becomes known, the interaction between neighbouring nuclei can be calculated by either point-charge [60] or Heitler-London simulation [55]. Another possibility develops where the total charge distribution, not of the point-charge type, becomes known. Simulations, based on this eventuality have become known as Density Functional Theory (DFT).

3.8 Density Functional Theory

DFT has led to a substantial simplification of quantum-chemical computations. Like the Hellmann-Feynman theorem it expresses the reasonable assumption of a reciprocal relationship between potential energy and electron density in a molecule. In principle this relationship means that all ground-state molecular properties may be calculated from the ground-state electron density $\rho(x,y,z)$, which is a function of only three coordinates, instead of a many-parameter molecular wave function in configuration space. The formal theorem behind DFT which defines the electronic energy as a functional of the density function provides no guidance on how to establish the density function $\rho(r)$ without resort to wave mechanics.

The main purpose of DFT is to simplify evaluation of the final term in the molecular electronic hamiltonian:

$$H = \sum_i \left[-\frac{1}{2}\nabla_i^2 + \sum_A \frac{Z_A}{r_{iA}} \right] + \frac{1}{r_{ij}}$$

where i and j refer to electrons and A refers to nuclei. Written in terms of an effectice potential ϕ, energy eigenvalues of this hamiltonian occur as solutions of the Schrödinger equation

$$\left[-\frac{1}{2}\nabla^2 + \phi \right] \psi_i(r) = \epsilon_i \psi_i(r)$$

with

$$\phi = v(R) + \int \frac{\rho(r)}{|R-r|} \mathrm{d}r + E_{xc}[\rho]$$

The three terms represent nuclear, coulomb and exchange-correlation potentials respectively. The third, problematic, term is written as a functional of the density. The same problem which occured in the Hartree-Fock simulation of atomic structure was overcome by defining the one-electron exchange potential with the Slater approximation for a uniform electron gas:

$$V_x = -3(3\rho/8\pi)^{\frac{1}{3}}$$

Adapted to molecular environments the same assumption leads to the *local density approximation*:

$$E_{xc}[\rho] = \int \rho(r)\epsilon_{xc}(\rho)\mathrm{d}r$$

where $\epsilon_{xc}(\rho)$ is the exchange-correlation energy per electron of the uniform electron gas of density ρ, and the density,

$$\rho(r) = \sum_{i=1}^{N} |\psi_i(r)|^2$$

DFT calculation starts from an assumed value of $E[\rho]$. The socalled Kohn-Sham orbitals ψ_i are expanded in a 'reasonably selected' truncated set of Gaussian functions representing atomic orbitals [58]. After iteration between $\rho(r)$ and ϵ_i to self-consistency the final values of the parameters are used to calculate E_0 and other one-electron properties.

Despite several uncertainties DFT has become the method of choice to aid in the interpretation of environmental effects, especially in the field of coordination chemistry.

Part II

Alternative Theory

Chapter 4

The Periodic Laws

4.1 Introduction

The most conspicuous failure of quantum physics, as a theory of chemistry, is the demonstrated inability to account in detail for the observed periodic order of the elements, the single most important feature of theoretical chemistry. The importance of this failure, if not completely ignored, is routinely underplayed in elementary chemistry texts, by statements such as [61]:

> "We need not dwell on these exceptions beyond noting that they occur."

A supposedly better informed source [51] states:

> ".... the 4s state, which has a higher energy than the 3d state in hydrogen, is depressed because of its low angular momentum, which causes its orbital to be large at small r, where it can feel the full nuclear attraction."

Still, these non-explanations[1] are generally considered sufficient to rationalize all discrepancies between observed and predicted electronic configurations.

Such enthusiasm is unfortunate in science. It provides little incentive for further analysis once a problem is declared solved. The structure of the periodic table is a problem of this type: as a chemical problem it refers to systems, conditioned by their environment, and not to an isolated entity such as a hydrogen atom. The energy spectrum of a hydrogen electron can therefore, by

[1] Note that the ns levels on the coinage metals are consistently at higher energy than the $(n-1)d$ levels.

definition, not account for the observed periodic table of the elements, while *ad hoc* adjustments only mask the deficiencies without addressing the fundamental problem. This problem is best appreciated on re-examination of periodic material systems, especially when the more representative sample, consisting of all stable, non-radioactive nuclides [62] is considered.

4.2 Nuclide Periodicity

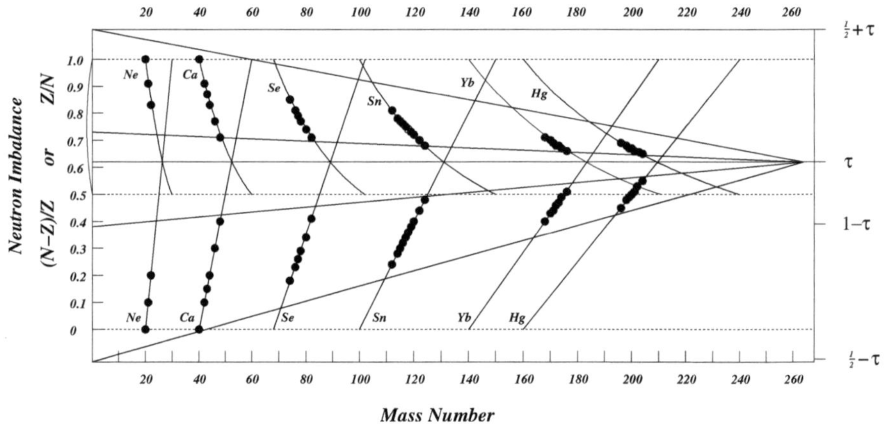

Figure 4.1: *Neutron imbalance as a function of mass number for the isotopes of selected elements*

Before the discovery of neutrons, the familiar empirical form of the periodic table was established as a function of atomic number, which represents less than half the matter that makes up an atom, except for hydrogen and ^{3}He. Later work has shown that nuclear stability is related to nuclear composition, characterized by a neutron imbalance that increases with mass number. For a nucleus of N neutrons and Z protons, neutron imbalance is represented either by the ratio $r = Z/N$ or the relative excess $N_x = (N-Z)/Z$. These parameters are plotted against mass number for selected elements in Figure 4.1.

In the excess plot the isotopes of a given element occur on a straight line through the points $(N_x, A) = (0, 2Z)$ and $(1, 3Z)$. In the ratio plot these isotopes lie on circular segments between the coordinates $(r, A) = (1, 2Z)$ and $(1/2, 3Z)$. The straight lines and circles intersect along a straight line defined by $r = N_x$, *i.e.*:

$$\frac{Z}{N} = \frac{N-Z}{Z} \quad \text{or} \quad Z^2 + NZ - N^2 = 0$$

This quadratic equation has solutions

$$Z = \frac{-N \pm \sqrt{N^2 + 4N}}{2} = -N\left(\frac{1 \mp \sqrt{5}}{2}\right) = N\tau$$

The negative quantity in brackets is an irrational number known as the golden ratio, $\tau = 0.61803....$ The solution $Z = -N(1.61803...) = -\Phi N$ defines $\Phi = 1/\tau$. The field of nuclide stability, as defined on both plots, converges to a point on the line $r = N_x = \tau$ at $A \simeq 267$, the maxinum possible mass number for nuclides, stable against β-type decay. By definition, this maximum,

$$A_{max} = Z_{max} + N_{max} = Z_{max}(1 + \Phi)\,,$$

fixing the corresponding $Z_{max} = 102$.

The heaviest non-radioactive nuclide, ${}^{209}_{83}\mathrm{Bi}$, is interpreted as the maximum, on this scale, in the gravitational field of the solar system. ${}^{204}_{102}\mathrm{No}$ is postulated to be the heaviest stable nuclide anywhere in the cosmos.

The distribution of naturally occurring stable nuclides is examined more closely in Figure 4.2. The total of 264 nuclides separate into four families

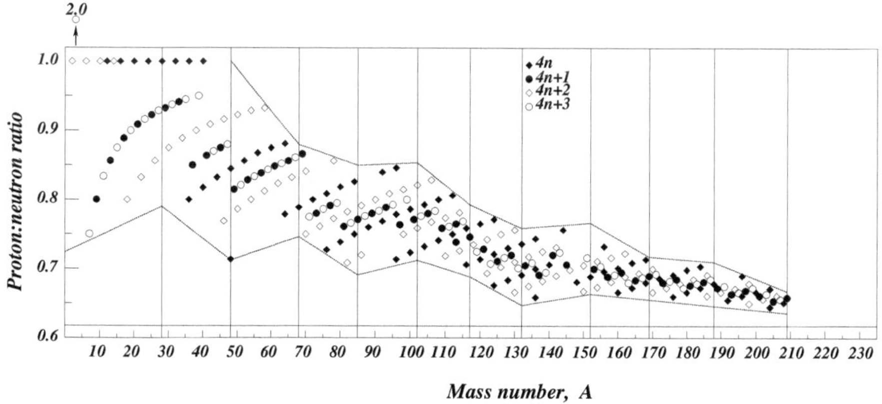

Figure 4.2: *Distribution of stable nuclides as a function of proton:neutron ratio and mass number*

defined by the modular series $A(\text{mod}4) \equiv 0 \rightarrow 3$. The two even series, $A = 4n$

and $A = 4n + 2$, consist of 81 members each, and the odd series consist of 51 members each. Each of the four distributions converge, separately, to $A = 267$. The field of stability, outlined by the converging straight lines in Figure 4.1, is better defined by the two zig-zag envelopes with common inflection points on 11 vertical hem lines that divide the nuclides into groups of 24.

The nuclides of each modular group are spread along 11 festoons that terminate in a radioactive nuclide at each end. Nuclides on the high-ratio side decay by positron emission or electron capture; those on the low-ratio side by β-emission. All nuclides with $A > 209$ decay by α-emission. The 81 naturally stable elements have, on average, 3 isotopes each. The predicted 100 elements on the cosmic scale, by the same reckoning, correspond to 300 isotopes.

4.3 The Number Spiral

The conclusions of the previous section are summarized well in terms of the number spiral with a period of 24, shown in Figure 4.3. The arrows mark eight radial directions where all prime numbers, except for 2 and 3, and

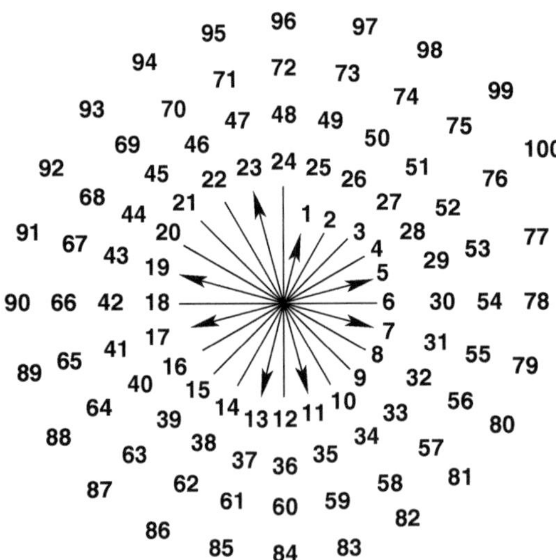

Figure 4.3: *Natural numbers arranged on a spiral of period 24. Arrows identify the eight radial prime-number directions*

their binary products are located. These directions are associated with the chemical elements, with an implied periodicity of 8, while the 24 numbers per cycle represent the nuclides. The sum of all numbers over any complete cycle is given by:

$$\sigma_{j+1} = \sum_{n=24j}^{24(j+1)} n = (2j+1)300\,,\ j = 0, 1, 2, \ldots$$

The sums $\sigma_i = a, 3a, 5a, 7a, \ldots,\ (a = 300)$ have coefficients that match the numbers of s, p, d, f electron pairs required to fill atomic sub-shells. The constant $a = 300$ corresponds to the maximum possible number of stable nuclides and $a/3$ to the maximum number of stable elements.

The relative energies of sub-levels and hence the order in which these are occupied by electrons cannot be predicted in detail, but this pattern is revealed when the periodicity of the nuclides is examined as a function of atomic number. This distribution of stable nuclides as a function of neutron imbalance is shown in Figure 4.4. To ensure that the grouping into sets

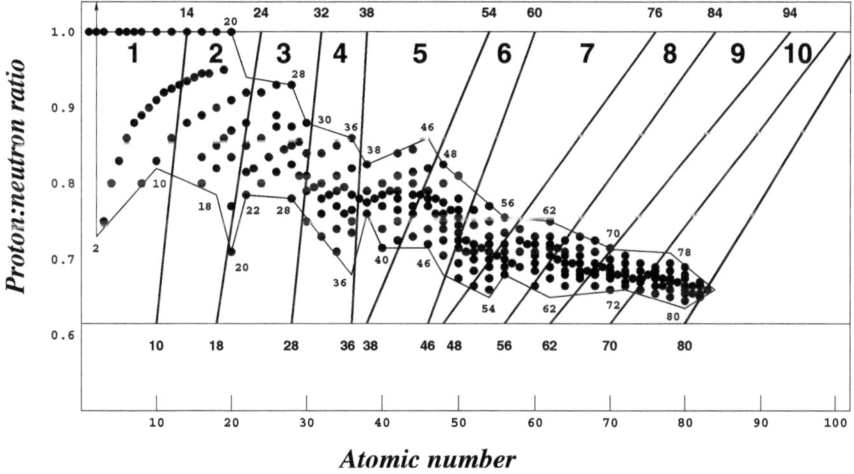

Figure 4.4: *Nuclide distribution as a function of atomic number*

of 24, which defines the periodicity, is undisturbed, the eleven hem lines are drawn at different angles with respect to the coordinate axes. The distribution converges to τ at $Z = 102$, as before, and the hem lines intersect the line $Z/N = \tau$ at points that define elemental periodicity at this limiting ratio. The points coincide with the empirically known atomic numbers at the completion of energy levels according to the conventional form of the periodic table of the elements.

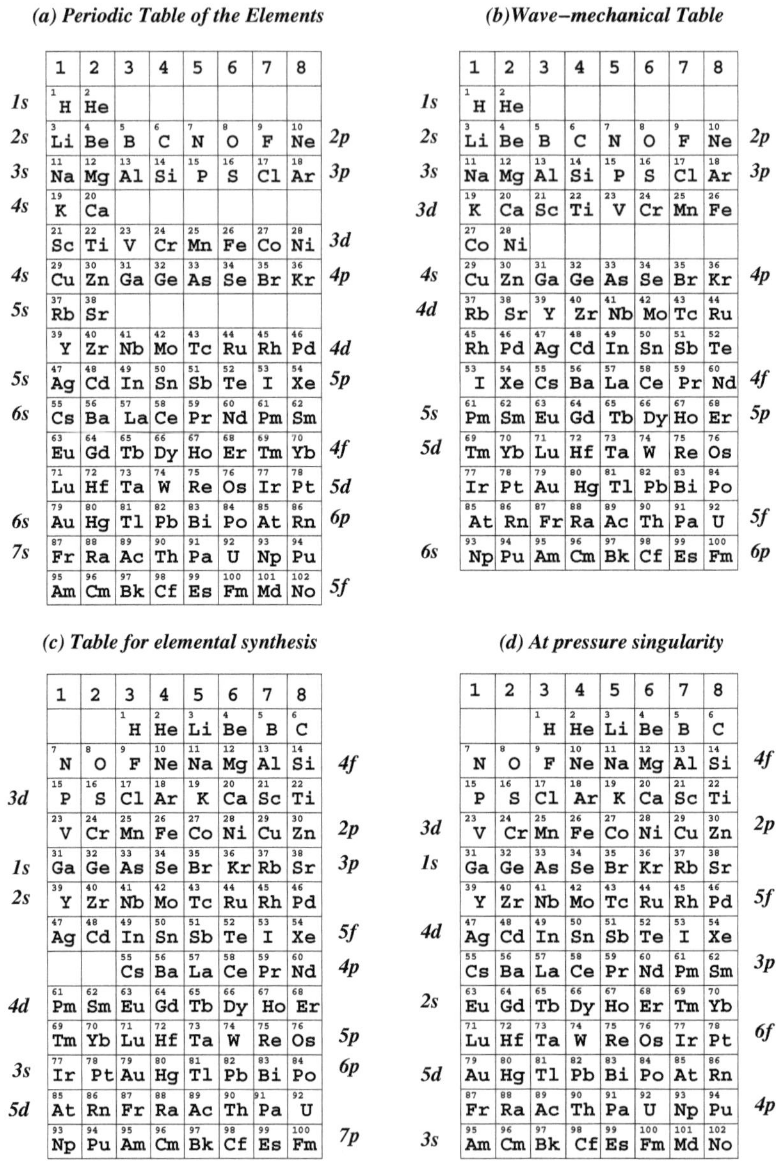

(a) Periodic Table of the Elements

	1	2	3	4	5	6	7	8	
1s	1 H	2 He							
2s	3 Li	4 Be	5 B	6 C	7 N	8 O	9 F	10 Ne	2p
3s	11 Na	12 Mg	13 Al	14 Si	15 P	16 S	17 Cl	18 Ar	3p
4s	19 K	20 Ca							
	21 Sc	22 Ti	23 V	24 Cr	25 Mn	26 Fe	27 Co	28 Ni	3d
4s	29 Cu	30 Zn	31 Ga	32 Ge	33 As	34 Se	35 Br	36 Kr	4p
5s	37 Rb	38 Sr							
	39 Y	40 Zr	41 Nb	42 Mo	43 Tc	44 Ru	45 Rh	46 Pd	4d
5s	47 Ag	48 Cd	49 In	50 Sn	51 Sb	52 Te	53 I	54 Xe	5p
6s	55 Cs	56 Ba	57 La	58 Ce	59 Pr	60 Nd	61 Pm	62 Sm	
	63 Eu	64 Gd	65 Tb	66 Dy	67 Ho	68 Er	69 Tm	70 Yb	4f
	71 Lu	72 Hf	73 Ta	74 W	75 Re	76 Os	77 Ir	78 Pt	5d
6s	79 Au	80 Hg	81 Tl	82 Pb	83 Bi	84 Po	85 At	86 Rn	6p
7s	87 Fr	88 Ra	89 Ac	90 Th	91 Pa	92 U	93 Np	94 Pu	
	95 Am	96 Cm	97 Bk	98 Cf	99 Es	100 Fm	101 Md	102 No	5f

(b) Wave–mechanical Table

	1	2	3	4	5	6	7	8	
1s	1 H	2 He							
2s	3 Li	4 Be	5 B	6 C	7 N	8 O	9 F	10 Ne	2p
3s	11 Na	12 Mg	13 Al	14 Si	15 P	16 S	17 Cl	18 Ar	3p
3d	19 K	20 Ca	21 Sc	22 Ti	23 V	24 Cr	25 Mn	26 Fe	
	27 Co	28 Ni							
4s	29 Cu	30 Zn	31 Ga	32 Ge	33 As	34 Se	35 Br	36 Kr	4p
4d	37 Rb	38 Sr	39 Y	40 Zr	41 Nb	42 Mo	43 Tc	44 Ru	
	45 Rh	46 Pd	47 Ag	48 Cd	49 In	50 Sn	51 Sb	52 Te	
	53 I	54 Xe	55 Cs	56 Ba	57 La	58 Ce	59 Pr	60 Nd	4f
5s	61 Pm	62 Sm	63 Eu	64 Gd	65 Tb	66 Dy	67 Ho	68 Er	5p
5d	69 Tm	70 Yb	71 Lu	72 Hf	73 Ta	74 W	75 Re	76 Os	
	77 Ir	78 Pt	79 Au	80 Hg	81 Tl	82 Pb	83 Bi	84 Po	
	85 At	86 Rn	87 Fr	88 Ra	89 Ac	90 Th	91 Pa	92 U	5f
6s	93 Np	94 Pu	95 Am	96 Cm	97 Bk	98 Cf	99 Es	100 Fm	6p

(c) Table for elemental synthesis

	1	2	3	4	5	6	7	8	
			1 H	2 He	3 Li	4 Be	5 B	6 C	
	7 N	8 O	9 F	10 Ne	11 Na	12 Mg	13 Al	14 Si	4f
3d	15 P	16 S	17 Cl	18 Ar	19 K	20 Ca	21 Sc	22 Ti	
	23 V	24 Cr	25 Mn	26 Fe	27 Co	28 Ni	29 Cu	30 Zn	2p
1s	31 Ga	32 Ge	33 As	34 Se	35 Br	36 Kr	37 Rb	38 Sr	3p
2s	39 Y	40 Zr	41 Nb	42 Mo	43 Tc	44 Ru	45 Rh	46 Pd	
	47 Ag	48 Cd	49 In	50 Sn	51 Sb	52 Te	53 I	54 Xe	5f
			55 Cs	56 Ba	57 La	58 Ce	59 Pr	60 Nd	4p
4d	61 Pm	62 Sm	63 Eu	64 Gd	65 Tb	66 Dy	67 Ho	68 Er	
	69 Tm	70 Yb	71 Lu	72 Hf	73 Ta	74 W	75 Re	76 Os	5p
3s	77 Ir	78 Pt	79 Au	80 Hg	81 Tl	82 Pb	83 Bi	84 Po	6p
5d	85 At	86 Rn	87 Fr	88 Ra	89 Ac	90 Th	91 Pa	92 U	
	93 Np	94 Pu	95 Am	96 Cm	97 Bk	98 Cf	99 Es	100 Fm	7p

(d) At pressure singularity

	1	2	3	4	5	6	7	8	
			1 H	2 He	3 Li	4 Be	5 B	6 C	
	7 N	8 O	9 F	10 Ne	11 Na	12 Mg	13 Al	14 Si	4f
	15 P	16 S	17 Cl	18 Ar	19 K	20 Ca	21 Sc	22 Ti	
3d	23 V	24 Cr	25 Mn	26 Fe	27 Co	28 Ni	29 Cu	30 Zn	2p
1s	31 Ga	32 Ge	33 As	34 Se	35 Br	36 Kr	37 Rb	38 Sr	
	39 Y	40 Zr	41 Nb	42 Mo	43 Tc	44 Ru	45 Rh	46 Pd	5f
4d	47 Ag	48 Cd	49 In	50 Sn	51 Sb	52 Te	53 I	54 Xe	
	55 Cs	56 Ba	57 La	58 Ce	59 Pr	60 Nd	61 Pm	62 Sm	3p
2s	63 Eu	64 Gd	65 Tb	66 Dy	67 Ho	68 Er	69 Tm	70 Yb	
	71 Lu	72 Hf	73 Ta	74 W	75 Re	76 Os	77 Ir	78 Pt	6f
5d	79 Au	80 Hg	81 Tl	82 Pb	83 Bi	84 Po	85 At	86 Rn	
	87 Fr	88 Ra	89 Ac	90 Th	91 Pa	92 U	93 Np	94 Pu	4p
3s	95 Am	96 Cm	97 Bk	98 Cf	99 Es	100 Fm	101 Md	102 No	

Figure 4.5: *The four periodic tables*

To emphasize the periodicity of 8, suggested by the number spiral, the periodic table is rearranged as shown in Figure 4.5. Closure of the eleven periods coincides with the completion of electronic sub-levels, except for atomic numbers 62 and 94 that split the f-levels into sub-sets of 6 and 8. Additional structure, in complete agreement with the experimentally known sub-level order of the elements, is revealed by the zig-zag profiles that define the field of nuclide stability in Figure 4.4.

4.4 Elemental Synthesis

It is no accident that golden-ratio sampling predicts the observed periodic table, whereas the hydrogen solution of Schrödinger's equation fails to do so. From observations in botany and astronomy the structure of natural objects appears related to the geometry of space-time, through the golden ratio, and it may be inferred that the assumed geometry of the Schrödinger simulation is a good, but inadequate, approximation. It probably means that the Schrödinger result may be recovered by sampling at a different ratio of $Z/N \neq \tau$. This possibilty is explored in Figure 4.6. In fact, extrapolation

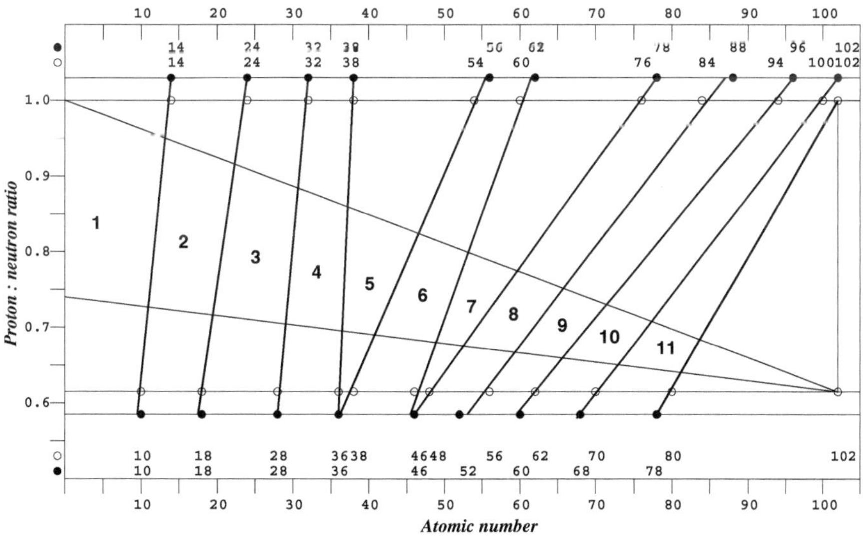

Figure 4.6: *Four periodic arrangements defined by extrapolation of the 11 hem lines*

of the hem lines to $Z/N \simeq 0.58$ predicts the same energy spectrum as the Schrödinger equation.

The difference between the observed and Schrödinger spectra is due to an environmental effect. The Schrödinger equation assumes coulombic interaction between the proton and electron of hydrogen, and nothing else, *i.e.* an essentially empty universe. As argued before, environmental pressure exists in the real world and has an effect on electronic levels, which can be simulated by uniform compression of the atom. The effect of such compression should become even more apparent on extrapolation of the hem lines to the ratio $Z/N = 1$, and it does.

The intersection of hem lines at this ratio points at an inversion of observed energy levels, such that $4f < 3d < 2p < 1s \ldots etc.$, predicted to occur under extreme pressure. Extrapolation to $Z/N > 1$ is interpreted as an approach to infinitely high pressure and complete inversion of the Schrödinger spectrum. Periodic arrangements corresponding to all special conditions, *i.e.* $Z/N = \tau$, $< \tau$, $= 1$ and > 1 are shown in Figure 4.5.

Recall the reciprocity between matter and curvature, implied by the theory of general relativity, to argue that the high-pressure condition at $Z/N = 1$ corresponds to extreme curvature of space-time caused by massive objects such as quasars, and the like. The argument implies that the Schrödinger solution is valid in empty, flat euclidean space-time, that $Z/N = \tau$ corresponds to the real world, $Z/N = 1$ occurs in massive galactic objects where elemental synthesis happens, and $Z/N > 1$ implies infinite curvature at a space-time singularity.

It has been demonstrated [62] that nuclear synthesis can be rationalized in terms of continued α-particle (^{4}He) addition, starting from the elementary units ^{n}He ($n = 2, 3, 4, 5$), to yield the four modular series of nuclides shown in Figure 4.2. By assumption this process happens under cosmic conditions where all stable nuclides consist of protons and neutrons in the ratio $Z/N = 1$. The even mass number series, $A = 4n$ and $A = 4n + 2$, result from the equilibrium chain reactions:

$$^4_2\mathrm{He} \overset{\alpha}{\rightleftharpoons} {}^8_4\mathrm{Be} \overset{\alpha}{\rightleftharpoons} {}^{12}_6\mathrm{C} \cdots + n\alpha \cdots \overset{\alpha}{\rightleftharpoons} {}^{200}_{100}\mathrm{Fm}$$

$$^2_1\mathrm{H} \overset{\alpha}{\rightleftharpoons} {}^6_3\mathrm{Li} \overset{\alpha}{\rightleftharpoons} {}^{10}_5\mathrm{B} \cdots + n\alpha \cdots \overset{\alpha}{\rightleftharpoons} {}^{202}_{101}\mathrm{Md}$$

which represent 100 single-isotope elements. On release of this equilibrium mixture, into more moderately curved space-time regions, most nuclides become unstable and decay radioactively by positron emission, which produces new elements and lowers the Z/N ratio, without affecting the mass number. This process results in a decay cascade that extends all the way down to a nuclide which is stable in the new environment. After n steps of decay the initial nuclide is converted into a stable daughter:

$$^A_{Z_0}\mathrm{X} \overset{-2n\beta^+}{\longrightarrow} {}^A_{Z_s}\mathrm{Y}$$

where $Z_0 = A/2$ and $Z_s = Z_0 - 2n$.

The initial ($Z/N = 1$) and final ($Z/N \simeq \tau$) states of the decay process are mapped in Figure 4.7. The upper frame shows the decay of even mass

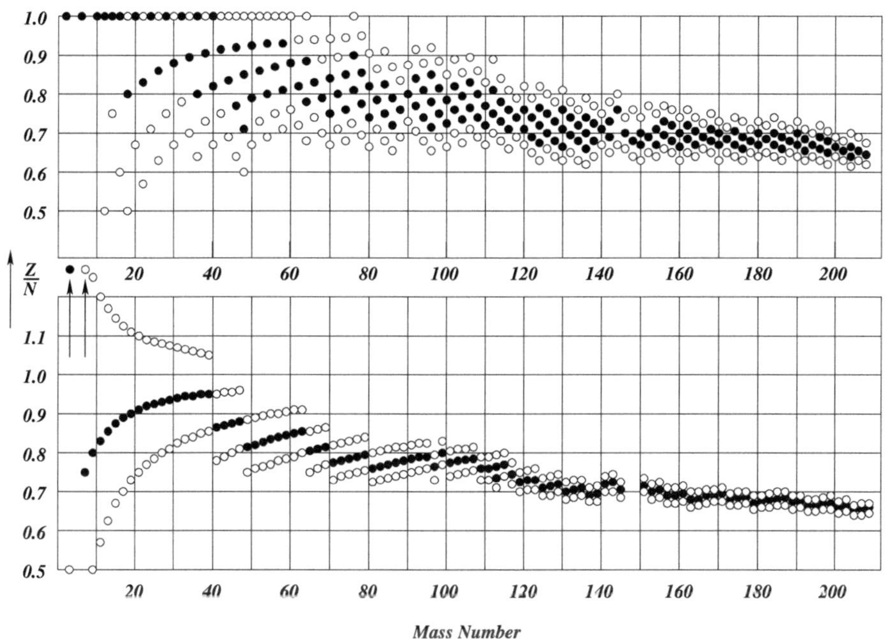

Figure 4.7: *Diagram to explain nucleogenesis*

number nuclides by the emission of two positrons per event. Emission of a single positron would lead to the formation of odd-odd nuclei, which is not observed, presumably due to an intrinsic instability of such nuclides. The last β^+ and first β^- unstable nuclides are shown as open circles for each mass number. The positions of α-unstable nuclides are left blank. As in Figure 4.1, the isotopes of a given element occur on obvious circular segments.

The very lightest nuclei ($A < 16$) are seen not to suffer any decay and others with $16 < A < 46$ experience only one stage of decay, consisting of two steps of positron emission and/or electron capture. The effect of the two-step process is that no odd-odd ($Z, N = 2n + 1$) nuclides, beyond ^{14}N, occur naturally.

Two different stable nuclides, related by two steps of positron decay, occur with the same mass number of 36:

$$^{36}_{18}\text{Ar} \rightarrow {}^{36}_{16}\text{S}$$

This is interpreted to indicate that ^{36}Ar is metastable in free space ($Z/N \leq \tau$) where it decays to ^{36}S. However, when incorporated into a new condensing galaxy ($Z/N \geq \tau$), enhanced gravity stabilizes the metastable nuclide, resulting in the occurrence of nuclide pairs with the same mass number and hence, elements with more than one isotope. Naturally occurring α-emitters are still in a metastable state. The same happens in about half of all cases – in four instances three stable nuclides with the same mass number are formed. The result is the formation of 81 stable nuclides from 50 starting products, in each of the even A series.

The situation is somewhat different for the two odd series of nuclides. Odd A implies that either N or Z be odd, *i.e.*:

$$A_o = N + Z_o \quad \text{or} \quad A_o = Z + N_o\,;\ (Z = N \pm 1)$$

To have $Z/N = 1$ for given A_o, both of the two different nuclides should be present. Each of the two odd series, $A_o = 4n \pm 1$, therefore consists of 100 isotopes of 50 elements. All told, the four series of nuclides, with $Z/N = 1$, consist of 100 elements and their 300 isotopes. Because the pairs of isotopes are interconvertible, on positron exchange, they produce the same decay product at $Z/N < 1$, always selecting $Z < N$. The isotope ^{3}He is the only surviving nuclide with $Z > N$. With one exception each odd-A pair decays to single nuclides, each series producing 51 nuclides. Hence, $2 \times 81 + 2 \times 51 = 264$, accounts for the observed number of stable isotopes of 81 elements.

From Figure 4.7 it may seem that the known nuclides are decay products of 309 isotopes of 102 elements. However, elements 43 and 61 and the nuclides of mass number 4, 5, 8, 147 and 149 do not feature, thus reducing the starting set to 300 isotopes of 100 elements.

The mechanism of nuclear synthesis by α-addition, followed by β^+ decay, as assumed here, should produce a final nuclide for each mass number, such that the next member in the decay chain has $Z/N < \tau$. The only nuclides that satisfy this requirement are ^{2}H, ^{3}He, ^{6}Li, ^{7}Li, ^{9}Be, ^{11}B, ^{12}C, ^{16}O, ^{18}O, ^{22}Ne and ^{48}Ca. Some borderline cases are ^{36}S, ^{46}Ca, ^{136}Xe, ^{198}Pt and ^{204}Hg. For all other even mass numbers decay invariably stops one step short of reaching τ, in line with the proposition that the limiting ratio inside galaxies exceeds τ.

The mechanism assumed here is equivalent to the previous proposal [62] of α-addition, with the added advantage of demonstrating the transition between two states defined by $Z/N = 1$ and τ respectively. This proposed synthesis of elements is more economical than the widely accepted big-bang scenario that requires special hypothetical conditions for the production of each nuclide. Unlike the big-bang mechanism, it also accounts for the demonstrated periodicity of nuclide stability and cosmic abundance [62], without further assumptions.

4.5 The Golden Parameter

The four different periodic tables account for the observed elemental diversity and provide compelling evidence that the properties of atomic matter are intimately related to the local properties of space-time, conditioned by the golden parameter $\tau = 1/\Phi$. The appearance of τ in the geometrical description of the very small (atomic nuclei) and the very large (spiral galaxies) emphasizes its universal importance and implies the symmetry relationship of self-similarity between all states of matter. This property is vividly illustrated by the formulation of τ as a continued fraction:

$$x = 1 + \frac{1}{1 + \frac{1}{1+\frac{1}{1+\frac{1}{1+\dots}}}} = 1 + \frac{1}{x}, \quad i.e.\ x = \Phi$$

The golden parameter is perhaps best known from botanical phyllotaxis and from its relationship to five-fold symmetry, as evidenced by trigonometric formulae such as;

$$\begin{aligned} \Phi &= 2\sin\left(\frac{3\pi}{10}\right) - 2\cos\left(\frac{\pi}{5}\right) \\ \tau &= 2\sin\left(\frac{\pi}{10}\right) = 2\cos\left(\frac{2\pi}{5}\right) \end{aligned}$$

These angles are all fractions of $3\pi/5 = 108°$, characteristic of a regular pentagon.

As a working hypothesis it is postulated that the natural curvature of space-time is a function of τ and π. The growth of any structure, or object, large or small, in free space, is guided by the same general geometrical template. Objects that grow in regions of high curvature, such as nuclides during synthesis, are of a different architecture, and may not necessarily be stable in free space. Some of the nuclides, all of which are synthesized in the ratio of $Z/N = 1$, are therefore unstable and decay radioactively on release into free space. Only 264 of the original 300 nuclides and only 81 of the original elements decay into stable entities.

The geometrical principles that underlie interactions inside atomic nuclei must also prevail in molecular space and the golden parameter is likely to surface again in problems of chemical bonding. The advantage is that once the role of the golden ratio has been identified the mathematical analysis of the situation is simplified enormously. Such is the case with the periodic classification of the elements.

4.6 Periodic Table of the Elements

The periodic table of the elements represents the classification that results on arranging the elements in order of increasing atomic number. This arrangement mirrors the distribution of primes within a natural-number spiral of period 24, well characterized in terms of a few simple concepts from number theory.

The sum over all numbers from 0 to k is given by the triangular number $k(k+1)/2$. The sum over numbers on each cycle of a periodic number spiral follows as:

$$\frac{1}{2}[lk.l(k+1)-(l-1)^2k(k+1)]=\frac{1}{2}(2l+1)k(k+1)\,,$$

where k and l are the period and number of the cycle and $2l+1$ is the difference between successive square numbers. On the nuclide cycle of 24 these numbers are interpreted as counting electrons: $\sum_l k(k+1)(2l+1)$; nuclides: $k(k+1)/2 = 300$, $(Z_{max} = 100)$; electron pairs per sub-shell: $2l+1 = 1,3,5,7$ for s,p,d,f sub-shells; and electron pairs per energy level: $\sum_{l=0}^{n-1}(2l+1) = n^2$.

By matching these occupation numbers to the observed periodic structure derived from Figure 4.4 the correct energy spectrum is indicated:

$1s$	2		$5s,4d$	38, 46
$2s,2p$	4, 10		$5s,5p$	48, 54
$3s,3p$	12, 18		$6s,4f(6)$	56, 62
$4s,3d$	20, 28		$4f$	70
$4s,4p$	30, 36		$5d,6s$	78, 80

The novel feature is that ns sub-levels become vacant towards the end of each $(n-1)d$ sub-level to make room for 10 d electrons and to be filled again next. In this way each transition series consists of only 8 elements and not 10 as implied by the Schrödinger spectrum.

The principal and azimuthal quantum numbers are directly defined as n and l respectively. The $2l+1$ multiplicity of sub-levels defines the allowed values of the magnetic quantum number m_l, on assuming the Bohr condition:

$$L_zY_l^{m_l}=m\hbar Y_l^{m_l}$$

Only those periodic aspects of electronic configuration that depend on electron spin and the exclusion principle remain unaccounted for.

4.6.1 Farey Fractions and Ford Circles

The most convincing derivation of periodic structure, using the concepts of number theory, comes from a comparison with Farey sequences. The Farey scheme is a device to arrange rational fractions in enumerable order. Starting from the end members of the interval [0,1] an infinite tree structure is generated by separate addition of numerators and denominators to produce the Farey sequences $\mathscr{F}_n$, of order n, where n limits the values of denominators in the sequence.

$\mathscr{F}_1$	$\frac{0}{1}$										$\frac{1}{1}$
$\mathscr{F}_2$	$\frac{0}{1}$					$\frac{1}{2}$					$\frac{1}{1}$
$\mathscr{F}_3$	$\frac{0}{1}$			$\frac{1}{3}$		$\frac{1}{2}$		$\frac{2}{3}$			$\frac{1}{1}$
$\mathscr{F}_4$	$\frac{0}{1}$		$\frac{1}{4}$	$\frac{1}{3}$		$\frac{1}{2}$		$\frac{2}{3}$	$\frac{3}{4}$		$\frac{1}{1}$
$\mathscr{F}_5$	$\frac{0}{1}$	$\frac{1}{5}$	$\frac{1}{4}$	$\frac{1}{3}$	$\frac{2}{5}$	$\frac{1}{2}$	$\frac{3}{5}$	$\frac{2}{3}$	$\frac{3}{4}$	$\frac{4}{5}$	$\frac{1}{1}$

A pair of fractions, adjacent to each other in any sequence, has the property of unimodularity, such that for the pair h/k, l/m the quantity $|hm - kl| = 1$.

Each rational fraction, h/k, defines a *Ford circle* with a radius and y-coordinate of $1/(2k^2)$, positioned at an x-coordinate h/k. The Ford circles of any unimodular pair are tangent to each other and to the x-axis. The circles, numbered from 1 to 4 in the construction overleaf, represent the Farey sequence of order 4. This sequence has the remarkable property of one-to-one correspondence with the natural numbers ordered in sets of $2k^2$ and in the same geometrical relationship as the Ford circles of $\mathscr{F}_4$.

Any chemist immediately recognizes the periodic table of the elements in this arrangement. Rearrangement into sets of 8, as shown, generates the natural periodic table of Figure 4.5. The familiar attempt to relate the periodic order of the elements to a sequential occupation of increasingly higher energy levels:

$$(1s) < (2s, 2p) < (3s, 3p, 3d) < (4s, 4p, 4d, 4f) < \cdots,$$

each accommodating $2n^2$ members, does not arise from $\mathscr{F}_4$.

The detail of secondary periods embedded within primaries, *e.g.* 2 within 8, 8 within 18, *etc.*, to generate substructures such as $2+6=8$, $8+10=18$,

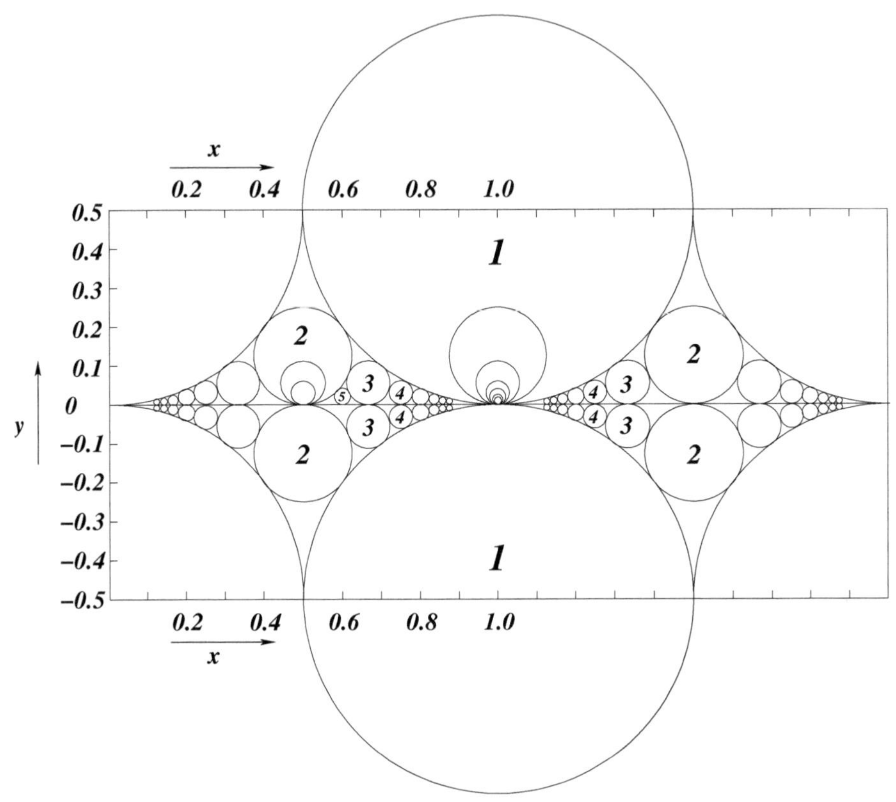

																																$2k^2$
87	88	89	90	91	92	93	94	95	96	97	98	99	100	101	102																118	
														37	38	39	40	41	42	43	44	45	46	47	48	49	50	51	52	53	54	
																								11	12	13	14	15	16	17	18	
																								1							2	2
																								3	4	5	6	7	8	9	10	8
														19	20	21	22	23	24	25	26	27	28	29	30	31	32	33	34	35	36	18
55	56	57	58	59	60	61	62	63	64	65	66	67	68	69	70	71	72	73	74	75	76	77	78	79	80	81	82	83	84	85	86	32
	2+6													6+2					8						2+6							

$18+14=32$, is correctly mapped by the Ford circles that represent non-primitive fractions. A nesting of Ford circles 1/1, 2/2, 3/3, 4/4 …, is shown by way of illustration.

The numerical values of primary, secondary and supplementary periods are $2k^2$, $2(k-1)^2$, $2(2k-1)$ for $k = 1, 2, 3, 4$ respectively. The substructure of the periodic table is fully accounted for.

Fibonacci Fractions

The rational fractions defined by successive Fibonacci numbers in the sequence:

$$\begin{array}{lcccccccccl} F_n & = & 1 & 1 & 2 & 3 & 5 & 8 & 13 & \ldots & (n = 1, 2, \ldots) \\ i.e. & & \frac{1}{1} & \frac{1}{2} & \frac{2}{3} & \frac{3}{5} & \frac{5}{8} & \frac{8}{13} & & \ldots & \end{array}$$

have the same unimodular property as the Farey sequences and the additional property of converging to the golden ratio, $\tau = 0.61803....$ The corresponding Ford circles, of diminishing size, now converge to τ and $1 + \tau$ rather than to 0 and 1.

The convergence follows the Fibonacci fractions which appear in the Farey tree structure that develops between the limits $\frac{1}{2}$ and $\frac{2}{3}$.

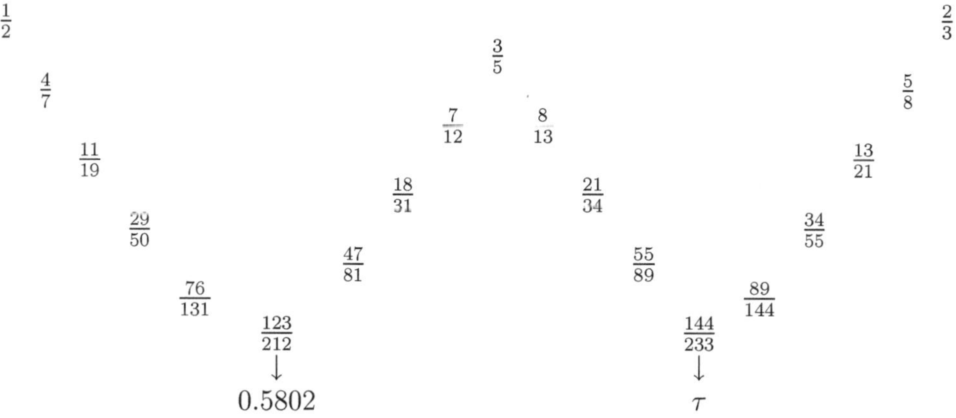

The branch above $\frac{3}{5}$ converges through $\frac{144}{233}$ to τ and the lower branch to 0.5802. The numerators of the latter unimodular series:

$$\frac{1}{2} \quad \frac{3}{5} \quad \frac{4}{7} \quad \frac{7}{12} \quad \frac{11}{19} \quad \frac{18}{31} \ldots$$

are recognized as the *Lucas* numbers, defined by

$$L_n = F_{n-1} + F_{n+1}$$

Alternatively, $F_n = 5(L_{n-1} + L_{n+1})$, with $L_0 = 2$. The simplest formulation of the converging Lucas fractions is therefore given by

$$\frac{L_n}{L_n + F_{n+1}} \quad , \quad n = 1, 2, \ldots$$

It follows that the two branches of the Fibonacci-Lucas tree describe Ford-circle periodicities that converge respectively to τ and 0.5802 as in Figure 4.6. Assuming that the two branches sprout from $\mathscr{F}_4$ and $\mathscr{F}_2$ respectively, different Ford-circle sequences can be identified to generate the observed periodicities:

$$2,8,8,18,18,32,32,50,\ldots \quad , \quad 2,2(2n^2\,,\,n=2,3\ldots)$$

and

$$2,(8,18,32),50,\ldots, \quad 2n^2 \text{ for } n=1,2,\ldots$$

Parentheses indicate nested circles. These alternatives match the periodic tables of Figure 4.5, derived at $Z/N=\tau$ and 0.58 respectively.

4.7 Electron Spin

All electrons, protons and neutrons, the elementary constituents of atoms, are fermions and therefore intrinsically endowed with an amount $\hbar/2$ of angular momentum, known as *spin*. Like mass and charge, the other properties of fermions, the nature of spin is poorly understood. In quantum theory spin is treated purely mathematically in terms of operators and spinors, without physical connotation.

In practice, spin is as real as mass and charge, and routinely measured spectroscopically by the techniques of electron and nuclear magnetic resonance. These measurements are done under widely different conditions and with minor interference between the two phenomena. As implied by the terminology, atomic spin is of two different types – separately associated with the nucleus and extranuclear electrons respectively. The theoretical challenge is how to describe these two independent rotations within the same body.

It is necessary to first understand the spin of a free fermion. Considered as an isolated dimensionless point particle, no conceivable mechanism can explain the physical origin of its magnetic moment. Even the rotation of a spherically symmetrical indivisible unit charge, associated with a wave structure in the aether, cannot have an intrinsic magnetic moment.

4.7.1 Spherical Rotation

The most convincing physical model [63] explains electron spin in terms of *spherical rotation*, another way of rotating a solid object, different from the well-known mode of rotation about an axis. It starts as a slight wobble, which, by continuous exaggeration of the motion, develops into a double rotation of the object, without rotating about an axis[2]. One advantage of this mode of rotation is that an object connected to a supporting frame by flexible strings does not get the strings tangled up during the spherical gyrations, in stark contrast with axial rotation. While the supporting strings of the spherically rotating object appear to wind up during the first half of the spherical rotational cycle, they unwind during the second half, and return to their original configuration on completion of the full cycle. The only effect on the strings is an undulation at half the frequency of the spherical rotation.

Should a fermion represent some special distortion, or knot, in the aether, spherical rotation allows it to move freely through the space-time continuum without getting entangled with its environment, which consists of the same stuff as the fermion. While rotating in spherical mode adhesion to the environment is rythmically stretched and relaxed as the fermion moves through space. This half-frequency disturbance of the wave-field, that supports the fermion in space, constitutes the effect observed as spin.

The peculiarity of spherical rotation is that rotation by 2π fails to return the rotating object to its initial orientation. Evidently there is an additional aspect to the state of orientation that needs to be taken into account. Two *versions* are said to be associated with each orientation. The quaternion operator

$$R = \cos(\theta/2) - i\sin(\theta/2)$$

for axial rotation, changes sign on rotation through odd multiples of $\theta = 2\pi$, which corresponds to spherical rotation of 2π. However, this sign alternation only becomes evident when the operation is performed on a column vector, known as a *spinor*.

In order to take the version into account spherical rotation is represented symbolically by

$$\begin{bmatrix} i & 0 \\ 0 & -i \end{bmatrix} \begin{bmatrix} 1 \\ 0 \end{bmatrix} \rightarrow \begin{bmatrix} i \\ 0 \end{bmatrix}$$

This operation represents a quarter turn on a great circle or rotation of π radians about an axis. An intermediate position is given by the spinor:

$$\begin{bmatrix} \cos\theta + i\sin\theta \\ 0 \end{bmatrix} = \begin{bmatrix} e^{i\theta} \\ 0 \end{bmatrix}$$

[2]This operation is the basis of the party trick in which the joker twirls a bowl of soup around its vertical axis without spilling. The bowl makes two full turns in executing the complete gyration.

where θ is the great-circle displacement, or a 2θ rotation of the core about an axis. The matrix that generates this position from the initial configuration is

$$\begin{bmatrix} e^{i\theta} & 0 \\ 0 & e^{-i\theta} \end{bmatrix} \begin{bmatrix} 1 \\ 0 \end{bmatrix} \rightarrow \begin{bmatrix} e^{i\theta} \\ 0 \end{bmatrix}$$

In a continuous process, which is a linear function of time, the operation that generates spin becomes:

$$\begin{bmatrix} e^{i\omega t} & 0 \\ 0 & e^{-i\omega t} \end{bmatrix} \begin{bmatrix} 1 \\ 0 \end{bmatrix} \rightarrow \begin{bmatrix} e^{i\omega t} \\ 0 \end{bmatrix}$$

The spin, measured along an axis, is inverted by the operation:

$$\begin{bmatrix} 0 & -1 \\ 1 & 0 \end{bmatrix} \begin{bmatrix} e^{i\omega t} \\ 0 \end{bmatrix} \rightarrow \begin{bmatrix} 0 \\ e^{i\omega t} \end{bmatrix}$$

Inversion of the time parameter, $t \rightarrow -t$, or angular velocity $\omega \rightarrow -\omega$, inverts the spin into $\{e^{-i\omega t}, 0\}$. This motion is generated from the initial configuration $\{1, 0\}$ by the operation:

$$\begin{bmatrix} e^{-i\omega t} & 0 \\ 0 & e^{i\omega t} \end{bmatrix} \begin{bmatrix} 1 \\ 0 \end{bmatrix} \rightarrow \begin{bmatrix} e^{-i\omega t} \\ 0 \end{bmatrix}$$

Spin, reversed in this sense, is interpreted as anti-spin, associated with an anti-fermion. The inverted antispin is $\{0, e^{-i\omega t}\}$.

A spinor system, such as

$$\begin{bmatrix} e^{i\omega t} & 0 \\ 0 & e^{-i\omega t} \end{bmatrix} \begin{bmatrix} \phi_1 \\ \phi_2 \end{bmatrix}$$

has space and time derivatives, in a Lorentzian frame, that satisfy Dirac's relativistic wave equation [63]. In a Galilean frame it leads to Schrödinger's equation [7].

4.7.2 Schrödinger's Equation and Spin

Consider a spinor that moves along x,

$$\begin{bmatrix} e^{i(\omega t - kx)} & 0 \\ 0 & e^{-i(\omega t - kx)} \end{bmatrix} \begin{bmatrix} \phi_1 \\ \phi_2 \end{bmatrix} = \begin{bmatrix} \phi_1 e^{+} \\ \phi_2 e^{-} \end{bmatrix},$$

in shorthand notation. Form the derivatives of the spinor:

$$\begin{aligned}\frac{\partial\phi}{\partial t} &= \frac{\partial}{\partial t}\left[\begin{array}{c}\phi_1 e^+\\ \phi_2 e^-\end{array}\right] = \left[\begin{array}{c}i\omega\phi_1 e^+\\ -i\omega\phi_2 e^-\end{array}\right] = i\omega\left[\begin{array}{c}\phi_1 e^+\\ -\phi_2 e^-\end{array}\right]\\ \frac{\partial\phi}{\partial x} &= i\left[\begin{array}{c}-k\phi_1 e^+\\ k\phi_2 e^-\end{array}\right] = -ik\left[\begin{array}{c}\phi_1 e^+\\ -\phi_2 e^-\end{array}\right]\\ \frac{\partial^2\phi}{\partial x^2} &= k^2\left[\begin{array}{c}\phi_1 e^+\\ -\phi_2 e^-\end{array}\right]\end{aligned}$$

It follows that

$$\frac{1}{i\omega}\frac{\partial\phi}{\partial t} = \frac{1}{k^2}\frac{\partial^2\phi}{\partial x^2}$$

with similar results for y and z. In three dimensions

$$-i\frac{\partial\phi}{\partial t} = \frac{\omega}{k^2}\nabla^2\phi$$

This expression resembles Schrödinger's equation

$$-i\hbar\frac{\partial\Psi}{\partial t} = \frac{\hbar^2}{2m}\nabla^2\Psi$$

The two equations become identical for $2m\hbar\omega = (\hbar k)^2$, *i.e.* for $\hbar\omega = E = p^2/2m$, such that $k = 2\pi/\lambda$ and $p = h/\lambda$, the De Broglie condition.

This result is interpreted to show that a region of the continuum which rotates in spherical mode, interacts inertially with its environment by generating a wave-like shroud that undulates at half the angular frequency of the core. The inertial mass is related to the angular frequency through Planck's constant. The angular momentum on the surface of a unit sphere is $L = m\omega$, and with the wavelength of the undulation, $\lambda = 2\pi$, $k = 1$, the spin angular momentum follows as $m\omega = \hbar/2$.

To transform the squared Schrödinger operator

$$S^2 = -i\hbar\frac{\partial}{\partial t} - \frac{\hbar^2}{2m}\nabla^2 \quad , \quad S^2\Psi = 0$$

into a spinor operator it is written in the linear form

$$S = -i\hbar A\frac{\partial}{\partial t} + \frac{iB\hbar\nabla}{2m} + C \quad , \quad S\Phi(spinor) = 0,$$

where A, B and C are square matrices. In terms of the identity matrix and the matrices that effect rotations of π radians about the coordinate axes (the components of B), the operator

$$S = A\left[\begin{array}{cc}1 & 0\\ 0 & 1\end{array}\right]\frac{\partial}{i\partial t} - \left[\begin{array}{cc}0 & i\\ i & 0\end{array}\right]\frac{\partial}{\partial x} - \left[\begin{array}{cc}0 & -1\\ 1 & 0\end{array}\right]\frac{\partial}{\partial y} - \left[\begin{array}{cc}i & 0\\ 0 & -i\end{array}\right]\frac{\partial}{\partial z} + C\left[\begin{array}{cc}1 & 0\\ 0 & 1\end{array}\right]$$

On substituting the derivatives obtained before, the Schrödinger equation reads:

$$\begin{aligned} S\Phi &= \left\{\hbar\omega A\begin{bmatrix} 1 & 0 \\ 0 & 1 \end{bmatrix} + \hbar k_x\begin{bmatrix} 0 & -1 \\ -1 & 0 \end{bmatrix} + \hbar k_y\begin{bmatrix} 0 & -i \\ i & 0 \end{bmatrix} + \hbar k_z\begin{bmatrix} -1 & 0 \\ 0 & 1 \end{bmatrix}\right. \\ &+ \left. C\begin{bmatrix} 1 & 0 \\ 0 & 1 \end{bmatrix}\right\} \times \begin{bmatrix} \phi_1 e^+ \\ -\phi_2 e^- \end{bmatrix} \end{aligned} \tag{4.1}$$

Choosing the matrices

$$A = \begin{bmatrix} 0 & 0 \\ 1 & 0 \end{bmatrix} \quad \text{and} \quad C = \begin{bmatrix} 0 & 2m \\ 0 & 0 \end{bmatrix}$$

the equation becomes

$$S\begin{bmatrix} \psi \\ \chi \end{bmatrix} = \left\{\begin{bmatrix} 0 & 2m \\ \hbar\omega & 0 \end{bmatrix} + \hbar\begin{bmatrix} \mathbf{k}\cdot\sigma_i & 0 \\ 0 & \mathbf{k}\cdot\sigma_i \end{bmatrix}\right\}\begin{bmatrix} \psi \\ \chi \end{bmatrix} = 0$$

$$\Longrightarrow \begin{bmatrix} \hbar\mathbf{k}\cdot\sigma_i & 2m \\ \hbar\omega & \hbar\mathbf{k}\cdot\sigma_i \end{bmatrix}\begin{bmatrix} \psi \\ \chi \end{bmatrix} = 0$$

In more familiar form

$$(\sigma\cdot\mathbf{p})\psi - 2m\chi = 0$$
$$E\psi - (\sigma\cdot\mathbf{p})\chi = 0$$

Elimination of χ leads to

$$E\psi = \frac{(\sigma\cdot\boldsymbol{p})^2}{2m}\psi$$

When shown that $(\sigma\cdot\boldsymbol{p})^2 = p^2$, the spinor form of Schrödinger's equation reduces to the normal complex form with E and p in operator notation:

$$\left(2mi\hbar\frac{\partial}{\partial t} + \hbar^2\nabla^2\right)\psi = 0$$

In an external electromagnetic field the appropriate operators are

$$E \leftarrow (\hbar i\partial/\partial t - V)$$
$$p \leftarrow (-\hbar i\nabla - e\boldsymbol{A}/c)$$

in terms of the scalar potential V and vector potential $\boldsymbol{A}$ of the field. Substitution of these variables, after some vector manipulation [15] leads to the Pauli equation:

$$i\hbar\frac{\partial\psi}{\partial t} = \frac{\hbar^2}{2m}\left(\nabla - \frac{ie\boldsymbol{A}}{\hbar c}\right)^2\psi + \frac{\hbar e}{2mc}(\sigma\cdot\boldsymbol{B})\psi + V\psi$$

in a magnetic field of strength $\boldsymbol{B}$. The term

$$\mu = \frac{\hbar e}{2mc}\sigma$$

represents the intrinsic magnetic moment due to spin.

4.7.3 The Spin Model

To understand the appearance of spin it is necessary to consider a fermion as some inhomogeneity in the space-time continuum, or aether. In order to move through space the fermion must rotate in spherical mode, causing a measurable disturbance in its immediate vicinity, observable as an angular momentum of $\hbar/2$, called spin. The inertial resistance experienced by a moving fermion relates to the angular velocity of the spherical rotation and is measurable as the mass of the fermion.

The spinor that describes the spherical rotation satisfies Schrödinger's equation and specifies two orientations of the spin, colloquially known as up and down: ($\uparrow$) and ($\downarrow$), distinguished by the allowed values of the magnetic spin quantum number, $m_s = \pm\frac{1}{2}$. The two-way splitting of a beam of silver ions in a Stern-Gerlach experiment is explained by the interaction of spin angular momentum with the magnetic field.

The components of the spinor $\phi\{e^+, e^-\}$ define standing waves for fermions and anti-fermions respectively; in each case by a combination of two spherical waves that respectively, converges to and diverges from $r = 0$, *e.g.*:

$$\begin{aligned} \Phi &= \frac{A}{r}\left\{e^{i(\omega t+kr)} \pm e^{i(\omega t-kr)}\right\} \\ &= Ae^{i\omega t}\left\{\frac{\cos kr}{kr} \text{ or } \frac{\sin kr}{kr}\right\} \end{aligned}$$

In the limit, $r \to 0$, $\cos kr/kr \to 0$, $\sin kr/kr \to 1$, $\Phi_0 \to 0$ or A. The amplitude of the standing wave is interpreted proportional to its charge. The wave packet with $\Phi_0 = 0$ is electrically neutral and represents a neutron. $\Phi_0 = A$ represents a wave packet with charge proportional to $\pm A$. In this case the trigonometric part is a spherical Bessel function that modulates the oscillatory part. The standing wave has the shape of the wave packet of Figure 2.10. The distance between the nodal points defines the de Broglie wavelength of the fermion and the rapidly oscillating part, with the Compton wavelength of $2\pi/k$, represents Zitterbewegung.

It is significant that this reconstruction indicates equal charges of opposite sign for electrons and protons, but different mass, linked to angular velocity by Planck's constant.

The product, or field intensity,

$$\Phi^*\Phi = A^2\left(\frac{\sin kr}{kr}\right)^2 = \frac{C}{r^2}$$

defines the force between charges, in line with Coulomb's law, except when $r \to 0$. The finite value, Φ_0, of the electric potential at $r = 0$ has been interpreted [64] as equivalent to renormalization in quantum electrodynamics which assumes an arbitrary cut-off of the Coulomb potential to avoid an unwanted infinity at $r = 0$.

4.7.4 Hund's Rule

Apart from free radicals, which contain unpaired electrons, spin is generally observed in chemical species associated with parallel valence-shell spin multiplets. An empirical rule, first formulated by Hund, states that "the first $(2l+1)/2$ electrons on a given sub-level have the same spin orientation. Addition of more electrons causes stepwise pairing of spins". This rule is a manifestation of the tendency of atoms to assume spherically symmetrical shapes [65]. It means that there is no resultant orbital angular momentum on any atom in the ground state. The reason for this lies with the $2l+1$ allowed values of m_l from $-l$ to l, which describe l balanced pairs with $\pm m_l$ and one state of zero L_z. Any odd number of electrons therefore occur as balanced pairs and a single electron with $m_l = 0$. Given this situation the exclusion principle demands normal multiplets (parallel spins) when a sub-level is less than half full and inverted multiplets when the sub-level is more than half full [65], as in Figure 4.8. The largest concentration of electron spin on

σ \ m_l	−2	−1	0	1	2
1			↑		
2	↑				↑
3	↑		↑		↑
4	↑	↑		↑	↑
5	↑	↑	↑	↑	↑
4	↑	↑	↑↓	↑	↑
3	↑	↑↓	↑	↑↓	↑
2	↑	↑↓		↑↓	↑
1	↑↓	↑↓	↑	↑↓	↑↓
0	↑↓	↑↓	↑↓	↑↓	↑↓

Figure 4.8: *Spin orientation in d-multiplets with up to 10 electrons in spherical atoms that obey the exclusion principle. The spin count is defined as* $\sigma = 2\sum m_s$.

an atom occurs in the valence shell of the lanthanide series with the general configuration $6s^2 4f^n$. This valence shell constitutes a spherical mantle that surrounds a much smaller and more compact neutral core, which carries no orbital angular momentum nor spin. As for a free fermion the electronic spin of the atom arises from spherical rotation of this valence mantle. The intriguing possibility that nuclear spin could manifest in a comparable fashion cannot be ignored. The demonstrated [62] relationship between surface proton excess, spin multiplets of atomic nuclei and the appearance of elemental superconductivity, confirms this as a plausible model.

4.8 Nuclear Structure and Spin

The convergence of nuclear composition, shown in Figure 4.2, suggests that a golden ratio of protons to neutrons imparts special stability to an atomic nucleus. Had this been the only factor all nuclides would be expected to approach this ratio as closely as possible. Assuming that nuclides are produced with the ratio $Z/N = 1$, positron emission would reduce this ratio stepwise, in moderately curved space, without change in mass number, until $Z/N \to \tau$ for all nuclides. As this is well known not to be the case, at least one additional factor must have an effect on the final distribution of nucleons and on where the decay process ends. An overall excess of neutrons has, for instance, been postulated to counteract catastrophic coulombic repulsion between protons, without severely screening the attraction on extranuclear electrons. An interplay of such factors is responsible for the observed region of stability observed in Figures 4.2, 4.4 and 4.7.

To demonstrate the importance of the golden ratio it is assumed that protons and neutrons occur in the nucleus on three-dimensional spirals of opposite chirality, and balanced in the ratio $Z/N = \tau$, about a central point. The overall ratio for all nuclides, invariably bigger than τ, means that a number of protons, equal to $Z - N\tau$, will be left over when all neutrons are in place on the neutron spiral. These excess protons form a sheath around the central spiral region, analogous to the valence-electron mantle around the atomic core. The neutron spiral is sufficient to moderate the coulomb repulsion while the surface layer of protons enhances the attraction on the extranuclear electrons.

The golden excess, calculated for all stable nuclides is shown in Figure 4.9.

By fixing the scale of mass number to proton excess at 8:1, the latter quantity is mapped on a 44×44 square lattice, gauged on atomic number. The isotopes of any element map on to straight lines perpendicular to OZ, with Ru on the diagonal. The relative proton excess, $1 - \tau N/Z$, varies

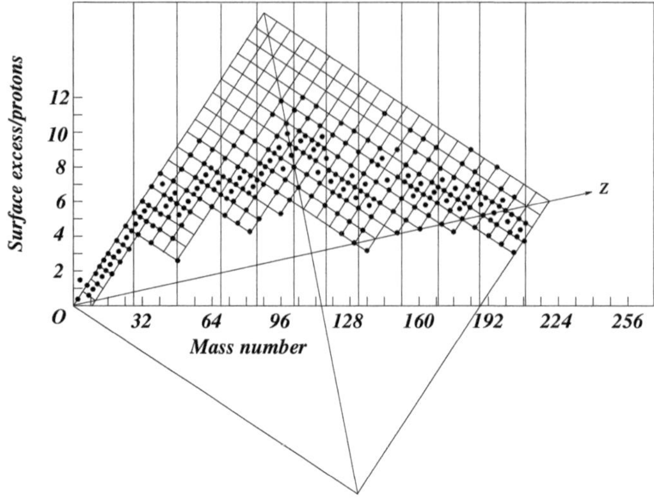

Figure 4.9: *Golden excess of protons on atomic nuclei as a function of mass number*

between $1 - \tau$ and 0 as Z/N varies from 1 to τ, and as a function of A displays the known periodicity, as shown in Figure 4.10.

The relevance of golden excess to nuclear spin follows from the demonstration [62] that within each nucleonic period of 24 the observed spins obey a nuclear equivalent of Hund's rule for $2l+1$ multiplets, with a few exceptions related to nuclear distortion. This observation is in line with central-field nuclear cohesion. Like valence-shell spin multiplets the spin of nuclei can be related to spherical rotation of the excess layer. The electron and nuclear spin functions are weakly linked and only show hyperfine interaction in an applied magnetic field. Strong correlation between surface excess and appearance of superconductivity has been interpreted [62] to mean that superconductivity is a nuclear magnetic effect, rather than a function of electron band structure. Superconduction develops in each periodic group only for those nuclides with surface excess less than the average between the extreme values for the group, meaning for those expected to exhibit normal spin multiplets.

4.9 Nucleon Periodicity

Like elemental periodicity, which is a function of electron configuration, an empirical periodic function for nucleons, known as the magic-number pattern, has been derived. The reason why this function has been less successful

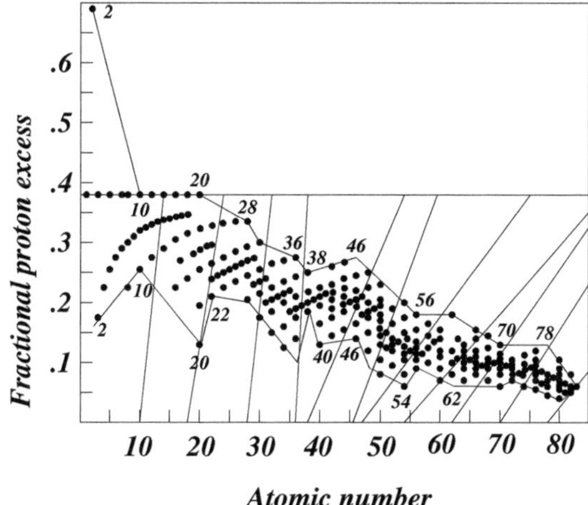

Figure 4.10: *Atomic-number periodicity derived from golden-ratio packing of nucleons.*

than its electronic counterpart, is because of the assumption that it applies equally to protons and neutrons. It provides a reasonable measure of nuclear properties for nucleon numbers less than 50. At higher mass number the assignment of magic numbers becomes increasingly arbitrary.

If, in a plot such as Figure 4.1, isotones, or nuclides with common neutron number, instead of isotopes, are singled out for emphasis, the characteristic circles and straight lines are interchanged, without affecting their points of intersection allong the line $r = N_x = \tau$. An alternative subset of nuclide periodicity, as a function of neutron number, therefore exists, and most likely relates to the magic-number pattern. This periodic function is defined in Figure 4.11.

Most of the empirical magic numbers are identified directly by the hem lines at $Z/N = \tau$. The hem lines, extrapolated to ratios of 1 and 0 also identify known magic numbers. The interesting conclusion to be drawn from this observation is that the nuclear periodic function remains constant under all cosmic conditions.

4.9.1 Farey and Ford Analysis

It is of interest to explore the possibility of an independent identification of magic numbers and the neutron spectrum by the relationship between neutron number and Farey sequences as mapped by Ford circles. Such an analysis presupposes the occurrence of periodic sequences of 32, 18, 8 and

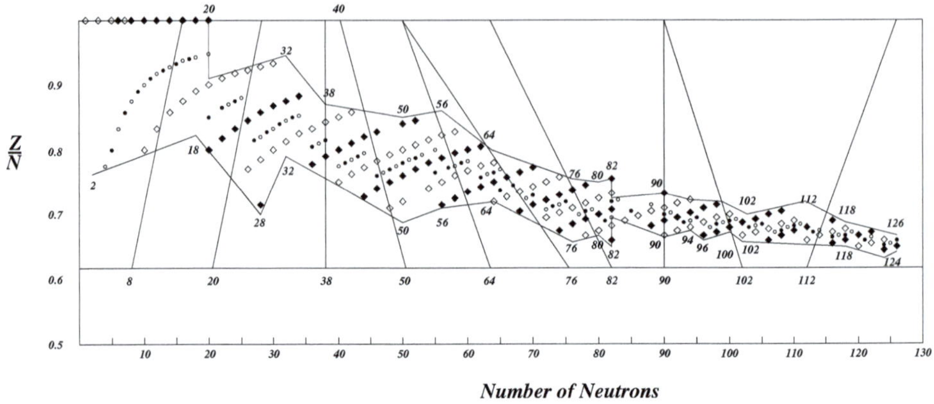

Figure 4.11: *Plot to demonstrate neutron periodicity*

2 ($2k^2$) nuclides. Guided by the important numbers from Figure 4.11 two patterns are recognized:

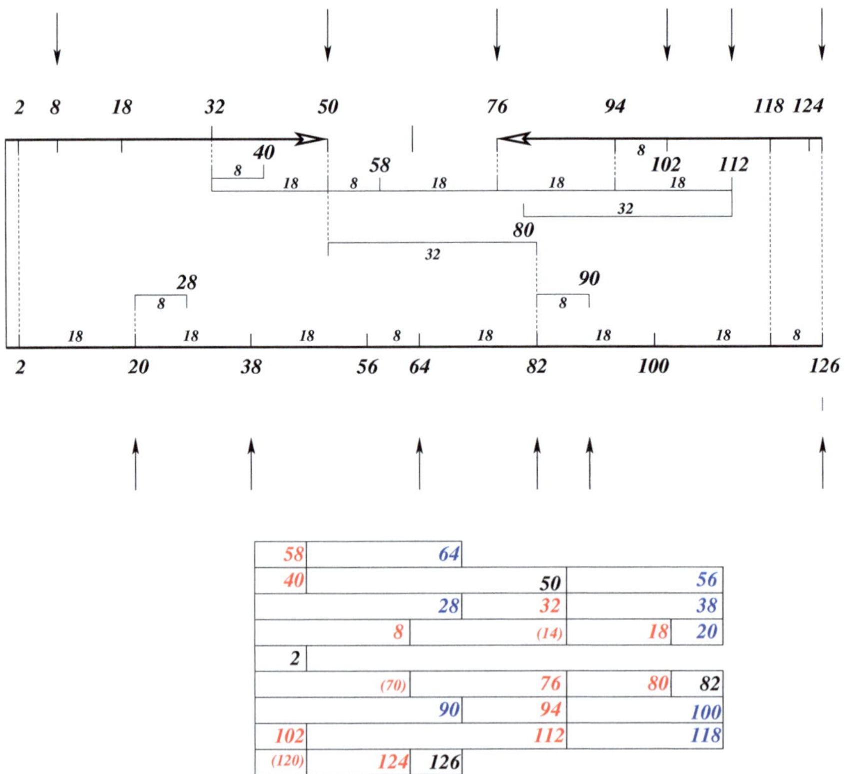

All of the primary and secondary sequences can be traced back to tangent Ford circles. The two independent patterns have common points at the four most significant, generally accepted, magic numbers: 2, 50, 82 and 126. The points at which the eleven hem lines intersect the golden ratio line are indicated by arrows. Ford circles from the Farey sequence $\mathscr{F}_5$ ($2k^2 = 50$) appear for the first time. When the two patterns are merged into a single construct, two-fold symmetry, as observed for elemental symmetry around atomic number 2, becomes apparent. A few numbers, in parentheses, have been added to emphasize the symmetry. Colour coding of the two independent patterns shows the complementarity of the two sequences. The numbers 32, 58 and 118, like the four magic numbers, fit equally well into both patterns.

Starting with the hemlines of Figure 4.2 mass-number periodicity in line with the Ford-circle model, and with near 2-fold symmetry around zero, is readily identified:

23	22	19	18	15	14	11	10	7	6	3	2	0	1	4	5	8	9	12	13	16	17	20	21
199	189		153					59	49	19		1		29	39					133	143	171	
200	201		154					60	50	20		2		30	40					134	144	172	
		163	155	125	117	93	85	61	51	21	11		3	31	41	69	77	101	109	135	145	173	181
208	198	170	162	132	124	100	92	68	58	28	18		10	38	48	76	84	108	116	142	152	180	188

Discrepancies occur in those cases where Figure 4.2 indicates hemlines at odd mass numbers: 85, 117, 169 and 209. The Ford reconstruction predicts 84, 116, 170 and 208 respectively. The central-field model that was used for nuclear-spin assignment before [62] requires values of 86, 118, 168 or 170, and 210 respectively. In all cases the observed values are, remarkably, the average of two theoretical predictions.

Of more importance is the packing mode of both nucleons and extranuclear electrons suggested by the inverse relationship with sets of Ford circles. The transform of tangent Ford circles is interpreted as concentric layers that fit together snugly in space-filling mode. An illustration of such a structure is provided by the stacking of objects in the asteroid belt [12]. Plotting the number of asteroids against their distances from the sun, measured in units of Jupiter's orbital period, gaps in the belt occur at fractional distances, which define a row of unimodular $\mathscr{F}_7$ fractions:

$$\frac{1}{3} \quad \frac{2}{5} \quad \frac{3}{7} \quad \frac{1}{2} \quad \frac{3}{5} \quad \frac{2}{3} \quad \frac{3}{4}$$

These fractions generate a set of tangent Ford circles, the transform of a set of concentric rings separated by narrow gaps. The rings of Saturn share this same geometrical pattern and it can be argued that Nagaoka's Saturnian atom (section 2.3.6) contains the information that produces the periodic table of the elements.

Figure 4.12: *Voyager 2 false-colour image of the rings of Saturn. Courtesy NASA/JPL–Caltech*

The details of nuclear structure depend on the interplay of three periodic functions, regulated by A, Z and N respectively. Only the A periodicity is of central-field type. The physical properties of nuclides, the subject of nuclear physics, are conditioned by the irregular coincidences of the three types of energy level and will not be pursued here any further. The effect of nuclear structure on chemistry is minimal.

4.10 Conclusion

It has been customary for many years to look at quantum mechanics for theoretical guidance in chemistry. In retrospect it is evident that little more than a qualitative picture has emerged from this pursuit. For a quantitative exposition of chemical effects the only reliable model remains the empirical. Nothing illustrates this claim better than the periodic table of the elements. In qualitative terms it is superficially structured according to the energy and angular-momentum levels of electrons, as calculated quantum-mechanically. Many tricks have been devised to reconcile the empirical periodic function with the Aufbau procedure predicted by wave mechanics. One such device is the deceptively simple scheme to specify the order of electronic sub-level occupation, shown in the diagram below. However, it is as misleading as it is simple. By this scheme, each transition series should comprise ten elements, contrary to the observation of the 8-member series Sc–Ni, Y–Pd and Lu–Pt.

There is no quantum-mechanical guidance to elucidate the relative stability of different nuclides or define the number of naturally occurring stable elements and their isotopes. Nuclear periodicity in terms of the Elsasser magic numbers remains an empirical scheme.

The alternative derivation of atomic periodicity, based on the distribution of prime numbers and elementary number theory, makes firm statements on all of these unresolved issues. The number spiral predicts periodicities of 8 and 24 for all elements and nuclides respectively; limits their maximum numbers, in terms of triangular numbers, to 100 and 300 respectively; characterizes electronic angular-momentum sub-levels by the difference between successive square numbers $(2l+1)$; and electron pairs per energy level by the square numbers themselves. In this way the transition series fit in naturally with the periodicity of 8. The multiplicity of 2, which is associated with electron spin, is implicit in these periodicity numbers.

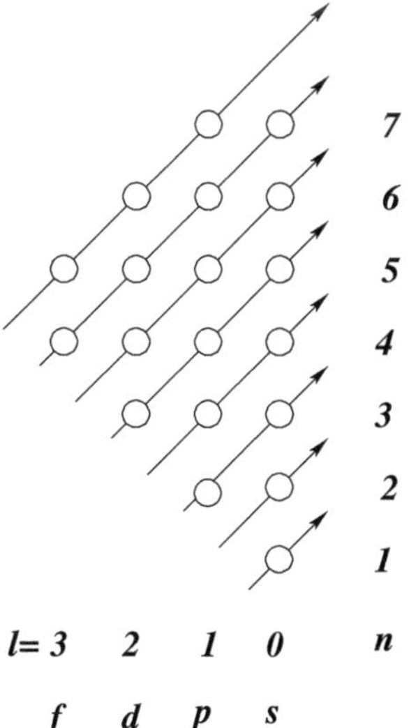

Details of the elemental and neutron periodicities follow directly as subsets of the 24-fold periodic function of the nucleons. The periodic ordinal numbers, derived in this way, define the stability limits of nuclides in terms of, either atomic number or neutron number, as shown in Figures 4.4 and 4.11 respectively.

Recognition of space-time curvature as the decisive parameter that regulates nuclear stability as a function of the ratio, Z/N, with unity and the golden mean, τ, as its upper and lower limits, leads to a consistent model for nucleogenesis, based on the addition of α-particles in an equilibrium chain reaction. This model is also consistent with the limitations imposed by the number spiral.

The natural appearance of nuclear magic numbers, and the golden-ratio limitation on nuclear distribution, indicate the development of an excess surface layer of protons, which correlates well with periodic variation of nuclear spin, and which may be an important parameter in the understanding of superconductivity.

Chapter 5

Chemical Interaction

The axioms that underpin any theory of chemical interaction were clearly stated by Kekulé, in the middle of the 19th century, as a theory of chemical affinity [16]. Restated in modern terminology:

- interatomic interaction is defined to be mediated by electrons;
- radicals that remain intact during chemical reaction are holistic molecular fragments;
- the way in which atoms stack together in molecules or radicals depend on the electronic configuration of the elements concerned;
- molecular shape – left undefined by Kekulé – can be ascribed to the minimization of orbital angular momentum.

To build a theory on these axioms it is necessary to have a clear understanding of the assumed nature of the electron and the conditions under which electron exchange between atoms becomes possible. These conditions will be taken to define an atomic *valence state.* The electronic configuration that dictates the mode of interaction between atoms of different elements will be interpreted to define the quantum potential energy of a valence electron in the valence state of an atom. This quantity will be shown to correspond to what has traditionally been defined empirically as the *electronegativity* of an atom.

5.1 The Valence State

Atoms in their energetic ground state should theoretically be unreactive since all of their electrons are in bound states and there is no feasible mechanism whereby any of these electrons can mediate interaction with another atom.

Electronic bound-state levels are inversely proportional to the square of an effective quantum number, $E \propto -1/n^2$, as shown on an arbitrary scale in the diagram.

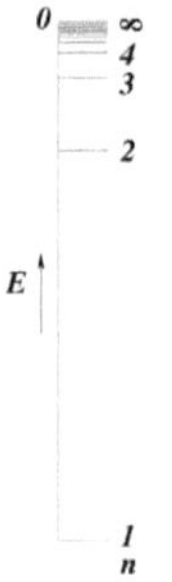

Near the ionization limit ($E \rightarrow 0$) the bound-state levels become increasingly closely spaced and the valence electron can be activated virtually continuously towards the zero energy level. At this level the electron becomes free to be exchanged between atoms. The process of activation can be simulated numerically very effectively as arising from increased hydrostatic pressure applied to the atom. Once decoupled from the core the energy of the valence electron, now in the valence state, is entirely in the form of quantum potential energy.

It is reasonable to assume that all atoms in a monatomic elemental sample under compression should reach the valence state simultaneously, at which point all valence electrons so released should spread across the entire set of anionic atomic cores to form a metal. Metallic Xe can, for instance, be obtained under pressure, but no systematic study of this effect has been conducted. In some cases more localized activation events, *e.g.* gas collision, leads, by the same mechanism, to the formation of smaller homonuclear molecules.

The physical nature of an electron in the valence state is not obvious. Although decoupled from the atomic core it remains associated with the core, because of environmental confinement. It therefore corresponds to unit negative charge confined to a sphere, described by the ionization radius of the atom concerned, but excluded from the core region of the atom.

Like its energy, the angular momentum of the valence electron also becomes spherically averaged. The electron rotates in spherical mode and generates a half-frequency wave field, known as *electron spin*, in its immediate environment, allowing the electron to couple with neighbouring charge fields and so initiate chemical change.

Activation by increased pressure, temperature or kinetic energy (collision), or some catalytic process, happens with conservation of angular momentum, (but not during photochemical activation). As the activated atom moves into an anisotropic molecular environment angular momentum vectors in projection along the Z-axis therefore reappear as for the free atom. Optimal alignment of all such vectors fixes the final molecular geometry.

On extending this idea to a quantitative study of the valence state, ionization radius, which is characteristic of each atom, is the important parameter. When using the Hartree-Fock-Slater method to calculate the ionization radii of non-hydrogen atoms the boundary condition is introduced by multiplying all one-electron wave functions by the step function:

$$S = \exp\left[-\left(\frac{r}{r_c}\right)^p\right] \quad , \quad \text{with } p >> 1$$

as part of the iterative procedure [53].

The quantum numbers n and l, which describe the energy levels of particular electrons, have a precise meaning only at the beginning of the iteration, where they refer to hydrogen-like one-electron wave functions. As the potential field varies with iterative cycling they loose this meaning and no longer represent sensible quantum numbers. They are retained as a book-keeping device, and not to indicate that the independent shell structure used as a starting configuration is preserved in the final self-consistent ground state of the compressed atom.

In standard HFS calculations all electrons with the same n and l labels are assigned similar energies at the average of the multiplet for that level. Since all electrons at the valence level are unlikely to reach the ionization limit simultaneously, the energy in excess over the ground state must be redistributed so as to promote a single electron towards ionization.

Two independent sets of ionization radii for all atoms have been calculated [53, 7] using exponential parameters $p = 20$ and 100, respectively. In both cases a clear periodic relationship appears, but the values at $p = 100$ are consistently lower. This observation reflects the steepness of the barrier that confines the valence electron, shown schematically in Figure 5.1.

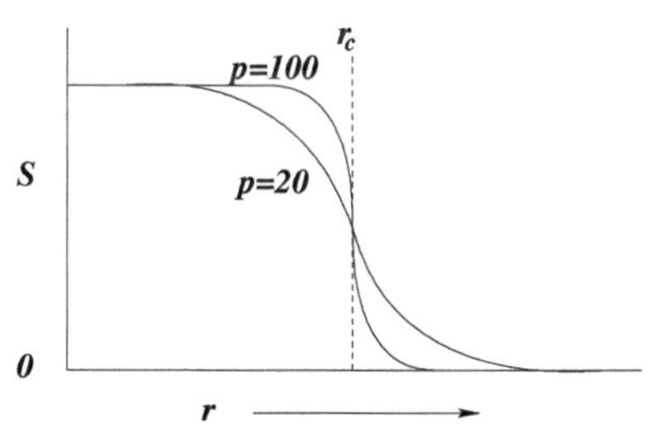

Figure 5.1: *Limiting radii of compressed atoms in relation to the exponential parameter, p*

In a chemical environment the barrier is unlikely to be infinitely sharp and there should be some optimum value of p that describes the situation best. The choice of $p = 20$ was established by comparison with independent chemical evidence [60] that relates to ionization radii. These radii are collated in the data table appended at the end of this chapter.

To show that the ground-state energy of an electron, decoupled from the atomic core, but confined to its ionization sphere, consists of quantum potential energy,

$$V_q = -\frac{h^2\nabla^2 R}{8\pi^2 mR} \tag{5.1}$$

equation (5.1) is rearranged into

$$\nabla^2 R + \frac{8\pi^2 m}{h^2} V_q R = 0$$

and, using eq. (3.34), V_q is equated with $E_0 = h^2/8mr_0^2$, such that:

$$\nabla^2 R + (\pi/r_0)^2 R = 0 \tag{5.2}$$

In one dimension (5.2) is the Helmholtz equation:

$$(D^2 + k^2)R = 0 \quad , \quad D \equiv \mathrm{d}/\mathrm{d}r \, , \, kr_0 = \pi \, ,$$

with solution $R = A \exp ikr \equiv A \cos kr + B \sin kr$. By definition, the radial wave function R is required to vanish at $r = r_0$, *i.e.*

$$R = A \cos \pi + B \sin \pi = 0 \quad \text{if } A = 0.$$

The most general solution of (5.2) is the Fourier sum of orthogonal functions:

$$R = \sum_k A_k e^{ikr} \tag{5.3}$$

where k has both negative and positive values. For closely-spaced energy levels the index k varies continuously and the set of coefficients A_k constitutes a Fourier integral,

$$R(r) = \frac{1}{\sqrt{2\pi}} \int_{-\infty}^{\infty} A(k) e^{ikr} \mathrm{d}r \tag{5.4}$$

The wave function, in turn, is the Fourier transform of $R(k)$, *i.e.*

$$A(k) = \frac{1}{\sqrt{2\pi}} \int_{-\infty}^{\infty} R(r) e^{-ikr} \mathrm{d}r \tag{5.5}$$

In three dimensions the spherically symmetrical Laplacian

$$\nabla^2 = \frac{\mathrm{d}^2}{\mathrm{d}r^2} + \frac{2}{r}\frac{\mathrm{d}}{\mathrm{d}r}$$

and (5.2) is solved by the spherical Bessel functions,

$$R = \sqrt{\frac{2}{\pi}} \cdot k \jmath_l(kr)$$

In the ground state

$$\jmath_0 = \frac{\sin(kr)}{kr} \, , \text{ and } E_0 = \frac{h^2}{8mr_0} \, , \text{ as before.}$$

The well-known [15]– (page 116) Fourier transform of the box function

$$f(x) = \begin{cases} a & \text{if} \quad |x| < \alpha \\ 0 & \text{if} \quad |x| > \alpha \end{cases}$$

is

$$\begin{aligned} g_k &= \frac{\alpha}{2\pi}\int_{-\alpha}^{\alpha} ae^{-ikx} = \frac{a\alpha}{\sqrt{2\pi}}\left[\frac{e^{-ikx}}{-ik}\right]_{-\alpha}^{\alpha} \\ &= \frac{a\alpha}{\sqrt{2\pi}}\frac{2\sin k\alpha}{k} \end{aligned}$$

For the special case $a = \sqrt{2\pi}/2\alpha$

$$g(k) = \frac{\sin k\alpha}{k\alpha} = j_0(k\alpha) \tag{5.6}$$

The most general solution to the wave equation of a spherically confined particle is the Fourier transform of this Bessel function, *i.e.* the box function defined by r_0. Such a wave function, which terminates at the ionization radius, has a uniform amplitude throughout the sphere, defined before (3.36) as

$$R(r) = \left(\frac{3c}{4\pi n}\right)\left(\frac{1}{r_0}\right)\exp\left[-\left(\frac{r}{r_0}\right)^p\right], \ p >> 1$$

The observation [66] that the energy, $E_g = h^2/8mr_0^2$, of a valence electron, decoupled from the core, but confined to the ionization sphere, consists entirely of quantum potential energy, has been interpreted to represent the electronegativity of an atom, also defined as the chemical potential of the valence state [67]

5.2 Electronegativity

The intuitive type of chemical reaction involves the participation of chemically dissimilar atoms, traditionally referred to as electropositive atoms reacting with electronegative atoms. The mutual affinity between such types is self-evident. The electronegativity series has been defined to arrange all elements in sequence, starting from the most electronegative at the highest value to the most electropositive, at the minimum.

The first known attempt to quantify electronegativity was due to Pauling, who devised an empirical scale, on the basis of thermochemical data. It relies on the simple idea that an *electrovalent* linkage between a pair of electropositive and electronegative atoms results from the transfer of an electron between the two:

$$e.g. \qquad \mathrm{Na{+}Cl} \rightarrow \mathrm{Na^+Cl^-}$$

whereas a *covalent* linkage between a homopolar pair requires equal sharing of two electrons:

$$e.g. \qquad \mathrm{Cl}\cdot + \cdot\mathrm{Cl} \rightarrow \mathrm{Cl{:}Cl}$$

All possible diatomic combinations, AB, correspond to situations between the two extremes. The larger the difference in electronegativity, $|x_A - x_B|$, the larger is the electrovalent component that stabilizes the linkage, and the smaller the covalent contribution. In thermochemical terms [68] – page 88:

> the values of the difference between the energy D(A-B) of the bond between two atoms A and B and the energy expected for a normal covalent bond, assumed to be the arithmetic mean or the geometric mean of the bond energies D(A-A) and D(B-B), increase as the two atoms A and B become more and more unlike with respect to the qualitative property that the chemist calls *electronegativity*, the power of an atom in a molecule to attract electrons to itself.

To hold for both definitions of the covalent mean it is assumed that, ideally

$$\frac{1}{2}\left(D_{AA} + D_{BB}\right) = \sqrt{D_{AA} \cdot D_{BB}}$$

or squared:

$$D_{AA}^2 + D_{BB}^2 - 2D_{AA} \cdot D_{BB} = 0$$

More generally:

$$\left(D_{AA} - D_{BB}\right) = 0 \text{ or } \Delta^2$$

For $A = B$ the difference in electronegativity is zero, and for $A \neq B$ the difference is defined as $\Delta = |x_A - x_B| = \sqrt{|D_{AA} - D_{BB}|}$. The well-known Pauling scale of electronegativities results from this definition of x_A on specifying dissociation energies in units of eV.

The only valid alternative, due to Mulliken, defines electronegativity as the mean of an atom's ionization potential and its electron affinity

$$\chi_M = \frac{1}{2}(I_P + E_A)$$

in units of eV. The relationship between the two scales, $\chi_M \simeq \chi_P^2$, is rarely emphasized and has caused massive confusion. Because of this confusion all of the numerous proposals that have been made can safely be ignored.

The problem has been resolved [69] by redefining electronegativity as the chemical potential of the valence state, calculated as the quantum potential of the valence electron, confined to its ionization sphere, *i.e.* $\chi^2 = h^2/8mr_0^2$, expressed in eV. Whereas χ corresponds to Pauling electronegativities, subject to simple periodic scaling, χ^2 corresponds to the Mulliken scale by the same type of operation. All of the many electronegativity scales in existence are simply related to the ionization radii, from which they ultimately derive.

Both of the χ_P and χ_M scales are empirical approximations based on incomplete experimental data. The theoretical definition of absolute electronegativity, $\chi = \sqrt{E_g} = \sqrt{\chi_M} \simeq x^{-\mu}\chi_P$ has been demonstrated to account for both empirical scales. The scale factor x varies with periodic shell and μ represents the number of valence energy-level vacancies.

A plot of ground-state confinement energies is shown in Figure 5.2 as a function of atomic number. The obvious periodicity corresponds in detail with the structure of the compact periodic table, based on number theory, Figure 4.5.

Electronegativies are listed in the data appendix, Table 5.5.

5.3 Covalent Interaction

In the same way that electronegativities determine the polarity of diatomic interactions, ionization radii should define the effective electronic charge clouds that interpenetrate to form diatomic molecules, as shown schematically in Figure 5.3. The overlap of two such spheres defines a lens of focal lengths fixed by the ionization radii, r_1 and r_2, at an interatomic distance $d = x_1 + x_2$.

The lens consists of two parts, each of which is generated as a solid of revolution by rotating the semi-circle $x^2 + y^2 = r^2$ about Ox. The resulting solid body has the volume

$$\begin{aligned}
v_1 &= \int_{x_1}^{r} \pi y^2 \mathrm{d}x \\
&= \int_{x_1}^{r} \pi \left(r^2 - x^2\right) \mathrm{d}x \\
&= \pi \left[r^2 x - \frac{1}{3}x^3\right]_{x_1}^{r} \\
&= \pi \left[\frac{2}{3}r^3 - r^2 x_1 - \frac{1}{3}x_1^3\right]
\end{aligned}$$

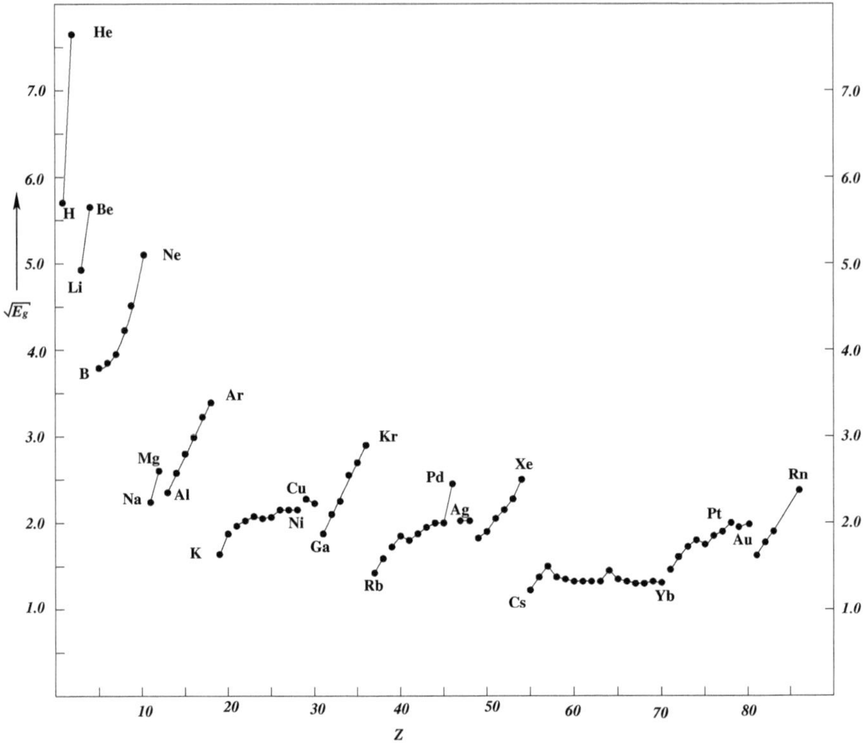

Figure 5.2: *Ground-state confinement energies E_g as a function of atomic number.*

In the problem of interest: $r_1^2 - x_1^2 = y^2$, $r_2^2 - x_2^2 = y^2$.

$$r_1^2 - r_2^2 = x_1^2 - x_2^2 \,,\; x_1 + x_2 = d \,,\; x_2 = d - x_1$$

Hence

$$\begin{aligned} r_1^2 - r_2^2 &= x_1^2 - d^2 + 2x_1 d - x_1^2 \\ &= 2x_1 d - d^2 = a \,(\text{say}) \\ x_1 &= \frac{d}{2} + \frac{a}{2d} \\ x_2 &= \frac{d}{2} - \frac{a}{2d} \end{aligned}$$

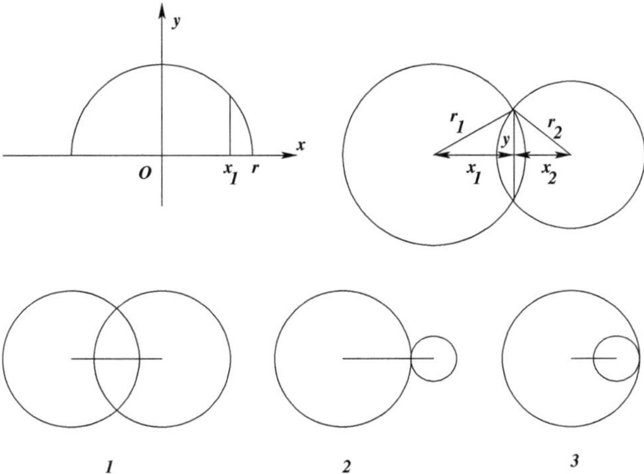

Figure 5.3: *Diagrams to explain the definition of point charges in the simulation of covalent interaction*

The overlap volume of the two spheres:

$$
\begin{aligned}
V_0 &= v_1 + v_2 \\
&= \pi\left[\frac{2}{3}\left(r_1^3 + r_2^3\right) - r_1^2\left(\frac{d}{2} + \frac{a}{2d}\right) + \frac{1}{3}\left(\frac{d}{2} + \frac{a}{2d}\right)^3 - r_2^2\left(\frac{d}{2} - \frac{a}{2d}\right) + \frac{1}{3}\left(\frac{d}{2} - \frac{a}{2d}\right)^3\right] \\
&= \pi\left[\frac{2}{3}(r_1^3 + r_2^3) - \frac{d}{2}(r_1^2 + r_2^2) - \frac{1}{4d}(r_1^2 - r_2^2)^2 + \frac{d^3}{12}\right] \qquad (5.7)
\end{aligned}
$$

For the special cases:

1.
$$r_1 = r_2 \quad : \quad V_0 = \pi\left(\frac{4}{3}r^3 - r^2 d + \frac{d^3}{12}\right) \qquad (5.8)$$

2.
$$d = r_1 + r_2 \quad : \quad V_0 = 0$$

3.
$$d \le r_1 - r_2 \quad : \quad V_0 = \frac{4}{3}\pi r_2^3$$

The valence electron that mediates interaction with another atom is considered to be distributed uniformly over the sphere of radius r_0 which surrounds the monopositive atomic core. On closer approach the atomic spheres start to overlap and bonding electrons are exchanged between the atoms. This polarization of the system is represented by point charges, related in magnitude to V_0, V_1 and V_2, by assuming that electron 1 is associated with nucleus 2 for a fraction $\varepsilon = V_0/V_1$ of the time, such that a charge of ε^+ appears on nucleus 1 and a charge of $\delta^+ = V_0/V_1$ on nucleus 2 [60]. The foreign electrons which move in the field of nucleus 1 for a time fraction δ can be represented by a charge δ^- in the overlap region, and likewise by a charge ε^- with respect to nucleus 2.

The total electrostatic energy of interaction is made up of four components:

1. Nuclear-nuclear repulsion amounting to $\delta\varepsilon/d$;
2. Electron-electron repulsion, given by $\delta\varepsilon/p$, where $p = r_1 + r_2 - d$, the thickness of the lens;
3. Attraction between nucleus 1 and electron 2, given by $\delta\varepsilon/b$, where $b = (r_1/r_2)(d/2)$, with $r_1 < r_2$.
4. Interaction between nucleus 2 and electron 1, given by $\delta\varepsilon/(d-b)$.

From the expression (5.7) for V_0 follows the formula for

$$\delta\varepsilon = \left[\frac{d^3}{16} - \frac{3d}{8}\left(r_1^2 + r_2^2\right) - \frac{3}{16d}\left(r_2^2 - r_1^2\right)^2 + \frac{1}{2}\left(r_1^3 + r_2^3\right)\right]^2 \left[\frac{1}{r_1 r_2}\right]^3 \tag{5.9}$$

When the smaller cloud is completely enclosed by the bigger ($r_2 > r_1 + d$) the extra polarization amounts to an additional attraction between point charges of $\pm[1 - r_1/(r_2 - d)]$ at a distance d. The total energy of interaction reduces to:

$$E = K\left[\delta\varepsilon\left\{\frac{1}{b} + \frac{1}{d-b} - \frac{1}{d} - \frac{1}{p}\right\} + X\right] \tag{5.10}$$

where K is a dimensional constant and $X = 0$, unless $r_2 > r_1 + d$, when $X = [1 - r_1/(r_2 - d)]^2/d$.

Homonuclear diatomic molecules are the simplest chemical systems that contain a chemical bond. In this special case the overlap formulae reduce to:

$$V_0 = \pi\left[\frac{4}{3}\pi r^3 - r^2 d + \frac{d^3}{16}\right] \tag{5.11}$$

$$\varepsilon = 1 - \frac{3d}{4r} + \frac{1}{16}\left(\frac{d}{r}\right)^3 \tag{5.12}$$

$$E = K\varepsilon^2\left[\frac{3}{d} - \frac{1}{2r-d}\right] \tag{5.13}$$

These can be expressed in dimensionless units of the variable $d' = d/r$:

$$\varepsilon = 1 - 3d'/4 + (d')^3/16 \tag{5.14}$$

$$E = K\varepsilon^2/r \left[\frac{3}{d'} - \frac{1}{2-d'}\right] \tag{5.15}$$

Setting $K = r$, the binding energy (E'), obtained in dimensionless *diatomic units*, and shown as a negative quantity, varies as a function of d/r as shown in Figure 5.4. The energy function remains attractive at all values of d' until

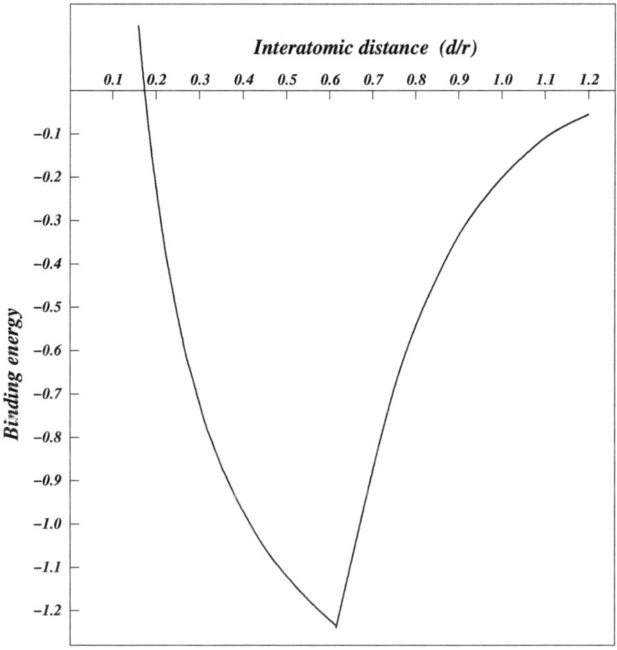

Figure 5.4: *Homonuclear diatomic energy curve in dimensionless units*

ε reaches its natural limit, which by definition corresponds to exactly one pair of electrons, the maximum allowed by the exclusion principle between the nuclei. An obvious turning point, by this argument, is expected to occur when $\varepsilon = \frac{1}{2}$, in which case each nucleus controls one half of the bonding pair. This sharing occurs when $d' = 0.695$. However, trial calculations show several diatomic molecules having $d' < 0.695$. As the bonding pair is spread over the total volume of overlap, the charge density in the internuclear region is interpreted to remain below the exclusion limit. Further calculations show that the minimum ratio for any diatomic molecule, $d/r = d'$ approaches the *golden ratio* τ.

5.3.1 The Diatomic Energy Curve

The point-charge density increases monotonically with decreasing d' until it reaches a maximum at the critical ratio of $d/r = \tau$. At this point $\varepsilon = 0.5513$ and

$$D' = 0.3\left[\frac{3}{\tau} - \frac{1}{2-\tau}\right] \tag{5.16}$$

which rearranges into

$$D' = 0.3\left[5.734 - \frac{1}{d'}\right] \tag{5.17}$$

that describes the energy for all $d' < \tau$. The total binding-energy curve therefore consists of attractive and repulsive parts that intersect at the point $d' = \tau$, $D' = 1.244$. The dissociation energy in familiar units is obtained as

$$D_c = KD'/r \tag{5.18}$$

with r in units of Å and the dimensional constant $K = 14.36$ or 1389 for units of eV or kJmol^{-1} respectively.

For heteronuclear bonds dimensionless distances $d' = d/R$, $r_1' = r_1/R$ and $r_2' = r_2/R$, with $R = \sqrt{r_1 r_2}$. Defining the ratio $r_1/r_2 = x$ the overlap formulae are:

$$\delta\varepsilon = \left[T_3 - \frac{3T_2}{16d'} - \frac{3}{4}T_1 d' + \frac{1}{16}(d')^3\right]^2 \tag{5.19}$$

$$T_1 = \frac{1}{2}\left[x + \frac{1}{x}\right] \tag{5.20}$$

$$T_2 = x^2 + \frac{1}{x^2} - 2 \tag{5.21}$$

$$T_3 = \frac{1}{2}\left[x^{3/2} + x^{-3/2}\right] \tag{5.22}$$

$$D' = \delta\varepsilon\left\{\frac{1}{d'}\left[\frac{x^2 - 2x + 4}{(2-x)x}\right] - \frac{1}{(1+x)/\sqrt{x} - d'}\right\} \tag{5.23}$$

An additional factor must be taken into account for bonds involving hydrogen, *i.e.* the depth to which the hydrogen sphere penetrates into the ionization sphere of the second atom. The relationship between dimensionless interatomic distance and dissociation energy, in this case, is summarized by the formulae:

$$D' = D'_1 + X' \tag{5.24}$$

$$D'_1 = x^3 \left\{ \frac{T_4}{d'} - \frac{1}{2\sqrt{x}} \right\} \tag{5.25}$$

$$T_4 = \frac{x^2 - 2x + 4}{x(2 - x)} \tag{5.26}$$

$$X' = \frac{1}{d'} \left[\frac{y - x}{y} \right]^2 \tag{5.27}$$

$$y = 1 - \sqrt{x}d' \tag{5.28}$$

According to these formulae polarization of the bonding density, as $r_1 \neq r_2$, causes displacement of the diatomic curve – to the right at large values of d' and to the left at small d'. For $0.75 \leq x < 1$ the binding curves stay very close to the homonuclear diatomic curve and merge with that as $d' \rightarrow \tau$. The curve for $x = 0.75$ is shown for comparison in Figure 5.5. Practically all known covalent bonds have ionization-radius ratios within these limits. For given $x = r_1/r_2$ the bonding curve intersects that for any system of larger x, as d' decreases. However, all curves coalesce as $d' \rightarrow 2$ and $E' \rightarrow 0$, the point at which the ionization spheres touch without interpenetration.

5.3.2 Generalized Covalence

The curves of Figures 5.4 and 5.5 show remarkable dependence on the golden ratio. The attractive and repulsive curves intersect where $d' = \tau$, at which point the diatomic dissociation energy $D' = 2\tau$. Again, the limiting curves (marked 1.0 and 0.25(H) in Figure 5.5) are maximally separated at $d' = 2\tau$, to merge again as $d' \rightarrow 2$.

The three major components of the combined covalency curve are readily simulated as circular segments as shown in Figure 5.6(a). The construction is performed in a coordinate system defined by the variables d' and D' measured on the same scale, as in Figure 5.4. The curves that reflect the energy of covalent interaction in dimensionless units as a function of dimensionless interatomic distance, start from two special points at $(d', D') = (2, 0)$ (A)and $(\tau, 2\tau)$ (B). A circle through B and centred at A describes the first (repulsive) segment of the general curve in the region $0 < D' \leq 2\tau$. A second circular (heteronuclear) segment over the region $\tau \leq d' \leq 2$ is centred at the point D $(2, 2.25\tau)$ – on the perpendicular bisector of the line AB. The third (homonuclear) segment, through $(2\tau, 0)$, is centred at $(2, 1.75\tau)$, on the line extended

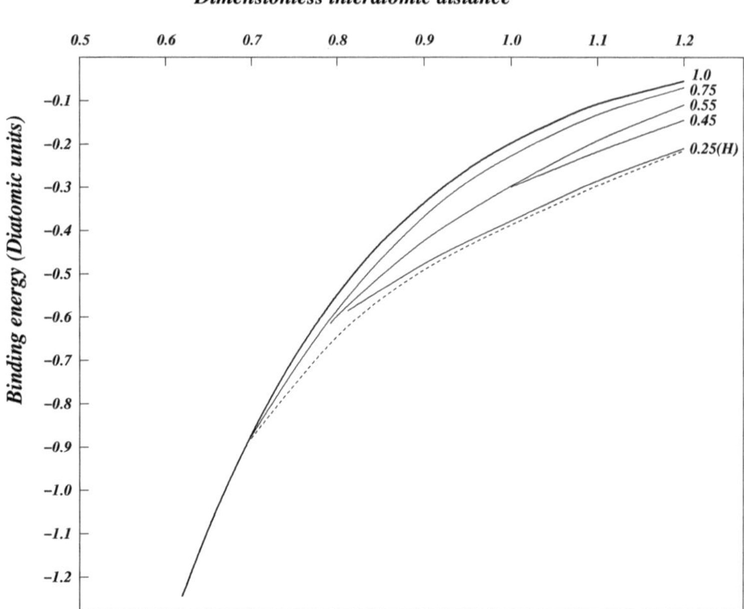

Figure 5.5: *Bonding curves at different radius ratios. A limiting curve is compounded from segments of the heteronuclear* ($x \leq 0.75$) *curves and the hydride curve at* $x = 0.25$.

from the origin through C. The two attractive segments merge at the point where $\varepsilon = 0.5$. The line that connects the points $(0, \tau)$ and D goes through this point. The entire construct fits into a golden reactangle and defines the topology of space-time that optimizes the covalent interaction between a pair of atoms as a function of their distance apart.

The principle embodied in Figure 5.6(a) is, not surprisingly, related to botanical phyllotaxis, or leaf arrangement on a central stem, and to the optimal spacing of points on the circumference of a circle. Fibonacci phyllotaxis has the effect of optimizing exposure of all leaves to sun and rain whilst growth angles related to the golden ratio lead to the most uniform distribution of points (leaves).

In a diatomic molecule it is of benefit to minimize the steric influence of those valence electrons not directly involved in mediating the interatomic interaction. The competing factors in this case are atomic size, interatomic distance, distribution of valence-electron density and the Pauli principle.

The polarization, caused by electron-pair covalent interaction, promotes

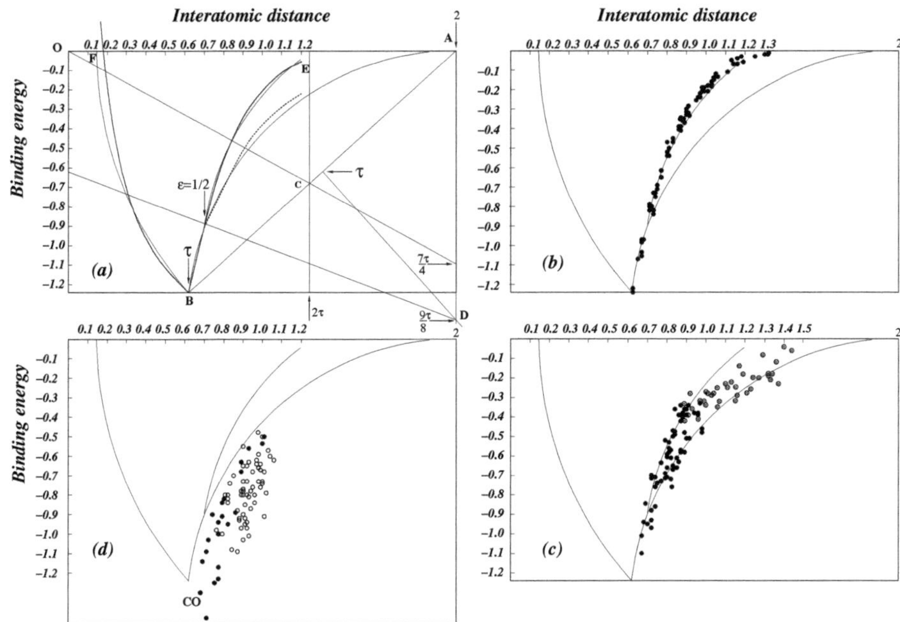

Figure 5.6: *(a) Geometrical construction of the golden-ratio circle segments that define the allowed (d', D')-field for covalent bonds. (b) Observed dissociation energies (D'_x) of homonuclear diatomic molecules. (c) Observed dissociation energies of heteronuclear diatomic molecules (black dots), including diatomic hydrides. (d) Observed dissociation energies of alkali and alkali-earth halides and oxides (open circles) and* CO*-type molecules and other dative bonds with substantial ionic bonding contributions. Calculated dissociation energies (at the same d') are invariably within the covalent field.*

more valence-shell electrons into the valence state. These electrons can either participate in the formation of more bonds, condense into lone pairs, or create a surface layer that screens the interaction between the atomic cores. When all additional valence electrons are associated with covalent bonds or lone pairs, the primary diatomic bond is known as *first order*. Screening of the nuclear repulsion by a pair of electrons (one per atom) allows the cores to move more closely together in an arrangement known as a second-order (or double) bond, and so forth.

The bonding parameters (d', D') of all covalent bonds, irrespective of order, lie within the field defined by the curves ABE. It is postulated that the field EBF represents covalent bonds under high pressure. The events at point F are open to conjecture.

The observation that bonds of all orders relate to the bonding diagram in equivalent fashion indicates that covalent bonds are conditioned by the geometry of space, rather than the geometry of electron fields, dictated by atomic orbitals or other density functions. The only special point related to electron density occurs at the junction of the attractive curves, where $\varepsilon = \frac{1}{2}$, indicating that one pair of electrons mediate the covalent interaction. It is interpreted as the limiting length (d'_1) for first-order bonds. It is of interest to note that all known first-order bonds have $d' > d'_1$. The covalence curve for the minimum ratio of $x = 0.18$ (for CsH) terminates at $d' = d'_1$.

Bonds with $\tau < d' < d'_1$ become possible because of nuclear screening (increased bond order), which causes concentration of the bonding pair directly between the nuclei. The exclusion limit is reached at $d' = \tau$ and appears as a geometrical property of space. The distribution of molecular electron density is dictated by the local geometry of space-time. Model functions, such as VSEPR or minimum orbital angular momentum [65], that correctly describe this distribution, do so without dictating the result. The template is provided by the curvature of space-time which appears to be related to the three fundamental constants π, τ and e.

5.3.3 Bond Dissociation Energy

By assuming atomic radii to be specified by ionization radii, r_0, the diatomic curves could, in principle, be used to calculate the dissociation energy of any covalent bond, irrespective of bond order, only from its interatomic distance. This promise is largely fulfilled, but the calculated energies are proportionately too low in virtually all cases. The obvious reason for this is that the calculation makes no allowance for steric interactions that occur in molecular environments. Because of this effect the (experimentally measured) interatomic distances used in the calculation do not refer to the strain-free situation modelled by ionization radii.

The problem is avoided by modification of the characteristic atomic radii to match the condition of strain that exists in molecules. As a typical example the variation of calculated dissociation energy with characteristic radius is shown in the table below for the C–C bonds in ethane, ethene and acetylene. The discrepancy approaches experimental uncertainty for all three bonds at the same value of $r_c = 1.85$ Å, which also models diatomic C_2 correctly.

Starting from ionization radii, r_0, and using experimentally measured values of dissociation energy and interatomic distance for homonuclear diatomic molecules, a self-consistent set of characteristic radii, suitable for the point-charge calculation of bonding parameters, of both homonuclear and

Table 5.1: *Dissociation energy as a function of characteristic radius*

C-C(Å): r_c(Å)	1.54	1.34 D($kJmol^{-1}$)	1.20
$r_0 = 1.60$	210	389	602
1.70	261	457	669
1.75	293	499	721
1.80	331	554	755
1.85	352	607	817
Expt.	368	598	836
d_0	1.36	1.20	1.09

heteronuclear interactions in molecules, has been generated and further updated by incorporating data pertaining to other covalent bonds. The calculated values are in general agreement with the set of empirical radii derived before [60]. An updated list of these characteristic radii are shown together with related data in the Appendix.

The comparison between ionization radii and characteristic radii is informative. All of the traditional non-metals have $r_x > r_0$. A number of traditional metals, which also have $r_x > r_0$, *i.e.* Li, Be, Na, Zr, Nb, Mo, La, Ta, W, Re and Os, are all known to form exceptionally stable diatomic molecules, compared to their congeners.

In order to test the point-charge method experimentally measured dissociation energy and interatomic distance are required for each chemical bond. Dissociation energies for most homonuclear diatomic molecules have been measured spectroscopically and/or thermochemically. Interatomic distances for a large number of these are also known. However, for a large number of, especially metallic diatomic molecules, equilibrium interatomic distances have not been measured spectroscopically. In order to include these elements in the sample it is noted that for those metals with measured r_e, it is found to be related, on average, to δ, the distance of closest approach in the metal, by $r_e = 0.78\delta$. On this assumption reference values of interatomic distance (d) become available for virtually all elements, as shown in the data appendix. In some special cases well-characterized dimetal bond lengths have also been taken into account for final assessment of interatomic distance.

By fixing the characteristic radii of all elements the dissociation energies and interatomic distances of diatomic covalent interactions are converted into dimensionless units and predicted to generate a set of points within the narrow band that defines covalence in Figure 5.6(a).

The plot of observed d' *vs* D' for homonuclear diatomics, shown in Figure 5.6(b), follows the theoretical covalence curve within experimental uncertainty.

In the case of heteronuclear diatomics there may be large differences in electronegativity, causing deviation from pure covalent interaction, either as dative bonds or ionic contribution to the bonding. Bonding plots, based on experimental measurement, are expected to distinguish between essentially covalent interactions and bonds with ionic character. This distinction is made obvious in Figures 5.6(c) and 5.6(d).

The diatomic halides and oxides of the alkali and alkaline earth groups must, by definition, have a considerable ionic contribution to their bonding. These diatomics, together with those of Al, Ga and In, are in fact found to lie outside the predicted field for covalent bonds, as shown in Figure 5.6(d). Molecules with dative bonds are expected along the borderline between covalent and ionic types, including several fluorides (of Sb, Si, Sn, Pb, Be and Ag) and chlorides (Si, Sn). They are arbitrarily grouped with the more ionic bonds.

Carbon monoxide is often identified as the molecule with the strongest known covalent bond, conventionally described as a σ-bond, augmented by two (p_x and p_y) π-bonds. The formula :C $\overset{\leftarrow}{\equiv}$ O:, with its two lone pairs, summarizes the many molecular-orbital, as well as valence-bond, descriptions that have been proposed as bonding schemes. A major consideration is the decrease in interatomic distance from 1.128Å to 1.115Å as the molecule ionizes into CO^+. Taken together with an observed increase in vibrational frequency this is interpreted as an increase in covalent bond strength. Compared with the observed dissociation energies of 1075 and 806 kJmol^{-1} for CO and CO^+ respectively, it means that, whatever the nature and order of the bond, there is a substantial ionic contribution to the CO bond.

This conclusion is substantiated by the calculated dissociation energies of the two molecules and the observation that, whereas the experimental point for CO^+ lies directly on the theoretical covalent curve, CO is well into the ionic region. It goes without saying that the same bonding pattern should repeat for all group14 – group16 diatomic molecules. This set includes the diatomic oxides, sulphides, selenides and tellurides of C, Si, Ge, Sn, and Pb, as well as Ti, Zr and Hf. With a few exceptions all of these occur in the ionic region, as shown in Figure 5.6(d). In addition, the oxides of periodic groups 3 and 5 occur in the same field.

Calculated dissociation energies do not reflect ionic contributions and invariably map into the covalent crescent. Ionic percentages are therefore estimated directly as the difference between calculated and observed dissociation energies.

5.3.4 The Quantum Model

Covalent interaction in diatomic molecules depends on the golden mean τ, the interatomic distance d' and the radius ratio $x = r_1/r_2$ of the constituent atoms, as summarized in Figure 5.6. The golden mean is a universal constant that matches the geometry and topology of space-time, the radius ratio is a known function of atomic number and d' relates to the optimal wave-mechanical distribution of valence-electron density in the diatomic system.

Interatomic distance is calculated by mathematical modelling of the electron exchange that constitutes a covalent bond. Such a calculation was first performed by Heitler and London using $1s$ atomic wave functions to simulate the bonding in H_2. To model the more general case of homonuclear diatomic molecules the interacting atoms in their valence states are described by monopositive atomic cores and two valence electrons with constant wave functions (3.36).

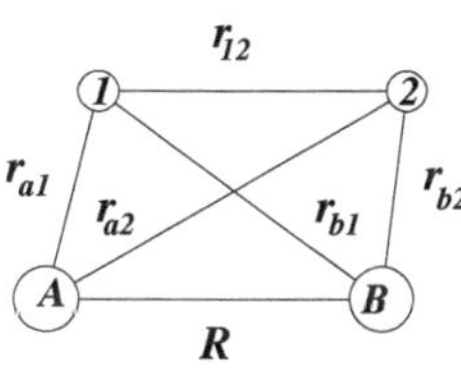

Figure 5.7: *Definition of electron and nuclear coordinates in the Heitler-London problem.*

The electron and core coordinates are labeled as shown in Figure 5.7. Assuming the cores to be clamped in fixed positions[1]the electronic motion is described by the Schrödinger equation (2.17)

$$H\psi = \left(-\frac{\hbar^2}{2m}\nabla^2 + V\right)\psi = E\psi$$

In atomic units[2] length is measured in units of $a_0 = (4\pi\epsilon_0)/me^2$, energy in (double) rydberg units $= 27.2$ eV, and the Hamiltonian operator is formulated as:

$$H = -\frac{1}{2}\left(\nabla_1^2 + \nabla_2^2\right) - \left(\frac{1}{r_{a1}} + \frac{1}{r_{a2}} + \frac{1}{r_{b1}} + \frac{1}{r_{b2}}\right) + \frac{1}{r_{12}} + \frac{1}{R} \tag{5.29}$$

The molecular wave function is defined by the combination of atomic wave functions:

$$\psi = \psi_a(1)\cdot\psi_b(2) + \psi_a(2)\cdot\psi_b(1)$$

which ensures equal sharing of exchanged electrons. At given internuclear distance the ground-state energy-eigenvalue solution is obtained by integra-

[1]This assumption does not imply a rigid linear structure of the diatomic molecule.
[2]Obtained by setting $m = \hbar = e = 4\pi\epsilon_0 = 1$.

tion of the differential equation over all coordinate space:

$$E = \frac{\int \psi H \psi \mathrm{d}\tau}{\int \psi\psi \mathrm{d}\tau}$$

This solution simplifies [70] to:

$$E = E_a + E_b + \left\{ \frac{1+S^2}{R} - (\epsilon_{aa} + \epsilon_{bb}) - S\,(\epsilon_{ab} + \epsilon_{ba}) + \right.$$

$$\left. \int \mathrm{d}\tau_1 \mathrm{d}\tau_2 \psi_a^2(1)\psi_b^2(2)/r_{12} + \int \mathrm{d}\tau_1 \mathrm{d}\tau_2 \psi_a(1)\psi_b(1)\psi_a(2)\psi_b(2)/r_{12} \right\} / \left(1+S^2\right) \quad (5.30)$$

in which

$$E_a, E_b - \text{energies of individual atoms;}$$

$$\epsilon_{aa} = \int \psi_a(1)\psi_a(1)(1/r_{b1})\mathrm{d}\tau$$

$$\epsilon_{bb} = \int \psi_b(1)\psi_b(1)(1/r_{a1})\mathrm{d}\tau$$

$$S = \int \psi_a(1)\psi_b(1)\mathrm{d}\tau$$

$$\epsilon_{ab} = \int \psi_a(1)\psi_b(1)(1/r_{a1})\mathrm{d}\tau$$

$$\epsilon_{ba} = \int \psi_a(1)\psi_b(1)(1/r_{b1})\mathrm{d}\tau$$

Because of the symmetry of the Hamiltonian integrals involving $\psi_a(1)$ and $\psi_a(2)$ are identical and calculated only once.

If the valence-state wave functions are written in their simplest, hard-sphere form,

$$\psi(r) = \sqrt{3c/4\pi n}/r_0\,,\ 0 < r < r_0\,,\ c = 0.46 \quad (5.31)$$

the Heitler-London integrals can be evaluated directly by summation over the atomic ionization spheres. An estimate of the optimal interatomic distance that minimizes the total energy is obtained by repeating the calculation at a series of fixed interatomic distances. A typical result, for the C-C interaction in dimensionless diatomic units, is shown in Figure 5.8. The agreement with experiment is precise. The calculated dissociation energy agrees with the point-charge result for the H_3C–CH_3 bond and in addition the correct equilibrium interatomic distance is predicted.

The two methods of calculation are complementary in the sense that the point-charge method does not distinguish between bonds of different order, but fails to predict interatomic distance. The Heitler-London calculation, without further modification, only applies to homonuclear single bonds.

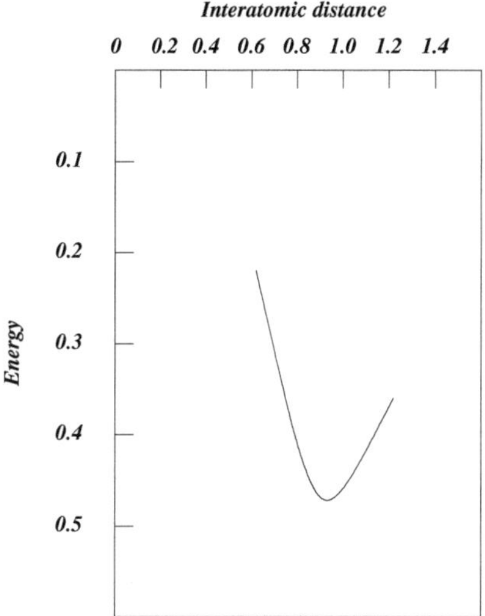

Figure 5.8: *Potential-energy curve of first-order C-C interaction, in dimensionless units, calculated by the Heitler-London method.*

5.3.5 Screening and Bond Order

In dealing with heteronuclear interactions it is necessary to compensate for electronegativity differences, which cause unequal sharing of bonding density. In the point-charge method such difference is automatically taken into account through the radius ratio $x = r_1/r_2$. It is logical to assume a related correction to Heitler-London results where the imbalance implies an effective core-charge product different from unity.

The relationship between nuclear charge and atomic wave functions is of the form $Z \propto \psi^{2/3}$. Core-charge imbalance could therefore be compensated for with an effective charge of $Z_e = x^{2/3}$ on one of the atoms. For p-block atomic pairs, this screening factor does indeed lead to the correct solution. For hydrides of p-block elements in periodic row n, compensating charge is defined as $Z_e = (k_n x/n)^{2/3}$, with screening constants $k_2 = 0.84$ and $k_3 = 0.70$.

The observation that point-charge simulation reproduces the exact same result as a Heitler-London calculation, but only for first-order bonds, confirms the previous conclusion that covalent interactions are mediated by the sharing of a maximum two electrons per bond, allowed by the exclusion principle. In view of this stipulation the conventional assumption, that several pairs of electrons contribute to the formation of high-order bonds, should therefore be re-examined[3].

An alternative interpretation is possible, by analogy with the correction for unequal sharing of bonding density, as a screening phenomenon. To understand the origin of this screening it is instructive to examine the changes in electronic arrangement that occur during chemical modifications which increase the C-C bond order:

$$H_3C-CH_3 \longrightarrow H_2C-CH_2 \longrightarrow H-C-C-H$$

In C_2H_6 the number of valence electrons per C atom matches the number of neighbouring ligands; in C_2H_4 there is an excess of one and in C_2H_2 an excess of two electrons per ligand.

In point-charge simulation this electronic rearrangement is of no immediate consequence except for the assumption of a reduced interatomic distance, which is the parameter needed to calculate increased dissociation energies. However, in Heitler-London calculation it is necessary to compensate for the modified valence density, as was done for heteronuclear interactions. The closer approach between the nuclei, and the consequent increase in calculated dissociation energy, is assumed to result from screening of the nuclear repulsion by the excess valence density. Computationally this assumption is convenient and effective.

To obtain the bonding curve for high order bonds it is only necessary to replace the factor $1/R$ in eq. (5.30) by k_ν/R, where k_ν is the screening constant for bonds of order ν. These screening constants are, not surprisingly, related to those for the p-block hydrides, *e.g.* $k_2 = (0.84)^2$. Screening constants for all bond orders are summarized by the linear relationship:

$$k_\nu = 1.295 - 0.295\nu \qquad (5.32)$$

[3]The formulation of spatially separated σ and π interactions between a pair of atoms is grossly misleading. Critical point compressibility studies show [71] that N_2 has essentially the same spherical shape as Xe. A total wave-mechanical model of a diatomic molecule, in which both nuclei and electrons are treated non-classically, is thought to be consistent with this observation. Clamped-nucleus calculations, to derive interatomic distance, should therefore be interpreted as a one-dimensional section through a spherical whole. Like electrons, wave-mechanical nuclei are not point particles. A wave equation defines a diatomic molecule as a spherical distribution of nuclear and electronic density, with a common quantum potential, and pivoted on a central hub, which contains a pith of valence electrons. This valence density is limited simultaneously by the exclusion principle and the golden ratio.

It is known from spectroscopic studies that bond-dissociation curves are simulated very well by the Morse function:

$$V(r) = D_e \left\{1 - \exp[-a(r - r_e)]\right\}^2 \tag{5.33}$$

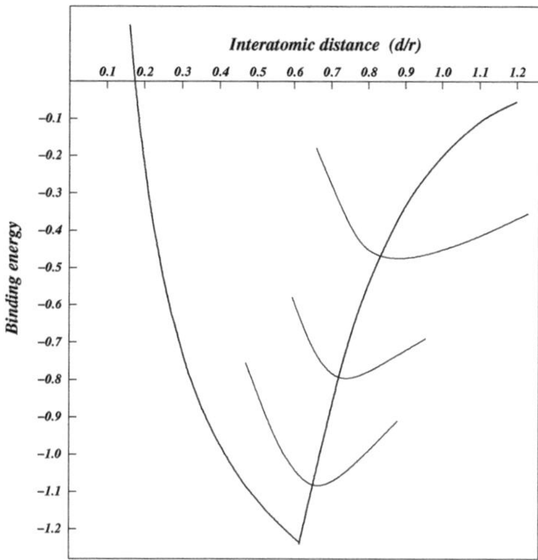

Figure 5.9: *Potential-energy curves of integral-order dicarbon bonds, calculated by the Heitler-London method for screened nuclei, superimposed on the point-charge covalency curve.*

The constant a relates to vibrational force constants (k_e), which means that a Morse curve for any single bond with known D_e, r_e and k_e can be calculated and, on modifying the internuclear repulsion by a factor k_ν/r, used to generate the corresponding potential-energy curves for all bonds of higher integral order. The exact procedure, with many examples, has been described in detail before [72, 73, 7]. A major advantage of the procedure is that the method is not restricted to integral bond orders and is especially effective to account for the variation in single-bond lengths adjacent to high-order bonds, in compounds such as R–CH=CH_2, CH_2=CH–CH=CH_2, R–C≡CH, *etc.*, all of non-integral order. Integral bond orders are conveniently identified along the general covalency curve, as shown for C–C interactions in Figure 5.9.

Because of second-neighbour interactions integral bond orders are the exception rather than the rule in all, but the simplest, molecules. Figure 5.9 suggests that bonds of different order should be related, also in point-charge simulation, by a simple formula like (5.32). However, point-charge simulation requires an interatomic distance for each bond order and modification of only the internuclear repulsion can therefore not predict the properties of other bond orders directly. Consequently, screening corrections to generate bonding parameters for higher order bonds are unlikely to be linear. Instead, effective screening constants, defined as

$$k_\nu = 2^{(1-\nu)/2}, \tag{5.34}$$

provide an adequate prediction of high-order parameters, based on first-order properties. This formula agrees with the linear function for first and second-order bonds, but deviates for orders 3 and 4. Dissociation energies for bonds of order ν are predicted correctly as:

$$D'_\nu = D'_1 + (\varepsilon/k_\nu)^2 \tag{5.35}$$

where ε is the calculated first-order nuclear charge.

5.4 Chemical Cohesion

The discussion of diatomic molecules has expanded imperceptibly to include general pairwise interatomic interactions, which may point at a robust three-dimensional molecular structure, in line with traditional chemical formulae. However, this is not an inevitable conclusion. The screening mechanism, which renders the Heitler-London method generally valid, provides a powerful counterargument. It shows that the interaction between any pair of atoms depends on all secondary interactions with other atoms within the same molecule. It is possible, in principle, to calculate, from the number and type of atoms in a molecule and the total number of valence electrons, the number of first-neighbour interactions with predictable dissociation energies. The total cohesive interaction in a molecule could even be calculated and minimized without predicting or assuming a definite three-dimensional structure, only based on average interatomic distances.

The question of molecular structure and shape is considered in the next chapter. It will be shown that the familiar molecular structure is a function of chemical history and the thermodynamic environment. It is emphasized, in particular, that experimental determination of molecular structure is strictly confined to the solid crystalline state.

It is also in the solid state that the nature of intramolecular interaction can be studied most sensibly. The firm distinction which is traditionally made between four types of interaction [74], shown in Figure 5.10, conceals

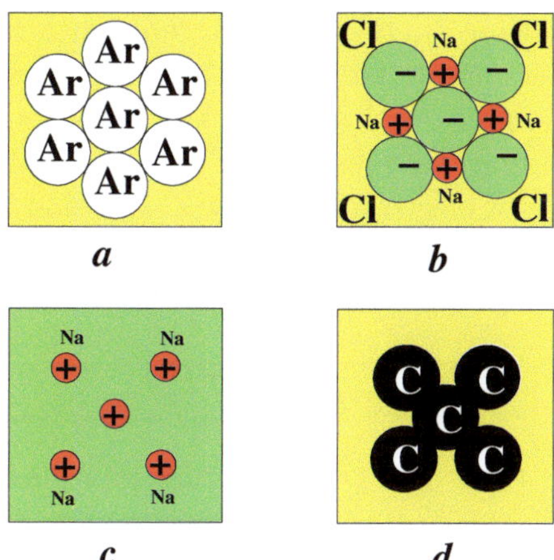

Figure 5.10: *Crystalline binding is traditionally ascribed [74] to four major types of interaction: (a) – Neutral atoms with closed electron shells held together by weak van der Waals forces associated with fluctuations in the charge distributions; (b) – Electrons are transferred from the alkali atoms to the halogen atoms, and the resulting ions are held together by attractive electrostatic forces between positive and negative ions; (c) – Valence electrons are given up by each alkali atom to form a communal electron sea in which the positive ions are dispersed; (d) – Neutral atoms appear to be bound together by the overlapping parts of their electron distributions.*

the commonality between all types of chemical interaction. It creates the impression that the so-called van der Waals, ionic, metallic and covalent types of chemical interaction are fundamentally different. However, this distinction is artificial and misleading. All interactions in Nature are of the same type and depend on the exchange of a mediating agent.

5.4.1 Interaction Theory

A simplified summary of electromagnetic interaction theory for chemists [7] shows that any interaction requires equal participation of an emitter and an absorber. The crucial argument is that emission without a receptive absorber and absorption in the absence of an emitter are equally unlikely events. What occurs in all cases is therefore best described as *transmission*. This means that emission, which causes propagation of energy as a function of time, is balanced by an inverse retrogression that runs from the absorber, backwards in time. If a signal, transmitted at time t_0 from the emitter, triggers an absorber response at time t_1, the return signal reaches the emitter at time t_0. The transaction is completed by this handshake, which is registered at time t_0 at the emission site and at t_1 by the absorber. In wave formalism it means that a standing wave exists for a period t_1 between emitter and absorber[4]. In the case of electromagnetic transmission this standing wave is called a *photon*.

There is nothing mysterious about the time-reversed signal. The wave equation

$$\nabla^2 U = \frac{1}{c^2}\frac{\partial^2 U}{\partial t^2}$$

is of second-order in time and therefore has solutions with time-inversion symmetry, known as retarded and advanced waves, respectively. The advanced solutions are usually ignored as physically unreal[5].

An accelerated charge (electron) is a source of both retarded and advanced spherical waves. An electron is, typically, accelerated by an incoming advanced wave from a distant absorber, which has in turn been stimulated by the retarded wave from the electron. Alternatively the electron could be excited by an incoming retarded wave from a distant emitter, responding to an advanced wave from the electron, now operating as an absorber. Either way, the ultimate transmission is not initiated at a single primary site. It is a concerted bootstrap process that involves two charges on an equal footing. It is colloquially stated that a photon is the agent that mediates an electromagnetic interaction.

Einstein's pioneering explanation of the photoelectric effect led to the perception of a photon as a particle that moves through the vacuum with constant speed c. Both of these conclusions are probably wrong[6]. The photon is not a particle but a standing (*i.e.* stationary) wave, as explained before. The interacting charges remain on the same relativistic world line and therefore effectively in contact.

Any charge centre, such as an electron, is both a source and a sink of electromagnetic radiation, with out-going and incoming spherical waves in balance. Should two charge centres be in different energy states, energy is transferred by the mechanism outlined above. In the equilibrium state each charge centre resembles the modulated standing wave packet of Figure 2.10.

---

[4]This transaction can last for an astronomical timespan.

[5]It is always dangerous to discard mathematically valid solutions on the basis of personal prejudice.

[6]The transaction mechanism explains the photoelectric effect even better than a particle model as the absorber in the metal can only be a single electron.

The interacting waves from myriads of charge centres constitute the electromagnetic radiation field. In particle physics the field connection between balanced charge centres is called a *virtual photon.* This equilibrium is equivalent to the postulated balance between classical and quantum potentials in Bohmian mechanics, which extends holistically over all space.

5.4.2 Cohesive Interaction

In the earlier analysis of covalency the interaction between atoms is assumed to consist of the interchange of valence electrons in a process analogous to the exchange of virtual photons between charge centres. Like the photon, an electron, in this situation, can be viewed in the Schrödinger sense as a standing wave between the nuclei. On describing this equilibrium charge distribution by an equivalent series of point charges, the covalent interaction is effectively defined by the local electromagnetic field. The same procedure lends itself to the analysis of all other interactions shown in Figure 5.10.

Ionic Interaction

The formation of electrovalent bonds between positive and negative ions is the simplest conception of chemical bonding. The first crystal structure ever determined by X-ray crystallography was that of sodium chloride, which unequivocally indicated a regular array of positive Na^+ and negative Cl^- ions. One unit cell is shown in Figure 5.11. The entire crystal consists of a single $(Na^+Cl^-)_n$ molecule. The total cohesive energy can be calculated with great precision by summing over all point-charge interactions on an infinite lattice.

In this instance the point-charge simulation is precise, but unlike the covalent case, the primary interaction is not shielded against secondary interference. Each ion is surrounded by six nearest neighbours of opposite charge, at a distance $d = a/2$, equivalent to an electrostatic coulomb interaction,

$$V = -\frac{6e^2}{4\pi\epsilon_0 d}\ \text{J}$$

with the elementary charge $e = 1.60219\times10^{-19}$C and d in meters. The twelve second neighbours with the same charge are at a distance of $\sqrt{2}d$, followed by a progression of shells with alternating charge type. The total interaction

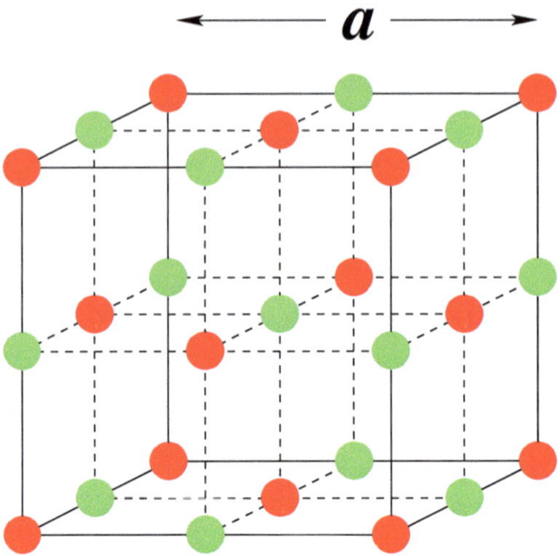

Figure 5.11: *The rocksalt structure.*

for an infinite crystal follows as:

$$
\begin{aligned}
V &= -\frac{e^2}{4\pi\epsilon_0 d}\left(6 - \frac{12}{\sqrt{2}} + \frac{8}{\sqrt{3}} - \frac{6}{2} + \ldots\right) \\
&= -\frac{e^2 A}{4\pi\epsilon_0 d}
\end{aligned}
$$

The infinite series converges to A, the so-called Madelung constant. The interaction for an Avogadro number of formula units consisting of n ions (in this case $n = 2$) each, amounts to the Madelung energy:

$$
\begin{aligned}
U &= -\frac{1}{2}\frac{ne^2 AN}{4\pi\epsilon_0 d} \ \mathrm{Jmol}^{-1} \\
&= \frac{KA}{d} \ \mathrm{kJmol}^{-1}, \ K = 1389
\end{aligned}
$$

The formula is converted into dimensionless units by defining a dimensionless distance based on a characteristic spacing for each compound. It is noted that the closest interionic distance can be specified as the sum of two ionic radii, $d = r^- + r^+$. Using the well established anionic radii for halide and chalconide ions [75], a conversion factor $R = \sqrt{r^+ r^-}$ is calculated in each case, and used to define the dimensionless distance $d' = d/R$, such that the Madelung energy,

$$U' = \frac{Z^2 A}{d'} = Z^2 D' \tag{5.36}$$

This formulation is valid for all crystals with the rocksalt structure (space group Fm$\bar{3}$m) made up of ions with charges of $\pm Z$. In fact, (5.36) is valid for any ionic crystal, providing A represents the Madelung constant for the relevant structure type. Madelung constants depend only on the geometrical arrangement of point charges and has the same value for isostructural ionic crystals. The Madelung constants for the three most common structure types[7] are shown in Table 5.2.

Table 5.2: *Madelung constants of common structures.*

Stucture	Space group	A
B1 Rocksalt	Fm3m	1.748
B2 CsCl	Pm3m	1.763
B3 Zincblende	F$\bar{4}$3m	1.638

The reduced Madelung energy (Setting Z=1 in (5.36)), *i.e.* A/d' is compared, as a function of d', for compounds with these three cubic structures, in Figure 5.12. The three structure types are clearly segregated as a function of energy, such that $A/d' < 0.77$ for zincblende structures, $0.77 < A/d' < 0.87$ for B1 crystals and $A/d' > 0.87$ for B2. The rationale for this discrimination is a minimum allowed radius ratio r^+/r^- for each structure type, as illustrated in Figure 5.13 that shows a section through the NaCl unit cell along a plane parallel to the cube face. As the radius ratio decreases a point is reached at which all ions are in close-packed contact. Any further reduction in the relative size of the cation would have no effect on the interionic distance or the Madelung energy. However, the cation will no longer be buttressed by its neighbours and would rattle around in its interstitial cavity between the anions. Simple geometry shows that the limiting situation occurs for $r^+/r^- = \sqrt{2} - 1 = 0.414$. The limiting ratio, derived for the CsCl structure, $r^+/r^- = \sqrt{3} - 1 = 0.732$ and for zincblende, $r^+/r^- = \sqrt{6} - 1 = 0.225$. These predicted limits are marked in Figure 5.12.

The predicted boundary between CsCl and NaCl structures, which are close in energy, corresponds well with the observed boundary. The predicted boundary between B1 and B3 structures is less convincing, but of the correct order. The more accurate discrimination, seen in Figure 5.12, takes

[7]Structure types are often distinguished in terms of their *Strukturbericht* symbols. Frequently encountered are A1 – Cu; A2 – W; A3 – Mg; B1 – NaCl; B2 – CsCl; B3 – ZnS (F$\bar{4}$3m).

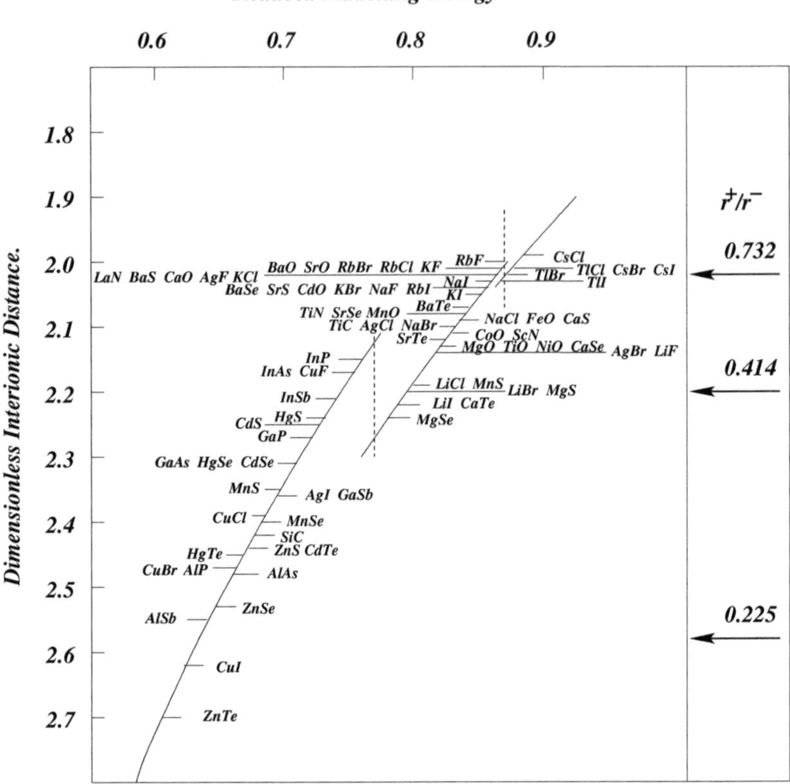

Figure 5.12: *Reduced Madelung Energy A/d' as a function of dimensionless interionic distance, for CsCl, rocksalt and zincblende crystals.*

the Madelung constant into account, together with the radius ratio. This criterion allows for the variation of ionic radii depending on differences in electronegativity.

Reduced Madelung energies are converted into familiar units by specifying actual charges (integer Z), such that

$$U = KZ^2 D'/R$$

The resulting energies are close to, but consistently higher than, experimentally measured lattice energies. An obvious interpretation of this result is that a degree of covalency occurs in nominally ionic crystals. Depending on electronegativity differences each compound should therefore have a characteristic charge $Z <$ integer n. The observation that lattice energies for most

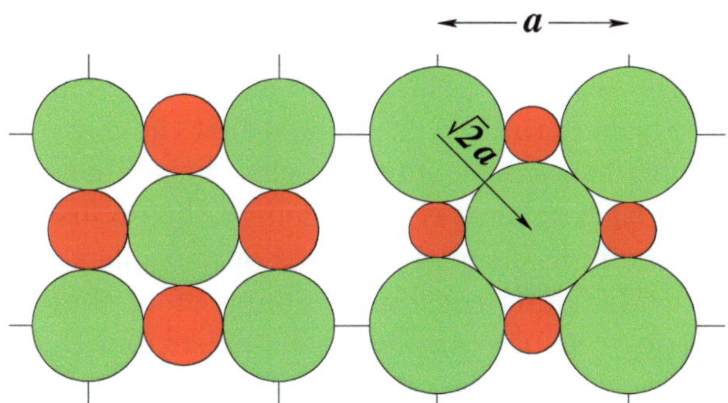

Figure 5.13: *The cube face of the rocksalt structure for different radius ratios. Structures with r^+/r^- less than the limiting ratio shown on the right-hand side are predicted to be unstable.*

diatomic structures are given, on average, by $U_c = 0.9U$, implies $Z = 0.95$, or a 5% covalency, which is not only a reasonable estimate, but also corresponds to the empirical Born repulsion formula:

$$U_c = U\left(1 - \frac{1}{n}\right)$$

The implied repulsion, $V = Br^n$, $n = 9$, is a purely *ad hoc* assumption, with dubious physical content.

The commonly stated rationale of a repulsive force, generated by overlapping charge clouds, is counter-intuitive in view of the accepted nature of covalency. Instead, the alternative formulation of ionic interaction, as in the point-charge covalency model, defines a general attractive curve for given Madelung constant, which terminates at a characteristic value of D', exactly as shown in Figure 5.9 for covalent interaction. As in the covalent case, the terminal point only depends on geometrical factors – in this case the radius ratio that limits the interpenetration of close-packed ionic spheres. Morse-like equilibrium curves for individual crystals appear centred on the attractive curve as shown schematically in Figure 5.9.

As in the covalent point-charge case the general curve, for given Madelung constant, specifies the binding energy for individual crystals according to their equilibrium interionic distances. It makes no allowance for variation of this distance. To estimate energy fluctuations, due to lattice vibrations, it is necessary to perform a Heitler-London type calculation, resulting in a Morse-like potential-energy curve, centred on the point-charge curve. The HL repulsive component is not the same as Born repulsion, which adjusts the equilibrium position along the point-charge (Madelung) curve. As in the covalence model the terminal point of the attractive point-charge curve is dictated by the exclusion principle. It is noted that the charges on both cation and anion are such as to ensure spherical closed-shell electronic structures. Interpenetration of such spheres is forbidden by the exclusion principle.

For crystals containing unequal numbers of multiple cations and anions, the definition of a Madelung constant becomes more complicated, but the Madelung (or lattice) energy can always be calculated by simple computerized methods [76]. In principle, a dimensionless interaction curve can be derived for any structure type.

Metallic Cohesion

Most theories of metallic cohesion are based on the Drude model which describes a metal as a regular array of cations immersed in a sea of electrons. This description contains elements of both the ionic and covalent models. Although metals lack the anions of ionic crystals it is not too difficult to imagine an equilibrium situation in which the supposed itinerant electrons are localized at the vacant, or interstitial, anion sites of corresponding ionic crystals. It is interesting to note that most metals crystallize with one of the three common structures: A1, A2 or A3 – *i.e.* cubic close packed (space group $Fm\bar{3}m$), body-centred cubic (space group Im3m) and hexagonal close packed (space group $P6_3/mmc$). The correspondence with binary ionic crystal types: B1, B2 and B3, is immediately obvious.

Modelling of the ionic crystalline salts has shown that, with an appropriate choice of ionic charge, the total ionic plus covalent contributions can be simulated in one step. A little experimentation shows that the same can be achieved in the case of metal crystals. However, knowing well that the covalent contribution can be substantial for metals such as Ta and W it is considered appropriate to make direct allowance for this effect. One possibility for such a simulation is shown in Table 5.3. The covalent contribution is arbitrarily assumed to be given by the dissociation energy of the diatomic molecules, calculated before. To get a convincing fit between calculated and observed [74] cohesive energies it is then only necessary to assume ionic charges of $\frac{1}{4}$ for metals of periodic group 1, of $\frac{1}{3}$ for group 11, of $\frac{2}{5}$ for the closed-shell metals of groups 2 and 10, and $\frac{1}{2}$ for all transition and lanthanide metals. The rationale behind these fractional charges remains obscure. The only systematic discrepancy, which could relate to experimental uncertainty, occurs towards the end of the lanthanide series. To accommodate the metals Al, Pb and Tl into the same scheme respective charges of $\frac{2}{5}$, $\frac{1}{3}$ and $\frac{1}{4}$ are assumed.

Table 5.3: *Structural details and cohesive energies of the metals with A1, A2 and A3 structures. Energies (in kJmol^{-1}) are listed as ionic and covalent contributions and total energies as calculated and observed [74] cohesive energies.*

Sym	*Struc*
Ion	*Cov*
Calc	*Obs*

Li	*A2*	**Be**	*A3*																		
87	*103*	*261*	*68*																		
190	*158*	*329*	*320*																		
Na	*A2*	**Mg**	*A3*																		
41	*74*	*185*	*15*																		
115	*107*	*200*	*145*																		
K	*A2*	**Ca**	*A1*	**Sc**	*A3*	**Ti**	*A3*	**V**	*A2*	**Cr**	*A2*	**Mn**		**Fe**	*A2*	**Co**	*A3*	**Ni**	*A1*	**Cu**	*A1*
33	*56*	*140*	*30*	*280*	*151*	*317*	*123*	*233*	*257*	*244*	*171*		*80*	*247*	*109*	*370*	*183*	*222*	*215*	*149*	*206*
89	*90*	*170*	*178*	*431*	*376*	*440*	*468*	*490*	*512*	*415*	*395*		*282*	*356*	*413*	*553*	*424*	*437*	*428*	*355*	*336*
Rb	*A2*	**Sr**	*A1*	**Y**	*A3*	**Zr**	*A3*	**Nb**	*A2*	**Mo**	*A2*	**Tc**	*A3*	**Ru**	*A3*	**Rh**	*A1*	**Pd**	*A1*	**Ag**	*A1*
31	*49*	*127*	*14*	*255*	*148*	*288*	*277*	*215*	*501*	*224*	*443*	*340*	*316*	*347*	*321*	*318*	*248*	*200*	*135*	*132*	*165*
80	*82*	*141*	*166*	*403*	*422*	*565*	*603*	*716*	*730*	*667*	*658*	*656*	*661*	*668*	*650*	*566*	*554*	*335*	*376*	*297*	*284*
Cs	*A2*	**Ba**	*A2*	**La**		**Hf**	*A3*	**Ta**	*A2*	**W**	*A2*	**Re**	*A3*	**Os**	*A3*	**Ir**	*A1*	**Pt**	*A1*	**Au**	*A1*
29	*48*	*90*	*38*		*255*	*290*	*295*	*196*	*480*	*224*	*500*	*339*	*376*	*339*	*393*	*318*	*324*	*198*	*308*	*132*	*232*
77	*78*	*128*	*183*		*431*	*585*	*621*	*676*	*782*	*724*	*859*	*715*	*775*	*732*	*788*	*642*	*670*	*506*	*564*	*364*	*368*

Ce	*A1*	**Pr**		**Nd**	*A1*	**Pm**		**Sm**		**Eu**	*A2*	**Gd**	*A3*	**Tb**	*A3*	**Dy**	*A3*	**Ho**	*A3*	**Er**	*A3*
233	*254*		*125*	*253*	*89*					*155*	*33*	*258*		*258*	*136*	*258*		*260*	*85*	*260*	
487	*417*		*357*	*342*	*328*				*206*	*188*	*179*		*400*	*394*	*391*		*294*	*345*	*302*		*317*
Al	*A1*	**Pb**	*A1*	**Tl**	*A3*											**Tm**	*A3*	**Yb**		**Lu**	
192	*139*	*109*	*97*	*67*	*68*											*263*	*52*	*222*	*41*	*266*	*150*
331	*327*	*206*	*196*	*135*	*182*											*315*	*233*	*263*	*154*	*416*	*428*

The relationship between cubic close-packed (ccp) structures and ionic compounds of type B1 is obvious. Interstitial sites with respect to metal positions are at fractional coordinates of the type $\frac{1}{2}00$ and equivalent to the ionic sites in B1. The Madelung constant of A1 type metals with interstitially localized free electrons is therefore the same as that of rocksalt structures. It is noted that the interstitial sites define the same face-centred lattice as the metal ions.

The situation for body-centred cubic metals (A2) is more complicated, but related to the ccp arrangement. As shown in Figure 5.14 a tetragonal face-centred unit cell can be constructed around the central axis of four contiguous body-centred cells. The interstitial points in the transformed unit cell define an equivalent face-centred cell, as before, and the same sites also define a body-centred lattice (shown in stipled outline) that interpenetrates the original A2 lattice. Each metal site is surrounded by six fcc interstices at an average distance d_6 – four of them at a distance $a/\sqrt{2}$ and two more at $a/2$. Referred to the averaged eight-coordination of a B2 lattice, the equivalent separation, $d_8 = \frac{4}{3}d_6 \simeq (\sqrt{3}/2)a$. The ionic lattice energies of Table 5.3 are

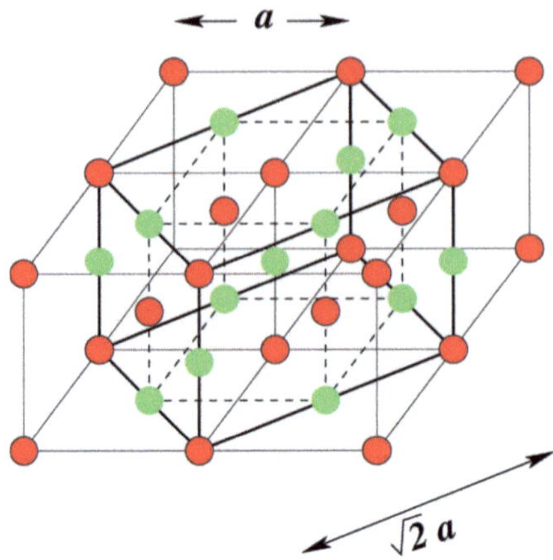

Figure 5.14: *Transformation from bcc to a face-centred tetragonal cell of double the volume.*

based on this average and the CsCl Madelung constant $A = 1.763$.

By the same argument a Madelung energy for A3 metals with interstitial electron density is approximated by that of an equivalent Wurtzite structure ($A = 1.641$). The same central motif occurs in all of the common metal structures, leaving equivalent sets of interstitial positions, as shown in Figure 5.15. To simplify the calculation, the charge separation is approximated in all cases as $d = 0.612a$, for metals with crystallographic axial ratios $1.57 \leq c/a \leq 1.62$. Zn and Cd, with $c/a \simeq 1.88$ are excluded.

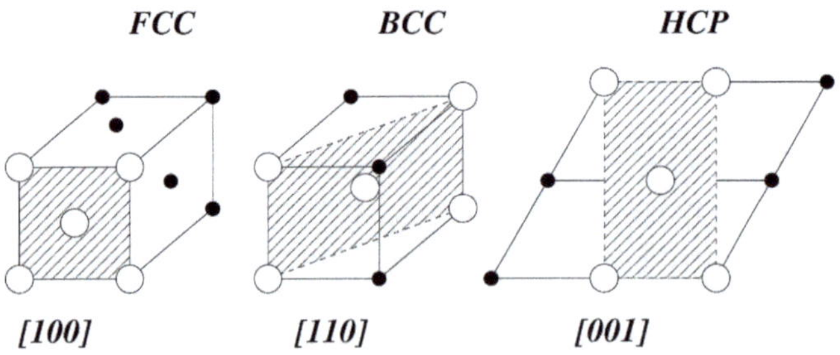

Figure 5.15: *Common motif in simple metal structures.*

Van der Waals Interaction

The cohesion in crystals of electrically neutral atoms, typified in Figure 5.10 by Ar, is ascribed to van der Waals interaction, generated by zero-point fluctuation of the electron density and its polarization effect on neighbouring atoms. The energy of interaction is given by the London formula, as a function of zero-point vibration frequency, between atoms at a distance d apart, as:

$$U \propto \frac{h\nu_0\alpha^2}{d^6},$$

or alternatively, in terms of first ionization potential, as:

$$U \propto \frac{I\alpha^2}{d^6}$$

The important factors in these formulae are the polarizability α, and the interatomic distance d. Noting that polarizability has the dimensions of a volume, the van der Waals interaction is seen to depend on the ratio of an atomic radius and an interatomic distance:

$$U \propto \left(\frac{R}{\sigma}\right)^6 I$$

This formula has been refined [77] for the calculation of the molar cohesive energy of noble gas crystals into:

$$U \simeq 7.22N\left[\frac{3\alpha^2 I}{4\sigma^6}\right] \tag{5.37}$$

where σ is the equilibrium interatomic distance in the solid. The empirical factor of 7.22 accounts for the twelve nearest-neighbour interactions, considered as six "full bonds" per atom, and some attraction to more distant neighbours. Using experimental values of α, I and σ [78], calculated cohesive energies compare well with experimental values [74], as shown in Table 5.4.

In order to avoid awkward explanation of the precise nature of this, so-called *dispersion* interaction, many authors label it a *quantum interaction without classical analogue*, content that it should remain a mystery. The similarity between van der Waals, metallic and ionic interactions in the solid state, argues against this dictum.

A section through the face centre of the cubic close-packed structure is shown in Figure 5.16. With randomly fluctuating electron density on all atoms it appears almost inevitable that some density should accumulate in

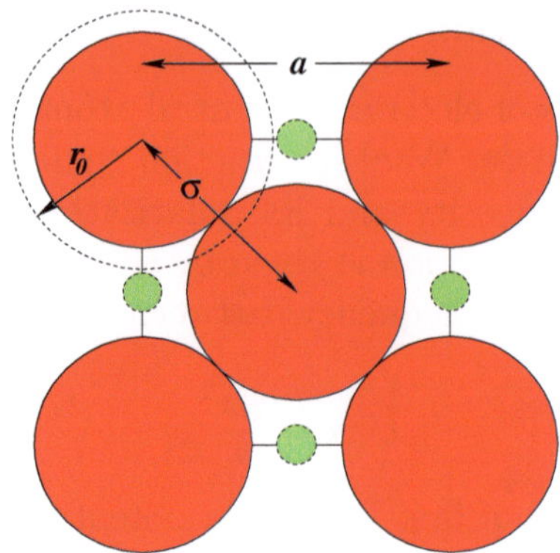

Figure 5.16: *Madelung simulation of the cohesion in inert-gas crystals.*

the octahedral interstices. As for a metal, the Madelung energy of such a crystal is estimated as:

$$U = \frac{2A}{a}(\delta Z)^2$$

where δZ is a fraction of the total Z (atomic number) electrons per atom that becomes interstitially localized. The factor δ must reflect the influence of the pulsating distorted charge cloud of radius r_0 (the ionization radius) at the distance $a/2$ from the centre. It should also depend on the polarization of an atom by a first heighbour, estimated before as $x = (R/\sigma)^6$. As an estimate of this factor the average (R/σ) per electron, calculated over the values of the four noble gases, $R/\sigma = 0.34$. Calculated cohesive energies, based on the formula

$$U = K\left(\frac{2A}{a}\right)\left(\frac{a}{2r_0}\cdot xZ\right) \tag{5.38}$$

are shown in Table 5.4. The calculated values are as convincing as the results

Table 5.4: *Details for the calculation of solid-state noble gas cohesive energies.*

									U(kJmol^{-1})		
	r_0(Å)	r_0^3	a(Å)	σ(Å)	α(Å^3)	R(Å)	R/σ	I(eV)	5.37	5.38	Expt.
Ne	1.20	1.73	4.462	3.16	0.39	0.73	0.23	21.56	1.71	1.15	1.88
Ar	1.81	5.93	5.316	3.76	1.62	1.18	0.31	15.76	7.72	2.32	7.74
Kr	2.12	9.53	5.810	4.11	2.46	1.35	0.33	14.00	9.16	8.07	11.2
Xe	2.49	15.44	6.350	4.49	3.99	1.59	0.35	12.13	12.01	15.71	16.0

obtained with the modified London formula (5.37). Having used empirical polarizabilities to estimate the x factor, it is intriguing to note the linear relationship between ionization spheres r_0^3 and cohesive energies, U. It accords with the interpretation of valence-state electron density flowing into interstitial sites and with the metallization properties of noble gases under pressure.

Covalent Crystals

Three of the four presumed types of chemical bond that occurs in the solid state have been reduced to the common basis of interaction between opposite charges localized at crystallographic lattice sites, apparently at variance with the pairwise covalency described before.

The diamond structure represents the archetype of a periodic covalent molecule and should resolve any paradox between the two models. For this purpose it is rewarding to re-examine the rocksalt structure as described in Figure 5.11. It is seen to derive from a CCP (A1) structure defined by the red spheres, with anions in the set of octahedral intersticial positions – green spheres. The crystallographic unit cell contains four metal-ion positions[8] and four octahedral intersticies. In addition there are two sets of four tetrahedral interstitial sites, as shown in Figure 5.17. In the zincblende structure the octahedral interstitial sites are vacant and anions are located in four equivalent tetrahedral sites.[9]

Diamond has the zincblende arrangement with C atoms at both cation and interstitial sites. The result is that each carbon atom in the crystal is tetrahedrally surrounded by four other equivalent carbon atoms in an infinite array. One tetrahedron centred on the zincblende anion position is shown in the diagram.

In order to describe the cohesion in the same terms as was done for metals and noble gas crystals, electron density is proposed to accumulate in the vacant interstitial octahedral sites and the remaining tetrahedral sites. This distribution is shown for two CC_4 tetrahedral units in Figure 5.17. The central carbon in the outlined tetrahedron is surrounded by an octahedron (in stippled outline) of interstitials defined by the set of vacant tetrahedral sites.

[8]Each of the eight sites at the corners of the cubic unit cell is shared between eight corner-sharing unit cells. Together they contribute one atom to the reference cell. Sites in the six cubic face centres are shared between two cells each and together contribute three atoms to the unit cell.

[9]In the fluorite (CaF_2) structure both sets of tetrahedral positions are occupied and the octahedral sites are empty.

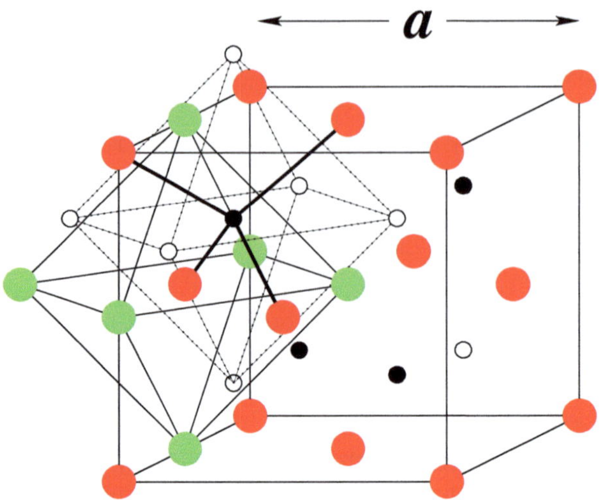

Figure 5.17: *The diamond structure.*

A second (green) octahedron, shown connected, surrounds a neighbouring C atom. Both arrangements are the same as the valence-density distribution in the CH_4 molecule as derived on the basis of orbital angular-momentum minimization [65] to be discussed in more detail in the next chapter.

The intuitive idea that the valence-density, which stabilizes a covalent interaction, is concentrated in a bond along the line between nuclei is, in fact, not supported by either of the models considered here. In the point-charge approximation, and the Heitler-London simulation, the exchange charge occurs in an overlap lens which is only *mathematically* equivalent to a point charge along the interatomic axis. It is nowhere implied that the limiting exclusion density only refers to the valence shell. The crystal model implies that, more likely, it refers to core density and to be interpreted as hard-sphere contact. The electrons that carry the interaction are extruded from this region and become concentrated in the interstitial positions of the crystal model which quenches the orbital angular momentum.

5.4.3 Conspectus

The four types of interaction that occur in the solid state have been shown to be dictated by a common principle. Pure covalent interaction only occurs in a few crystals, such as diamond, Si, Ge, silicon carbide and gray, or, α-Sn. However, it is the dominant interaction in most small molecules, including diatomic molecules, in which the three-dimensional distribution is easily overlooked. More careful consideration shows that all chemical interactions are electromagnetic in nature and mediated by the exchange of virtual photons between segregated charges.

Interatomic distances are determined by steric factors, of which the most important is the exclusion principle that depends directly on the geometry of space-time, observed as the golden ratio. Bond order depends on the ratio between the number of valence electrons and the number of first neighbours, or ligands, and affects interatomic distances by the screening of internuclear repulsion.

Intramolecular charge distribution depends on the number of valence electrons and electronegativity differences. In some extreme cases it resembles the interactions, traditionally considered to define van der Waals, metallic, ionic or covalent bonding. The true interaction in all cases however, contains elements of all four extremes.

The dominant interaction in molecules is covalent. The geometrical arrangement of atomic nuclei in molecules is mainly known from diffraction experiments in the solid state, and the persistence of such structures in the liquid, solution and gas phases is inferred from magnetic resonance and other spectroscopic studies. Apart from exceptional, nominally structureless, molecules to be discussed later on, most of these molecules may be considered as clusters of touching spheres, which represent atomic cores, and overlapping valence densities.

A perspective drawing of a cyclohexane molecule is shown in Figure 5.18 – overlapping first ionization spheres of the carbon atoms define the outer perimeter. It is obvious that concentration of the valence density on the con-

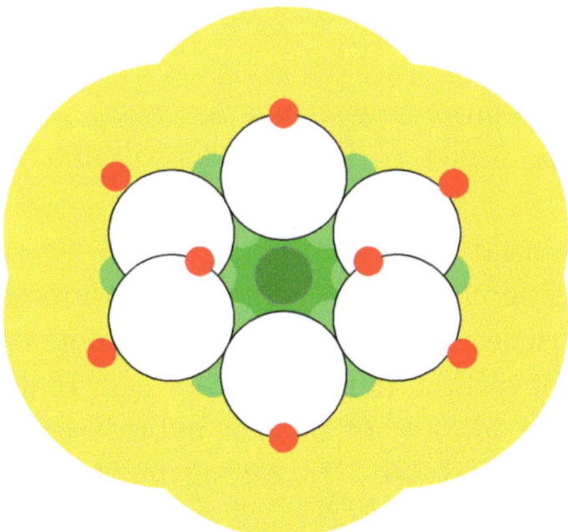

Figure 5.18: *Valence density distribution in cyclohexane*

necting lines between nuclei is highly unlikely. The colour-coded overlapping densities suggest that increased density appears in regions that correspond to the interstitial sites of regular lattices.

Although this suggestive drawing does not constitute a proof, it supports the idea that covalent interaction involves all atoms in a molecule. The idea of covalency as a pairwise electronic exchange is only a convenient bookkeeping device and the point-charge simulation over pairs of atoms should be interpreted as an additive contribution to a holistic intramolecular electromagnetic exchange, as discussed for the diamond structure.

The previous conclusion is in serious conflict with crystallographic charge-density analyses, which seem to indicate charge accumulation at the mid-points of all covalent bonds [79]. However, this result is contingent on several assumptions that need further consideration.

Deformation density, as originally conceptualized represents the difference between crystallographically observed electron density and calculated densities of the spherical atoms, which consitute the so-called promolecule. The effects of vibrational displacement, represented by ellipsoids in Figure 5.19, and ignored when defining a promolecule by spherical atoms, are most likely

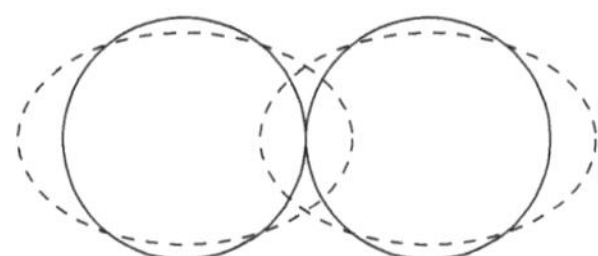

Figure 5.19: *Normal stretching vibration of a connected pair of atoms effectively smears out the spherical electron density into ellipsoids*

to show up between neighbouring atoms as excess difference density. However, despite this bias, depleted densities are observed in several instances.

To negate this observation, the conviction that covalent interaction mandates an excess bonding density in all cases, prompted the formulation of aspherical atomic densities to reflect the requirements of bonding theory. By multipole expansion of atomic densities, based on real spherical harmonics, in line with traditional models of orbital hybridization, the mandated deformation densities are retrieved. Increased flexibility of the model by the introduction of scaling parameters further ensures the elimination of any discrepancies with the theory. However, it is debatable whether this exercise proofs anything other than the power of well-chosen parameters to improve the fit between incompatible data sets.

The use of real spherical harmonics is particularly bothersome. It has been demonstrated convincingly that the notion of geometrical sets of oriented real atomic angular momentum wave functions is forbidden by the exclusion principle. The use of such functions to condition atomic densities therefore cannot produce physically meaningful results. The question of increased density between atoms must be considered as undecided, at best.

Table 5.5: APPENDIX: *Summary of ionization radii* (r_0)*, characteristic radii* (r_x)*, and Mulliken radii* (r_M) *in Å units; valence-state* (χ_v) *and Mulliken* (χ_M) *electronegativities in units of* $\sqrt{eV}$*; interatomic distance* (r_e) *or its estimate* (d)*; measured* (D_x) *and calculated* (D_c) *dissociation energies, in kJmol*$^{-1}$*, for diatomic molecules*

Z	Sy	r_0	χ_v	χ_M	$r_e, (d)$	d/r_x	D_x	D_c	r_x	r_M
1	H	0.98	6.25	6.25	0.741	0.96	436	432	0.77	0.98
2	He	0.30	8.18	(0)						
3	Li	1.25	4.91	1.73	2.673	0.99	110	103	2.70	3.54
4	Be	1.09	5.63	(0.8)	(1.94)	1.14	59	68	1.70	
5	B	1.62	3.79	2.07	1.590	0.86	290	300	1.85	2.96
6	C	1.60	3.83	2.50	1.312	0.71	596	614	1.85	2.45
7	N	1.56	3.93	(1.2)	1.098	0.67	857	830	1.62	
8	O	1.45	4.23	2.75	1.208	0.80	498	496	1.51	2.23
9	F	1.36	4.51	3.23	1.412	1.03	159	162	1.37	1.90
10	Ne	1.20	5.11	(1.7)						
11	Na	2.73	2.25	1.69	3.079	1.03	75	74	3.00	3.63
12	Mg	2.35	2.60	(0)	(2.78)	1.32	9	15	2.10	
13	Al	2.61	2.35	1.79	2.466	0.95	133	139	2.60	3.43
14	Si	2.40	2.56	2.18	2.246	0.77	310	301	2.90	2.91
15	P	2.20	2.79	2.37	1.893	0.67	485	488	2.81	2.59
16	S	2.05	2.99	2.49	1.889	0.71	425	427	2.66	2.46
17	Cl	1.89	3.24	2.88	1.988	0.86	243	241	2.30	2.13
18	Ar	1.81	3.39	(3.0)						
19	K	3.74	1.64	1.56	3.905	1.04	57	56	3.74	3.93
20	Ca	3.26	1.88	1.75	(3.43)	1.18	15	30	2.90	3.50
21	Sc	3.13	1.96	1.83	(2.79)	0.89	163±21	151	3.13	3.35
22	Ti	3.01	2.04	1.86	(2.52)	0.97	118	123	2.60	3.30
23	V	2.95	2.08	1.91	(2.28)	0.81	269	257	2.80	3.21
24	Cr	2.98	2.06	1.93	(2.43)	0.90	155	171	2.70	3.18
25	Mn	2.94	2.09	(0.2)	(3.00)	1.02	81	80	2.95	
26	Fe	2.87	2.15	2.01	(2.16)	1.03	118	109	2.10	3.05
27	Co	2.85	2.15	2.07	(2.18)	0.91	167	183	2.40	2.96
28	Ni	2.86	2.14	2.10	(2.17)	0.87	204	215	2.50	2.92
29	Cu	2.85	2.15	2.12	2.220	0.87	201	206	2.55	2.89
30	Zn	2.78	2.21	(0.2)	(2.32)	1.16	29	34	2.00	
31	Ga	3.29	1.86	1.79	(2.12)	1.01	138	129	2.10	3.43
32	Ge	2.94	2.09	2.14	(2.40)	0.80	274	299	3.00	2.87
33	As	2.62	2.26	2.30	2.103	0.72	382	375	2.92	2.67

Z	Sy	r_0	χ_v	χ_M	$r_e, (d)$	d/r_x	D_x	D_c	r_x	r_M
34	Se	2.40	2.56	2.43	2.166	0.75	331	325	2.90	2.52
35	Br	2.28	2.69	2.75	2.281	0.88	194	198	2.59	2.23
36	Kr	2.12	2.89	(2.4)						
37	Rb	4.31	1.42	1.53	(4.31)	1.05	46	49	4.10	4.01
38	Sr	3.83	1.60	1.69	(4.20)	1.22	16	14	3.43	3.63
39	Y	3.55	1.73	1.81	(3.09)	0.87	159±21	148	3.55	3.39
40	Zr	3.32	1.85	1.84	(2.50)	0.73	298	277	3.30	3.33
41	Nb	3.30	1.80	1.96	(2.10)	0.63	513	501	3.35	3.13
42	Mo	3.21	1.86	1.98	(2.21)	0.65	436	443	3.40	3.10
43	Tc	3.16	1.94	1.98	(2.35)	0.74	330	316	3.16	3.10
44	Ru	3.13	1.98	2.05	(2.31)					2.99
45	Rh	3.08	1.99	2.07	(2.45)	0.80	236	248	3.08	2.96
46	Pd	2.49	2.46	2.11	(2.39)	0.96	136	135	2.49	2.91
47	Ag	3.04	2.02	2.11	(2.51)	0.90	163	165	2.80	2.91
48	Cd	3.02	2.03	(0)	(2.59)	1.30	11	16	2.00	
49	In	3.55	1.73	1.74	(2.83)	0.98	100	104	2.90	3.52
50	Sn	3.26	1.88	2.06	(2.44)	0.87	195	192	2.80	2.98
51	Sb	3.01	2.04	2.20	(2.52)	0.74	299	293	3.40	2.79
52	Te	2.81	2.18	2.34	2.557	0.77	258	252	3.30	2.62
53	I	2.60	2.35	2,60	2.666	0.91	153	147	2.92	2.36
54	Xe	2.49	2.50	(1.5)						
55	Cs	4.96	1.24	1.48	4.47	1.04	44	48	4.30	4.14
56	Ba	4.48	1.37	1.64	(4.30)	1.08		38	4.00	3.74
57	La	4.13	1.48	1.74	(3.00)	0.73	247±21	255	4.10	3.52
58	Ce	4.48	1.37	1,80	(3.18)	0.71	243±21	254	4.48	3.41
59	Pr	4.53	1.35	1.79	(3.17)	0.91	130±29	125	3.50	3.43
59	Pr	4.53	1.35	1.79	(3.17)	0.91	130±29	125	3.50	3.43
60	Nd	4.60	1.33	(1.0)	(3.16)	0.99	84±29	89	3.20	
61	Pm	4.56	1.34	(1.1)						
62	Sm	4.56	1.34	(1.0)						
63	Eu	4.60	1.33	1.81	(3.47)	1.16	34±17	33	3.00	3.39
64	Gd	4.22	1.45	(0.5)						
65	Tb	4.59	1.34	(0.7)	(3.07)	0.90	131±25	136	3.40	
66	Dy	4.56	1.34	(0.7)						
67	Ho	4.63	1.32	(0.4)	(3.03)	1.01	84±30	85	3.00	
68	Er	4.63	1.32	(0.3)						
69	Tm	4.62	1.33	1.90	(3.00)	1.11	54	52	2.70	3.23
70	Yb	4.66	1.32	1.76	(3.38)	1.13	21±17	41	3.00	3.48

Table 5.4 (Continued)

Z	Sy	r_0	χ_v	χ_M	$r_e, (d)$	d/r_x	D_x	D_c	r_x	r_M
71	Lu	4.24	1.45	1.70	(2.99)	0.88	142 ± 33	150	3.40	3.61
72	Hf	3.83	1.60	(0.1)	(2.46)	0.70		295	3.30	
73	Ta	3.57	1.72	1.98	(2.23)	0.67	390 ± 96	480	3.59	3.10
74	W	3.42	1.79	2.08	(2.14)	0.62	486 ± 96	500	3.45	2.95
75	Re	3.38	1.81	2.00	(2.38)	0.68	386 ± 96	376	3.50	3.07
76	Os	3.37	1.82	2.18	(2.33)	0.67	415 ± 77	393	3.50	2.81
77	Ir	3.23	1.90	2.29	(2.36)	0.73	361 ± 68	324	3.23	2.68
78	Pt	3.16	1.94	2.35	(2.48)	0.75	308	308	3.40	2.61
79	Au	3.14	1.95	2.40	2.472	0.81	226	232	3.05	2.56
80	Hg	3.12	1.97	(0.9)	(2.61)	1.31	17	16	2.00	
81	Tl	3.82	1.61	1.75	(2.96)	1.06	63 ± 17	68	2.80	3.50
82	Pb	3.47	1.77	2.21	(3.05)	0.98	87	97	3.10	2.78
82	Pb	3.47	1.77	2.21	(3.05)	0.98	87	97	3.10	2.78
83	Bi	3.19	1.92	2.03	2.660	0.83	197	204	3.19	3.02
84	Po	3.14	2.06	2.27	(2.91)	0.83	186	187	3.50	2.70
85	At	3.12	2.16	(8.3)*						
86	Rn	3.82	32.31	(1.9)						

Chapter 6

Structure Theory

6.1 The Structure Hypothesis

The molecular concept has become so central in chemistry that understanding of chemical events is commonly assumed to consist of relating experimental observations to micro events at the molecular level, which means changes in molecular structure. In this sense molecular structure is a fundamental theoretical concept in chemistry. As the micro changes are invariably triggered by electron transfer, the correct theory at the molecular level must be quantum mechanics. It is therefore surprising that a quantum theory of molecular structure has never developed. This failure stems from the fact that physics and chemistry operate at different levels and that grafting the models of physics onto chemistry produces an incomplete picture.

Although the physics model may give a reasonable qualitative account of chemical concepts, such as chemical cohesion, it fails at the quantitative level, because essential factors are ignored. The most important factor is the environment. The free atom of physics represents a universe, completely empty, except for a solitary atom. Such an atom can never explain chemical effects, which occur because of the interaction of an atom with its environment. When the total environment is taken into account one deals with the familiar classical macro world. Between the two extremes is chemistry and it is important to know whether to describe chemical entities, like molecules, in classical or non-classical terms.

This problem relates to the question of where the so-called classical/quantum limit occurs. It has already been shown that Schrödinger's equation:

$$i\hbar\frac{\partial\Psi}{\partial t} = \left(\frac{\hbar^2}{2m}\nabla^2 + V\right)\Psi \tag{6.1}$$

separates into space coordinates, x and time coordinates, t, on assuming

$\Psi(x,t) = f(t) \cdot \psi(x)$. Substitution of

$$\frac{\partial\Psi}{\partial t} = \frac{\partial f(t)}{\partial t} \cdot \psi(x) \quad ; \quad \frac{\partial^2\Psi}{\partial x^2} = f(t)\frac{\partial^2\psi(x)}{\partial x^2}$$

into 6.1 gives:

$$i\hbar\frac{1}{f(t)}\frac{\mathrm{d}f(t)}{\mathrm{d}t} = \frac{\hbar^2}{2m}\frac{1}{\psi(x)}\frac{\mathrm{d}^2\psi(x)}{\mathrm{d}x^2} + V(x)$$

The left-hand side is a function of time only and the r.h.s. of coordinates only. This is only possible when both sides are equal to the same *separation constant*, E. Thus

$$\frac{\mathrm{d}f(t)}{f(t)} = \frac{iE}{\hbar}\mathrm{d}t$$

On Integration :

$$\begin{aligned} \ln f(t) &= iEt/\hbar + A \\ f(t) &= A\exp(iEt/\hbar) = A\exp(iS/\hbar) \end{aligned}$$

where the product $Et = S$ is known as the action of the sytem. When this solution is substituted into (6.1), the real part of the equation rearranges to:

$$\frac{1}{2m}(\nabla S)^2 + V - \frac{\hbar^2\nabla^2 A}{2mA} + \frac{\partial S}{\partial t} = 0$$

which has a form close to the Hamilton-Jacobi equation that describes the action (trajectory) of a classical system by:

$$\frac{1}{2m}(\nabla S)^2 + U + \frac{\partial S}{\partial t} = 0$$

The only difference between the classical and quantum formulations resides in the additional potential-energy term $\hbar^2\nabla^2 A/2mA$, known as the quantum potential, V_q. In the classical case $V_q = 0$. A quantum-mechanically stationary state occurs when $V_q = k$, a constant independent of x, *i.e.*

$$\nabla^2 A + \frac{2mA}{\hbar^2}k = 0$$

This expression reduces to Schrödinger's amplitude equation ($A = \psi$) by equating $k = E - V$, the eigenvalues of kinetic energy:

$$\nabla^2\psi + \frac{2m}{\hbar^2}(E - V)\psi = 0$$

The importance of the quantum potential lies therein that it defines the classical limit with $V_q \to 0$, or more realistically where the quantity $h/m \to 0$, which implies $h/p = \lambda \to 0$. It means that quantum effects diminish in importance for systems with increasing mass. Massless photons and electrons (with small mass) behave non-classically, and atoms less so. Small molecules are at the borderline, and macromolecules approach classical behaviour. When the system is in an eigenstate (or stationary state) of energy E, the kinetic energy $(E - V = k)$ is by definition equal to zero.

Returning to the idea of molecular structure, it is implied that isolated small molecules have no rigid structure or shape, and acquire this property only in condensed phases. The best documented example of such behaviour is the NH_3 molecule, but the phenomenon is certainly more general. The important implication is that shape is an attribute of molecules with appreciable mass. There is no need for shape when dealing with the small isolated entities of atomic physics, and for this reason it does not feature in traditional formulations of quantum theory.

Apart from detail, reformulation of quantum theory to be consistent with chemical behaviour, requires the recognition of molecular structure. In this spirit, it may be introduced as an essential assumption, or emergent property, without immediate expectation of retrieving the concept from first principles. Medium-sized molecules, especially in condensed phases, are assumed to have a characteristic three-dimensional distribution of atoms, which defines a semi-rigid, flexible molecular frame. The forces between the atoms are of quantum-mechanical origin, but on a macro scale, are best described in terms of classical forces.

6.1.1 Mechanical Simulation

It is a feature of classical mechanics not to enquire into the nature and origin of forces, but simply to quantify them in terms of suitable numerical parameters, such as the gravitational constant. Treating chemical forces between atoms in the same way, does not mean that they are of non-quantum origin. Whatever the nature or complexity of the interaction, an empirical polynomial function that describes the potential energy correctly, can in principle always be found. This aim is achieved mechanically by introducing a small number of so-called transferable force-field parameters.

Practical procedures are essentially of two types: purely empirical procedures, and chemical or valence-force methods. Empirical procedures model intramolecular cohesion only in terms of pairwise interactions, assuming potential energy of different forms for different regions, such as the bonding region, first and second neighbour regions and the non-bonded region. The valence method models deformation of an idealized structure in terms of chemically meaningful operations like valence bond and angle deformations. This approach appeals to chemists, but at present it remains empirical. It assumes that, in the absence of external forces, any bond or angle has an intrinsic characteristic value, determined only by the local charge distribution. Any distortion away from this equilibrium arrangement involves an amount of work that depends on the force constants, which resist the deformation. These force constants depend on bond order and chemical potential, and in practice angle deformation occurs more readily than bond stretching.

The simplest assumption is that bond deformation is proportional to the force, according to Hooke's law:

$$F = k(x - x_0) = -\frac{\partial V}{\partial x}$$
$$V = -\tfrac{1}{2}(x - x_0)^2 = \tfrac{1}{2}k(\Delta x)^2$$

For angle deformation, by analogy, $V_\theta = \frac{1}{2}k_\theta(\Delta\theta)$, *etc.* This assumption is clearly valid only for small displacements and cannot be used to model the rupture of a chemical bond. A Morse potential would be more appropriate in such application. To model the formation of a bond, an even more complicated potential, that takes activation effects into account, is required. However, most applications, known as *molecular mechanics* are less ambitious and have as their final objective only the modelling of the three-dimensional molecular structure. It assumes that the *strain* in a molecule is made up of the sums for various modes of distortion, *e.g.*:

$$U = E_r + E_\theta + E_\phi + E_\delta + E_{r\theta} + E_{\theta\theta} + E_{nb} + E_q + \ldots \tag{6.2}$$

ϕ and δ represent torsional and out-of-plane distortions, $r\theta$ and $\theta\theta$ indicate mixed modes, nb is for non-bonded van der Waals interaction and q for coulombic terms. From a knowledge of the various potential functions the total strain energy is calculated and minimized as a function of atomic coordinates.

The force constants and intrinsic parameters are assumed to be transferable, at least within families of related compounds and can therefore be optimized in terms of large volumes of experimental data. As the data banks grow, force fields become ever more precise and eventually should enable the optimization of any approximate structure to the desired degree of accuracy. From a theoretical perspective there is an understandable desire to explain molecular structure in terms of basic theory, rather than empirical modelling. It is therefore important to remember that the parameters of a valence force field are not arbitrary mathematical functions, but are defined in terms of well-known molecular parameters. In theory, at least, these parameters could therefore be derived from first principles.

6.1.2 Charge Density

The only remaining problem is calculation of the electron-density function, which cannot be done classically. However, for molecules in condensed phases the influence of the environment introduces another simplification. It has been shown that valence-state wave functions of compressed atoms are simpler, than hydrogenic free-atom functions. Core levels are largely unaffected and a nodeless valence-state wave function, which allows chemical distortion of electron density, can be defined. We return to this topic at a later stage.

6.2 Angular Momentum and Shape

Free atoms are spherically symmetrical, which implies conservation of their angular momenta. Quantum-mechanically this means that both L_z and L^2 are constants of the motion when $V = V(r)$. The special direction, denoted Z, only becomes meaningful in an orienting field. During a chemical reaction such as the formation of a homonuclear diatomic molecule, which occurs on collisional activation, a local field is induced along the axis of approach. Polarization also happens in reactions between radicals, in which case it is directed along the principal symmetry axes of the activated reactants. When two radicals interact they do so by anti-parallel line-up of their symmetry axes, which ensures that any residual angular momentum is optimally quenched. The proposed sequence of events is conveniently demonstrated by consideration of the interactions between simple hydrocarbon molecules.

Like the C atom, the simple hydrocarbons in their quantum-mechanically free state are all spherically symmetrical, presumably with C surrounded by a shell of hydrogen density at a characteristic distance, which depends on the chemical composition. The activation process breaks down the spherical symmetry, evidenced by the appearance of symmetry axes, as shown in Figure 6.1. The two units of angular momentum carried by the $C(2s^2 2p_{xy}^2)$ atom remain quenched, as pictured by charges that circulate in opposite sense, above and below the symmetry plane of each molecule. The formation of higher hydrides can, in all cases, be understood through the approach of an activated hydrogen atom along the molecular symmetry axis. The hydrogen density in CH_2 and CH_3 remains delocalized in annular shells.

When the elementary hydrocarbons combine to form dimers the relative orientation of the monomers is dictated by the vectorial quenching of the an-

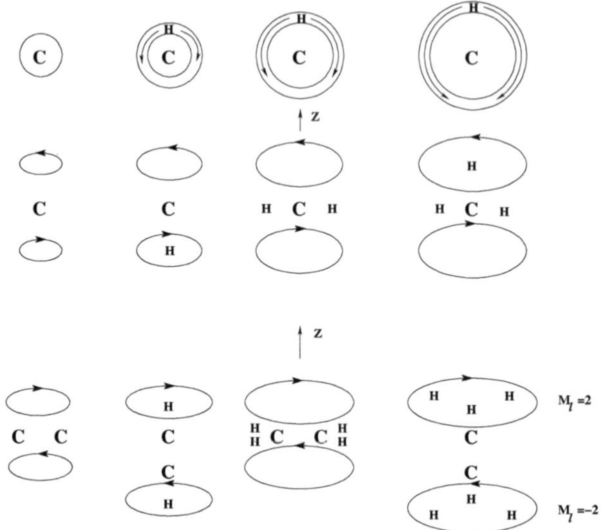

Figure 6.1: *Angular momentum vectors and hydrogen positions relative to C in small hydrocarbon molecules*

gular momentum. The formation of diatomic C_2 is, for instance, represented in this scheme by:

$$2\mathrm{C}(m_l = \pm 1) \rightarrow \mathrm{C}_2(M_l = \pm 2)$$

The same relationship is shown graphically for $(CH)_2$, C_2H_4 and C_2H_6 in Figure 6.1. At this stage there is no internal evidence to support the placement of hydrogen atoms, chosen to correspond with experimentally known solid-state structures.

6.2.1 Small Molecules

The hope of understanding the concept of molecular structure quantum-mechanically would obviously be at its most realistic for the smallest of molecules at the absolute zero of temperature. However, under these conditions completely different pictures emerge for the molecule in, either total isolation, or in a macroscopic sample. In the latter case the molecule appears embedded in a crystal, which is quantum-mechanically described by a crystal hamiltonian with the symmetry of the crystal lattice. The isolated molecule has a spherically symmetrical hamiltonian. The two models can obviously not define the same quantum molecule.

The central-field radial part of Schrödinger's equation (2.18) has the form:

$$-\frac{\hbar^2}{2m}\left[\frac{\mathrm{d}^2}{\mathrm{d}r^2}+\frac{2}{r}\frac{\mathrm{d}}{\mathrm{d}r}-\frac{l(l+1)}{r^2}\right]R+\left[E-V(r)\right]R=0$$

which is independent of m_l, the only quantum number that defines a vector quantity. In a non-central field the wave function $\psi(r,\theta,\varphi)$ is not separable in the same way and the energy eigenvalues depend on all quantum numbers n, l, m_l. Therein lies the difference between an isolated molecule with an essentially spherically symmetrical hamiltonian and a molecule that interacts with its environment. In the case of a free molecule the hamiltonian describes the density (nuclear and electron) as radial distribution functions. Such a function cannot generate a geometrical structure. In a non-central field the angular-momentum vectors defined by m_l impart specific geometrical features to the radial distribution function and generate a molecular structure, related to the Born-Oppenheimer approximation. The resulting structure is not only a minimum-energy arrangement, but derives in the first place from a minimum in orbital angular momentum.

In quantum-mechanical practice calculations are initiated by assuming molecular structures based on crystallographic results. The purpose of further optimization is not so much to obtain an improved geometrical description of the nuclear framework, but rather to refine the electron-density distribution function. A correct procedure could expand atomic electron densities as an infinite series of the orthonormal set of surface harmonics, which, for obvious reasons, is never attempted. However, in theory, the coefficients of the basis functions that minimize the total energy should define the desired charge distribution. In practice the series is truncated, on grounds of chemical intuition, to a small basis set of *real* surface harmonics. The intuition simply redefines the hybridization model in a basis set of real functions, selected to reproduce a favoured charge distribution. Although rarely admitted, the exercise therefore amounts to a biased simulation of both *assumed* nuclear and electron distributions.

An unbiased simulation may use a truncated basis set that represents the lowest complex surface harmonics of the atomic valence shell on a Born-Oppenheimer framework with the correct relative atomic masses. For small molecules, of less than about fifteen atoms, the nuclear framework could perhaps even be generated computationally without assumption. The required criterion is the optimal quenching of angular momentum vectors. The derivation of molecular structure by the angular-momentum criterion will be demonstrated qualitatively for some small molecules.

The valence shell of oxygen in the water molecule, $\mathrm{O}(2s^2 2p_z^2 2p_{xy}^2)$, in a polar environment, has a C_∞ rotation axis along Z, as shown for C in

Figure 6.1, predicting the same spherical structure of CH_2 for H_2O in a field-free environment. In a polarizing environment the excess valence density is displaced along Z to create what is known as a $2p_z$ lone pair, shown schematically in Figure 6.2. In ammonia the $N(2s^2 2p_z^1 2p_{xy}^2)$ valence shell

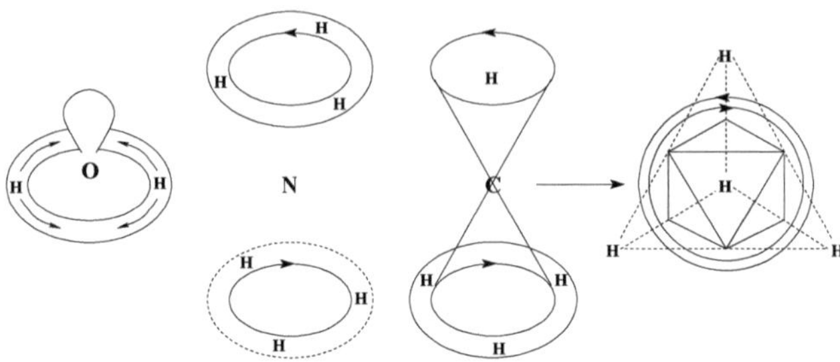

Figure 6.2: *Structures of water, ammonia and methane that quench orbital angular momentum*

has no lone pair in the NH_3 molecule and the polar directions along $\pm Z$ are equally likely. The annular distribution of hydrogen density could be in a plane perpendicular to Z on either side of N, defining the well-known non-classical structure of ammonia.

To understand the structure of methane it is necessary to assume that the four hydrogen atoms occupy symmetry-equivalent positions relative to C. The Z-axis can be defined along any of the four C-H directions. In each case it is necessary that the angular momentum vectors, represented by circulating charges, should cancel, as shown in one direction in Figure 6.2. The eight equivalent circles, so defined, intersect along the bisectors of the six H–H connecting lines, to define a variable octahedron. This conclusion that valence charge should accumulate in symmetrical disposition with respect to the hydrogen atoms is at variance with conventional bonding theory, but consistent with the covalent interactions as envisaged for crystalline diamond, Figure 5.17.

In earlier discussions of small-molecule structures, as shaped by angular-momentum vectors [7], the traditional practice of labelling electrons in chemical bonds, was followed. At the time it was considered necessary in order to render the argument more palatable to the tradition-minded, but it probably caused more confusion than instruction. In fact, such partitioning is totally unnecessary.

All electrons in a molecule are to be treated as equivalent. The angular momentum, carried by some electrons becomes a shared property of the total molecular valence shell. For first-row elements the simple octet valence rules can be used directly to predict the excess density per ligand, the number of lone pairs and the bond order for each interaction.

Each first-neighbour interaction is assumed to share a pair of electrons. In symmetrical dimers the odd electrons, not involved in covalent interaction, remain associated with their atoms of origin to form couples that screen the repulsion between neighbouring nuclei. Any remaining excess of valence electrons constitute symmetrical lone pairs. By way of illustration, consider diatomic C_2, derived from $C(2s^2 2p^2_{xy})$, the configuration dictated by the exclusion principle. The single covalent interaction leaves 3 electrons per atom unaccounted for. The two remaining odd electrons clearly constitute a separated screening couple, defining bond order 2, with an s^2 lone pair on each C. In N_2 the unpaired electrons of $N(2s^2 2p^2_{xy} 2p^1_z)$ form the bond. By Hund's rule the p^2_{xy} couple has parallel spins and cannot form a lone pair. It means that the remaining 4 electrons comprise two screening couples (bond order 3) and two lone pairs per N_2 molecule. The p^2_{xy} configuration on each oxygen atom generates one bond and a screening couple, together with four lone pairs to yield a paramagnetic O_2 molecule with bond order 2.

The same simple scheme gives the correct description of the dimeric hydrocarbons. Acetylene has 10 valence electrons and three bonds, H–C·C–H, leaving two odd couples, predicting a bond order of 3. Ethylene with 12 electrons, 5 bonds and 1 odd couple has bond order 2. Ethane is saturated.

The low-valence small molecules of the higher-period p-block elements follow the same valence rules as the first-period compounds. The higher-valence compounds, which involve the participation of unoccupird d-levels, have more complicated structures, but are commensurate with the angular-momentum argument.

Despite the completely different approach to chemical interaction, which has been followed here, the conventional standard symbols which are used to define the connectivity in covalent molecules, can also be applied, without modification, to distinguish between interactions of different order. However, each linkage pictured by formulae such as $H_3C{-}CH_3$, $H_2C{=}CH_2$, $HC{\equiv}CH$, represents the sharing of a single pair of electrons with location unspecified. The number of connecting lines only counts bond order and may be established from the classical valence rules, *e.g.* v(C,N,O,F)=(4,3,2,1). Special symbols are used for non-integral bond orders, as in the symbol for benzene:

The valence arguments used here predict the same bond orders as the conventional hybridization models. It is of interest to note that, on the basis of one odd electron per C–C interaction, uniform orders of $1\frac{1}{2}$ is predicted in a benzene ring, without resort to the awkward concept of π-bonding. Final elimination of this concept mandates an alternative explanation of non-steric barriers to rotation.

As will be shown, even-order bonds are conformationally rigid. Enhanced activity at higher-order sites relates to the increased number of odd electrons available for bonding.

6.2.2 Conformational Rigidity

Minimization of angular momentum on formation of an ethylene molecule requires anti-parallel alignment of orbital angular momentum vectors, perpendicular to the molecular plane, as shown in Figure 6.1. Twisting the molecule around the C-C axis decouples the vectors and generates residual angular momentum. Resistance against such torsion is known as the *barrier to rotation*, which is responsible for the conformational rigidity manifest in geometrical isomerism and the planarity of aromatic systems, including ethylene. The torsional barrier represents the kinetic energy required to initiate charge rotation on a circle of radius r, calculated [80] as:

$$T = \frac{M_l \hbar^2}{2mr^2}$$

Equating $T = 270\text{kJmol}^{-1}$, the experimental estimate of the torsional barrier, and solving for $r = 1.65\text{Å}$, gives an eminently reasonable radius for a circulating electron with $M_l = 2$. Compared to acetylene and ethane (Figure 6.1), these molecules should not have torsional barriers to rotation about their molecular axes, but perpendicular to that. This explains the linear structure of acetylene and the molecular axis with free rotation in ethane. The π-electron model predicts an enhanced barrier to rotation about triple bonds. It is significant to note that dimetal compounds with triple bonds have no geometrical isomers [81].

6.2.3 Molecular Chirality

Optical activity is one of the best known, but least understood, phenomena of organic chemistry. It is observed as the ability of certain substances to interact with linearly polarized light by rotating the plane of polarization. Linear polarization means that the electromagnetic radiation vectors oscillate in fixed orthogonal planes that intersect along the propagation vector.

A linearly vibrating vector can be considered as made up of two equal vectors that rotate at the same rate in opposite sense, as shown schematically in Figure 6.3. The resultant electric vector for two circularly polarized waves with $\theta_r = \theta_l$ is of the form $E = E_0 \cos\theta$. More generally, in complex

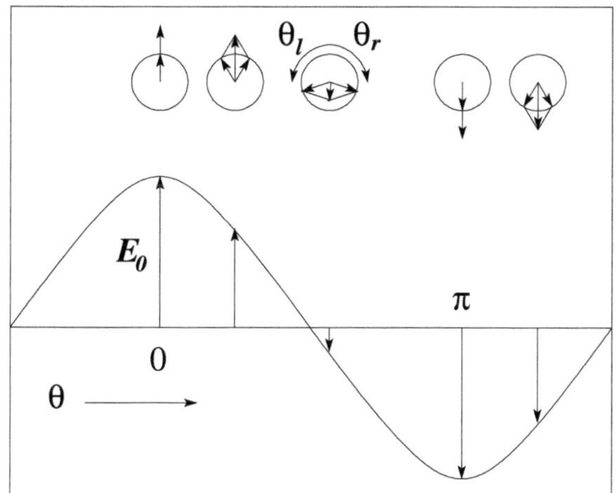

Figure 6.3: *The resultant of two equal components that rotate about the axis, defined by the propagation vector, defines a linear vibration*

notation[1], the two components are $E_{r,l} = E_0 \exp(\pm i\theta)$. A rotating vector that moves along the rotation axis describes a helix.

The rotating electric vectors generate equal, but opposite magnetic fields that cancel exactly, without any effect of most molecules in their path. However, any molecule with non-zero orbital angular momentum and wave function $\psi = A\exp(\pm iM_l\varphi)$, has a magnetic moment that would interact with one of the components of a plane polarized wave. When such interaction occurs the two rotating vectors no longer propagate at the same rate through a molecular sample and emerge out of phase. Their resultant is still a linear vector, but now rotated through an angle with respect to the original plane of polarization.

It is known from experience that only chiral molecules have the ability to rotate the plane of polarization. This observation suggests that molecular chirality is a manifestation of residual orbital angular momentum, and *vice versa*. When an achiral molecule is placed in a magnetic field the molecular magnetic vectors are aligned in the diection of the field and not along the local molecular symmetry directions. The molecule therefore acquires a

---

[1]

$$E_r + E_l = E_0(\cos\theta + i\sin\theta + \cos\theta - i\sin\theta)$$

magnetic moment and residual angular momentum [82], with the ability to interact with the magnetic vector of a polarized light beam. This well-known phenomenon is known as the *Faraday effect.*

The relationship between chirality and optical activity is conveniently demonstrated by a series of substituted methanes.

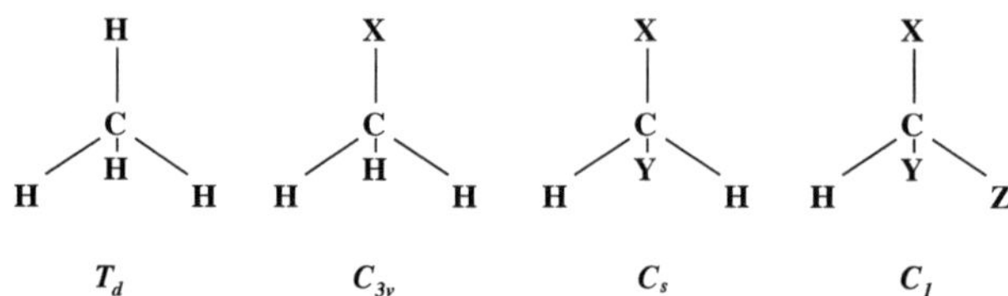

Quenching of the orbital angular momentum in methane (symmetry T_d) has been demonstrated in Figure 6.2. Replacing one hydrogen atom by some ligand X lowers the molecular symmetry to C_{3v}. It modifies the distribution of valence density but the angular momentum vectors remain balanced. Substitution of another hydrogen by ligand Y lowers the symmetry even further, but the two unique ligands stay in a mirror plane with the central C atom. The molecular symmetry remains sufficient to quench the orbital angular momentum. The vectors become disaligned only when this last element of symmetry disappears and the molecular symmetry reduces to C_1. At this stage angular momentum is no longer quenched, ($L_z \neq 0$), the molecular quantum number M_l is non-zero and polarized photons interact with the resulting magnetic moment. The plane of polarization is affected differently by enantiomers with respective positive and negative quantum numbers M_l. The enantiomers have identical molecular hamiltonians and energies – they only differ in angular momentum eigenstates. Decoupling of angular momentum vectors happens whenever a chiral centre, here defined in terms of four dissimilar ligands in tetrahedral relationship, occurs in a molecule.

An alternative modern explanation of optical activity, without mention of orbital angular momentum [83] states:

> "Classically, a molecule is optically active when in an electronic transition there is a helical movement of charge density.... A characteristic of a helix is that it corresponds to a simultaneous translation and rotation and so (...) optically active molecules are those in which a transition is simultaneously both electric dipole (charge translation) and magnetic dipole (charge rotation) allowed.... As an alternative general statement, one can say that optically active molecules do not have any S_n axis, where n can assume any value ($n = 1$ corresponds to a mirror plane and $n = 2$ to a centre of symmetry)."

This statement is hardly more informative than any popular interpretation (*e.g.* [84]) of van't Hoff's original proposal that the valences of carbon are directed towards the corners of a regular tetrahedron.

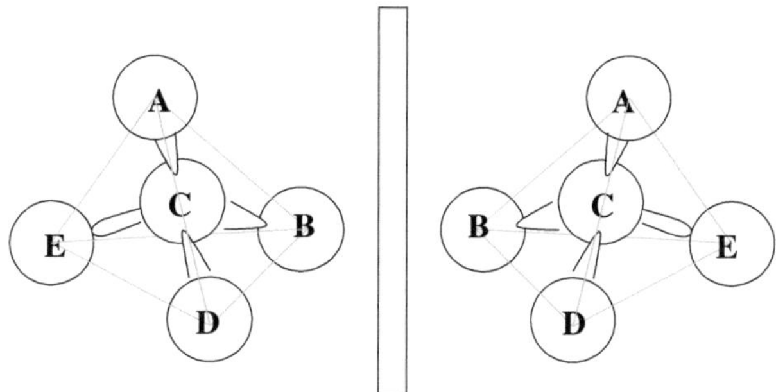

If the carbon atom has four different groups attached to it, then there are two different ways of arranging these groups. The two arrangements give rise to two types of molecule, related to one anther as nonsuperimposable mirror images, The two structures are enatiomers, and differ in the direction in which they rotate plane polarized light. Each structure is optically active.

These models offer no explanation of the Faraday effect.

6.3 Molecular Modelling

The relationship between orbital angular momentum and stereoisomerism argues convincingly for a pivotal role of angular momentum vectors in shaping the structures of small and medium-sized molecules. Especially satisfying is the prediction of a flat rigid structure with uniform bond lengths for the aromatic benzene molecule, arising from $6 \times p_{xy}^2$ pairs and six odd electrons. However, the failure to account for the non-planar conformation of cyclo-octatetraene (COT), and the alternating bond lengths observed in conjugated polyenes shows that some important consideration, presumably energy related, is clearly being overlooked. The successful free-electron model for aromatic and conjugated systems will be examined for possible clues.

6.3.1 Free-electron Modelling

The special properties of aromatic compounds and conjugated systems are related to the regular alternation of single and second-order C–C bonds along molecular chains. Each second order bond is associated with a balanced charge circulation and a couple of odd electrons with paired spins. The conclusion follows from the fact that ethylene is not a biradical. The properties of both aromatics and conjugate polyenes are traditionally ascribed to overlap of π-orbitals of p_z atomic type.

Quantum-mechanically delocalization may be likened to the appearance of a standing wave along the entire length of the chain. In one dimension such a wave obeys the wave equation:

$$-\frac{\hbar^2}{2m_e}\frac{\mathrm{d}^2\psi}{\mathrm{d}x^2} = E\psi$$

Setting $k^2 = 2m_eE/\hbar$, the resulting Helmholtz equation has solutions of the type:

$$\psi = a\sin kx + b\cos kx$$

In order to restrain the wave to the line segment that corresponds to the molecular chain it is necessary to introduce the boundary conditions, $\psi(0) = \psi(L) = 0$, to ensure that the wave vanishes at $x = 0$ and $x = L$, where L is the length of the chain[2]. Since $\cos 0 = 1 \neq 0$, the first condition requires $b = 0$ and hence that

$$\psi(x) = a\sin kx \quad (a \neq 0)$$

The condition $\psi(L) = 0$ next requires $\sin kL = 0$, *i.e.* $kL = n\pi$, with n a positive integer. The eigenvalues follow as

$$E_n = \frac{n^2h^2}{8m_eL^2}$$

and the wave function

$$\psi_n = a\sin\left(\frac{n\pi x}{L}\right), \quad n = 1, 2, 3, \ldots$$

Note that the condition $\psi = 0$ is not allowed, as that would imply $\psi = 0$, everywhere. The ground-state or *zero-point* energy $E_1 = h^2/8mL^2 \neq 0$.

The constant k, known as the *wave vector* defines the momentum of the delocalized electron, $p = k\hbar$, such that $E_x = p^2/2m$, or in terms of the de Broglie wavelength, $k = 2\pi/\lambda$. In summary:

$$E_n = \frac{n^2h^2}{8m_eL^2} = \frac{p^2}{2m_e} = \frac{1}{2m_e}\left(\frac{h}{\lambda}\right)^2$$

[2]This model is popularly known as the particle-in-a-box problem.

i.e.

$$\lambda^2 = \frac{4L^2}{n^2} \quad \text{or} \quad L = \frac{n\lambda}{2}$$

The electron along the polyene chain therefore behaves as a standing wave with an integral number of half wavelengths fitting into the line segment, such that $\psi_n(0) = \psi_n(L) = 0$.

Should the ends of the line segment, $x = 0$ and $x = L$ be joined to form a ring, the wave functions for odd n are not continuous at the joint.

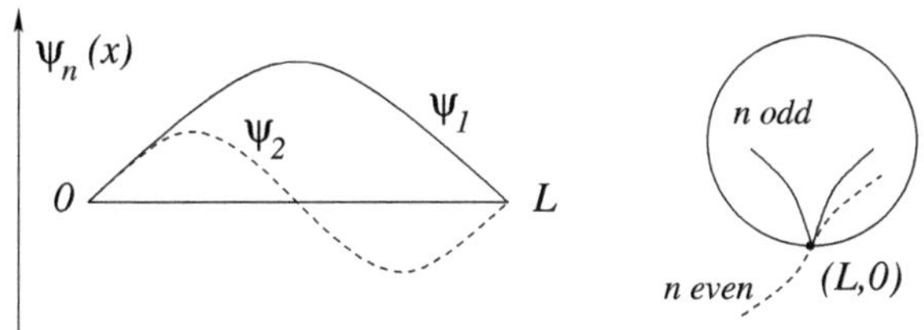

Only even functions of the linear case remain acceptable in the cyclic system and hence the quantum condition becomes $L(= 2\pi r) = n\lambda$, for all n. The eigenvalues are:

$$E_n = \frac{h^2}{2m_e\lambda^2} = \frac{h^2}{8m_eL^2}(2n)^2 = \frac{n^2h^2}{2m_eL^2}, \quad n = 0, \pm1, \pm2, \ldots \tag{6.3}$$

The different signs for quantum numbers $n \neq 0$ indicate that the electron can rotate either clockwise or counterclockwise, with either positive and negative angular momentum. Symbolically, the angular momentum,

$$\pm pr = \pm\frac{hr}{\lambda} \quad \text{or} \quad pr = \frac{nhr}{2\pi r} = n\hbar, \quad n = 0, \pm1, \pm2, \ldots$$

as required. Each energy level ($n \neq 0$) is doubly degenerate as the energy depends on n^2 and therefore is independent of the sense of rotation. The quantum number $n = 0$ is no longer forbidden because the boundary conditions no longer apply. In the cyclic case $n = 0$ implies infinite λ, *i.e.* ψ_0 =constant, $E_0 = 0$. The kinetic energy in the stationary state is zero, as in the Bohmian interpretation.

As the condition $\psi(0) = 0$ no longer applies the constant $b \neq 0$ and the wave functions for cyclic systems have the more general form, consisting of sine and cosine components. In the case of a four-membered ring the standing wave patterns, composed of sine and cosine components, as shown in Figure 6.4, will be identical, irrespective of the point at which the ring was closed originally. The zero energy level is non-degenerate. The predicted energy spectrum, assuming a non-degenerate zero level, has been

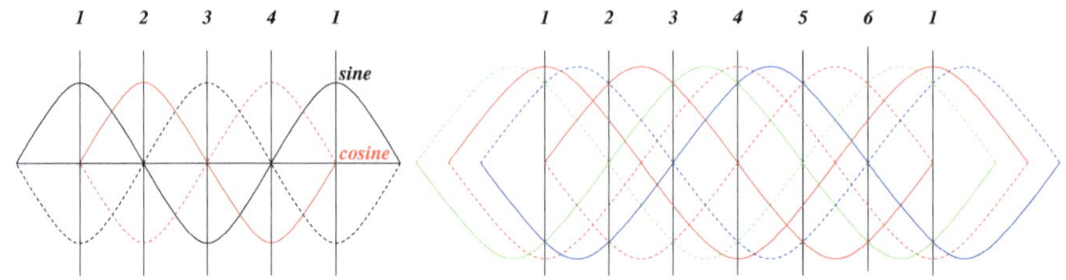

Figure 6.4: *Comparison of wave patterns on four and six-membered rings*

interpreted to mean that closed-shell configurations occur for cyclic systems with $2, 6, 10, \ldots, (4m+2)$ free electrons. In the traditional interpretation of aromaticity each carbon atom is considered to contribute one delocalized p_z electron, such that only those systems with $4m+2$ atoms have closed-shell arrangements. Despite this plausible characterization of aromatic systems, the predicted orbital degeneracy does not explain the orbital angular momentum of the free electrons: A p_z electron has, by definition, zero component of o-a-m along Z; the formula $(4m+2\,,\, m=1)$ makes provision for $8\hbar$; while six carbon atoms carry $12\hbar$. This discrepancy seems to indicate $m=3$, and hence 14 free electrons for benzene, instead of the assumed 18.

This final discrepancy is resolved by the degeneracy of the zero level of equation 6.2. There are three ways, rather than one, of closing a six-membered ring, as shown in Figure 6.4, giving rise to three different standing waves with the same principal quantum number and energy. The thing is that a standing wave, such as the one shown in red, has different sine and cosine components when sampled at carbon positions 1, 2, or 3.

The same difference can be shown to distinguish between cyclic systems with $4m$ and $4m+2$ carbon atoms respectively. The $n=0$ energy level has degeneracies of m and $2m+1$, for these respective systems. It is apparent that the zero level of $4m$ cyclics cannot accommodate more than $2m$ itinerant electrons, which means that the $2m$ odd couples cannot be delocalized to yield an aromatic system. The best known example of a non-aromatic conjugated cyclic molecule is cyclo-octatetraene, shown in Figure 6.5.

Cyclic systems with $4m+2$ carbon atoms are characterized by the aromatic index m. At the lowest level $(m=0)$ ethylene formally corresponds to a 2-membered ring. The electronic structure shown in Figure 6.1 is consistent with aromatic electron distribution. Benzene $(m=1)$ is the prototype of properly cyclic aromatics. The molecular o-a-m conserves the sum of atomic angular momenta. The odd couples with the screening function are delocalized around the ring, with zero o-a-m.

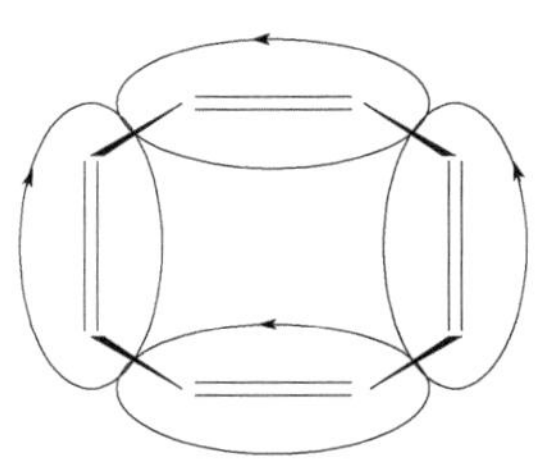

Figure 6.5: *Tub-like structure of COT*

It is of interest to see how the predicted aromatic electron configurations compare with observed electronic spectra. The wavelength for $n \rightarrow n+1$ transition follows from:

$$\Delta E_{(n+1,n)} = \frac{h^2(2n+1)}{2m_e(2\pi r)^2} = \frac{hc}{\lambda}$$

The radius of charge circulation follows from:

$$r^2 = \frac{h\lambda(2n+1)}{8m_e\pi^2 c}$$

Long wavelength absorption bands for ethylene ($n = 2$) and benzene ($n = 3$) are observed at 176 nm and 260 nm respectively, solving for $r = 1.64$Å and 2.36 Å respectively. This result is to be compared with $r = 1.65$Å calculated from the observed barrier to ethylene rotation (section 6.2.2).

The molecular structures and charge circulation circles for ethylene and benzene are compared to scale in Figure 6.6(a). In both cases the match is close enough to conclude that the circles pass through the hydrogen positions.

Aromatic Systems

The first member of the $4m+2$ series (ethylene, $m = 0$) is atypical in being noncyclic[3] and of bond order 2. The second, and first cyclic, member (benzene, $m = 1$) is considered the prototype of aromaticity. The o-a-m circulation, on a circle through the hydrogen positions, represents the vectorial addition of three angular-momentum vectors centred on the conjugated double bonds and tangent to the resultant circle. Remarkably, these circles intersect at the midpoints of the conjugated single bonds and at the centre of the cyclic system, as shown in Figure 6.6(b).

The component circles, by definition, have the same radius as the inscribed circle of the carbon hexagon, *i.e.* $r = 1.25$Å– somewhat smaller than the previously calculated o-a-m circle of ethylene. Using $r = 1.25$Å, the recalculated absorption wavelength, $\lambda = 146$nm, is within the range (145 – 180 nm) of the first intense absorption band of ethylene in the vacuum UV [85].

[3] Any straight line with an extra point at infinity may be considered a circle with infinite radius.

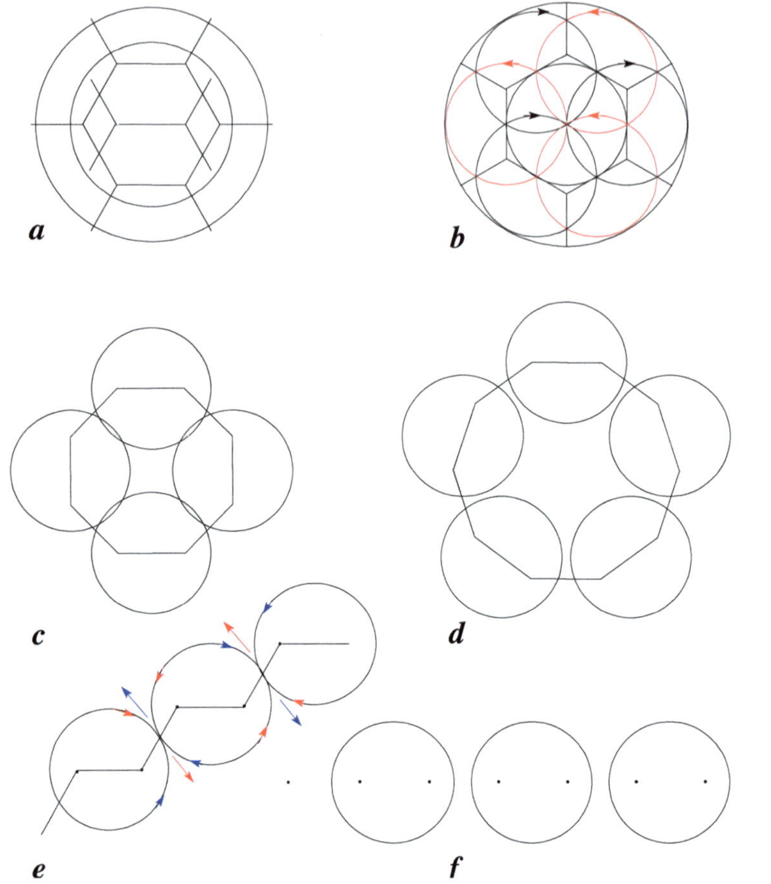

Figure 6.6: *Charge circulation figures – see text*

It is of interest to examine the aromaticity of the next possible monocyclic aromatic system, $C_{10}H_8$. The decagon with $1\frac{1}{2}$ order C–C bonds is shown in Figure 6.6(d) with superimposed o-a-m unit circles. As these circles do not intersect, delocalization of o-a-m is highly unlikely. Experimentally the compound is found to be non-aromatic, cyclodecapentaene. Whereas the effective C–C–C aromatic angle increases with carbon number there is no possibility of monocyclic aromatics beyond benzene. Cyclo-octatetraene (COT) (Figure 6.6(c)) is eliminated by the 4m+2 rule.

The case of a hypothetical linear aromatic chain ($m = \infty$) is also shown in Figure 6.6(f). Delocalization of o-a-m is ruled out. However, the alternative all-*trans* zig-zag arrangement of planar ethylene units, connected by first-order C–C bonds (Figure 6.6(e)), allows o-a-m vectors to interact, not by creating ring currents, but by linking up the entire molecule into a single planar structure with balanced o-a-m. The single bonds are somewhat shortened by partial screening and acquire a mild torsional barrier (21 kJmol^{-1}) to rotation, considerably less than the barrier in ethylene. Conventional theory has no convincing explanation of this stabilization of all-*trans* arrangements of polyenes with coplanar C atoms. It is interesting to note that *cis–trans* interconversion in retinal, a polyene, regulated by the intermediate barrier to rotation, is the function that enables colour vision [86].

The final conclusion is very close to the traditional organic chemist's view of aromaticity, being an exclusive property of fused six-membered rings, in which a chain of $4m + 2$ conjugated carbon atoms can be fitted in along the outer periphery of the molecule. In the majority of cases, called *cata condensed*, all carbon atoms are in the periphery and never shared between more than two rings. The general composition is $C_{4m+2}H_{2m+4}$. For $m > 2$ structural isomerism occurs, with $2, 5, 9, \ldots$ isomeric cata-condensed systems, consisting of three, four, five, *etc.*, fused six-membered rings.

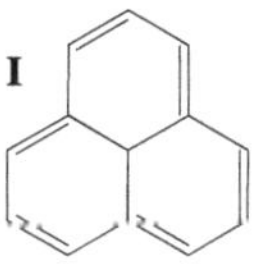

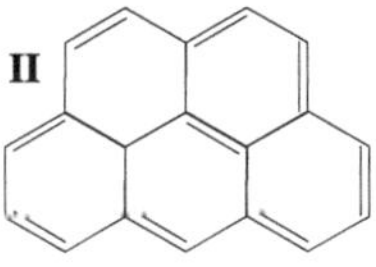

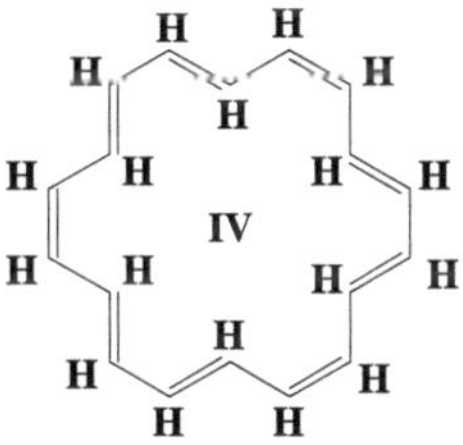

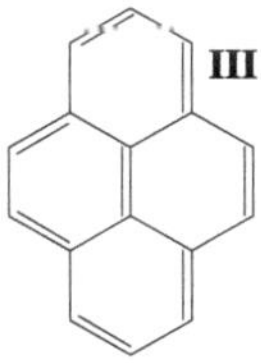

Three benzene rings can also fuse together in an arrangement of composition $C_{13}H_9$, (I). In this case the peripheral ring, $C_{12}H_9$, is conjugated, but not of $4m + 2$ type. For an open ring with $4m$ atoms the zero energy level has been shown to be m-fold degenerate. However, the presence of the central C atom causes further differentiation of the peripheral set and doubles the zero-level degeneracy. The compound becomes aromatic by using the $6 + 4m\,(m = 3)$ energy levels. The conventional formulation cannot account for the aromatic nature of this *peri-condensed* compound with an odd number of carbon atoms.

Further odd-carbon aromatics, $C_{17}H_{11}$ ($C_{16}H_{11}$ periphery), with four fused rings, and six $C_{21}H_{13}$ ($C_{20}H_{13}$) isomers of five condensed rings, follow the same pattern. The compound (II), $C_{19}H_{11}$, with two peri-condensed rings (3 central C atoms) has a $C_{16}H_{11}$ periphery and is aromatic as explained before.

Pyrene (III) is an example of a peri-condensed aromatic with two central carbon atoms and a $4m+2$ periphery. There are three isomeric five-ring aromatics ($C_{20}H_{12}$ – $C_{18}H_{12}$) of the same type.

Nuclear Magnetic Resonance Many textbooks mention charge circulation, inferred from NMR measurements in a magnetic field, applied perpendicular to aromatic planes, as evidence for the delocalization of p_z electrons. The only mechanism whereby such a ring current can be induced is by

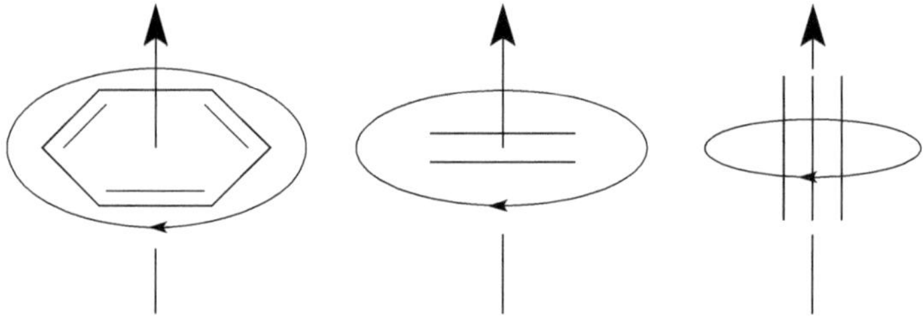

Figure 6.7: *Ring currents in benzene, ethylene and acetylene, as inferred from NMR shielding effects*

coupling to a molecular magnetic moment, which immediately disqualifies p_z electrons with their zero component of angular momentum in the field direction.

Similar experiments indicate ring currents parallel and perpendicular to second and third order bonds, respectively, as shown in Figure 6.7. These observations support the structure-generating role of o-a-m and refute the conventional hybridization model that ties up the p_{xy} electrons (with o-a-m, and implicated by the exclusion principle) in the formation of sp^2 hybrids. In the alternative interpretation the magnetic field simply decouples the anti-parallel o-a-m vectors of p_{xy} pairs to generate ring currents.

The conjugated polyene, [18]annulene, IV, is a potential $4m+2$ aromatic system. Because of steric repulsion between hydrogen atoms inside the ring, the molecule is distorted away from planarity. Nevertheless, its NMR shielding effects indicate an induced ring current in the mean molecular plane, once more in line with the decisive role of o-a-m. All evidence derived from the exclusion principle, conformational rigidity, aromaticity, electronic spectra and NMR shielding is therefore consistent with the alternative picture, and at variance with the conventional model.

6.3.2 The Jahn-Teller Model

The JT theorem defines the relationship between symmetry and stability of non-linear molecules and the origin of three-dimensional molecular structure as a function of o-a-m.

The way to understand the first step in the formation of a molecule is to consider a given atom as surrounded by a number of non-interacting secondary atoms, or ligands. The energy and o-a-m of the central atom is affected by the presence of the *coordination shell* of ligands, within the demands of the relevant conservation laws. Their effect can be simulated by recalculation of the electronic energy and o-a-m of the central atom in the modified symmetry environment, defined by the distribution and nature of the ligands.

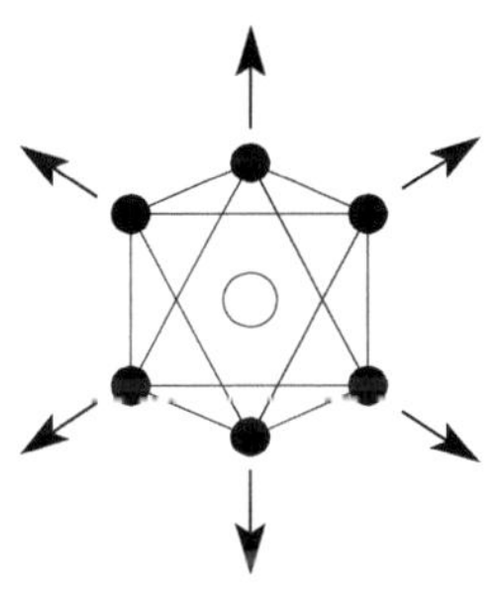

Figure 6.8: *Symmetrical deformation of an octahedron*

In the case of a free atom the exclusion principle dictates spherical symmetry through the balance of electronic angular momenta (section 4.7.4). This special symmetry is destroyed when the atom finds itself in a chemically reactive environment. Eventually the reference atom is surrounded by a number of ligands commensurate with electronegativity and size factors. The tendency to keep o-a-m in balance results in a symmetrical arrangement of ligands around the central atom.

Symmetry alone cannot dictate the final geometry of the coordination shell. Suppose that the central atom is surrounded by six equivalent ligands in octahedral arrangement. Totally symmetrical displacement of the ligands, as indicated in Figure 6.8 gives rise to a continuous set of configurations consistent with the octahedral symmetry. Among all of these configurations there is one that minimizes the total energy of the system and decides the equilibrium arrangement of the molecule. However, this configuration is not nenessarily stable against all other types of nuclear displacement. The JT model explores the likelihood of such displacements occurring.

An important special case arises when a valence electron of the central atom is on a degenerate energy level. Breaking the symmetry by some displacement, not of the totally symmetrical type, may lift the degeneracy and lower the total energy. Although such displacement lowers the total energy, sufficient symmetry must remain to quench the o-a-m. This process is recognized to mirror the arguments used when discussing the distortion in substituted methanes (section 6.2.3). The energy required to decouple angular momentum vectors is sufficient to prevent the formation of chiral structures by JT distortion.

The original JT analysis examined the modification of central-field energy eigenvalues as a function of the representations of the important molecular symmetry groups. The results agree [65] with the heuristic analysis of o-a-m conservation.

6.3.3 Molecular Mechanics

Molecular mechanics as a minimization of strain energy makes a rigid distinction between steric and electronic effects. Electronic effects are introduced in the form of empirical constants such as characteristic bond lengths and angles, the corresponding force constants, torsional rigidity of even-order bonds, planarity of aromatic systems and the coordination symmetry at transition-metal centres. These constants are accepted, without proof, to summarize the ensual of electronic interactions and used without further optimization.

Any molecular parameter which, in a trial structure, has a value at variance with the characteristic electronic standard, adds to the strain energy. It is considered a steric effect and subject to optimization. At convergence the actual molecular structure is recovered, providing that all empirical constants had been specified correctly.

Because of the empirical specification of its electronic interaction parameters molecular mechanics is routinely discounted as devoid of theoretical underpinning. This conclusion is certainly too harsh. Apart from connectivity, the electronic basis of all other force-field concepts has already been demonstrated:

Bond length: The Heitler-London method allows the calculation of all first-order diatomic interactions using valence-state wave functions as defined in terms of characteristic ionization radii.

Bond order: Any chemical bond is stabilized by a pair of shared electrons. When there is an excess of valence electrons over bonding pairs, bond dissociation energy is increased by the screening of nuclear repulsion, and the bond length is contracted accordingly, in discrete steps.

Valence angle: The arrangement of secondary ligands around a central atom is conditioned by the symmetry that quenches o-a-m and fixes the valence angles. It has been shown to generate a tetrahedral arrangement for methane. The planarity imposed on ethylene predicts a distorted trigonal arrangement at each carbon atom. The precise geometry is obtained by minimizing the repulsion between charge centres, located at ligand positions in trigonal array. A computerized procedure of general utility for the estimation of bond angles by this procedure has been described [87].

Torsional rigidity: Bonds of even order are sterically rigid because of a barrier to rotation, representing the energy needed to generate the o-a-m that occurs in the twisted system, and which can be calculated directly.

Aromaticity: Extended planar fragments, such as aromatic systems, are due to the same law of o-a-m conservation. The amount of energy needed for out-of-plane distortion is in principle calculable by the same procedure as before.

Steric interaction: Non-bonded or van der Waals interaction has been demonstrated to be no different from, albeit weaker than, both covalent and ionic interactions. It has in fact been demonstrated [88] that all pairwise interactions in a molecule are correctly simulated by the point-charge model of section 5.3.

All terms in equation (6.2) are therefore accounted for. It remains to show that the essential concepts of *force constant* and *bond parameter free of strain* also have well-defined meaning in the theory of chemical cohesion.

Force Constants

An unexpected feature of Table 5.1 is the remarkable similarity between the energies calculated from the characteristic radius r_c and those calculated from the ionization radius r_0, for the same interactions, but with bond orders increased by unity. It means that the steric factor which is responsible for the increase in bond order (*i.e.* screening of the internuclear repulsion) is also correctly described by an adjustment to r_0 to compensate for modified valence density. Calculating backwards from first-order $D_0 = 210$ kJmol^{-1}, an effective zero-order C–C bond length of 1.72 Å is obtained.

Noting the difference

$$\Delta E = D_0 \left(\frac{r_c - r_0}{r_c}\right) \tag{6.4}$$

in energy, calculated for the same interaction from characteristic and ionization radii, it becomes possible to calculate harmonic force constants for deformation from these observations.

To compensate for experimental uncertainty it is assumed that the length of isolated electron-pair bonds are calculated from observed dissociation energies using $r_0 = 1.60$ for carbon, and tabulated as d_0 in Table 5.1. Differences in bond length, corresponding to the energy differences ΔE, are assumed given by

$$\Delta d = (1.72 - 1.54;\ 1.54 - 1.36;\ 1.36 - 1.20)\ \text{Å}$$

for bond orders of 1, 2 and 3 respectively. Starting from dissociation energies $D = 352$, 607 and 817 kJmol^{-1} as calculated from r_c, and the assumed Δd, force constants are calculated as

$$k_r = \frac{\Delta E}{(\Delta d)^2 \times 301} \text{mdyneÅ}^{-1} \quad (\text{or Ncm}^{-1}) \tag{6.5}$$

The calculated values of 4.88, 8.44 and 14.33 are in excellent agreement with the literature values of 4.50, 8.43 and > 12.16 Ncm^{-1} for k_1, k_2, k_3 respectively [78].

The effective ionization and characteristic radii for the heteronuclear C–H interaction, $R_e = \sqrt{r_C \cdot r_H}$, are used to calculate $\Delta E = 460 \times 0.01/1.2 = 3.83$ kJmol^{-1}. Using the same dissociation energy an isolated bond length of 1.03 Å is calculated, compared to the observed 1.08 Å. Using this difference to calculate

$$k_r = \frac{3.83}{(0.05)^2 \times 301} = 5.09\ \text{Ncm}^{-1},$$

compared to the spectroscopic value of 4.83 Ncm^{-1}. The variability in reported values of aliphatic C–H bond lengths (1.06 – 1.10 Å), dissociation energies (425 – 556 kJmol^{-1}) and harmonic force constants (4.48 – 5.57 Ncm^{-1}) [78] substantiates the calculation.

Bonds free of strain

The stretching of an isolated bond to the observed length of an actual bond is caused by a combination of steric and electronic factors. Take ethane as an example:

$$\begin{array}{ccccc} d_e & & d_0 & & d_{obs} \\ & \textit{electr.} & & \textit{steric} & \mathrm{H}\quad\mathrm{H} \\ \mathrm{C:C} & \longrightarrow & \vdots\mathrm{C:C}\vdots & \longrightarrow & \mathrm{HC:CH} \\ & & & & \mathrm{H}\quad\mathrm{H} \\ \textit{isolated} & & \textit{strain–free} & & \textit{actual} \end{array}$$

The stretching force constant, as calculated by stretching a second order into a first-order bond, $\Delta d = d_1 - d_2$, is not affected by the mode of stretching. In molecular mechanics however, only the steric component of the stretch is considered, $\Delta d_s = d_{obs} - d_0$, and its calculation requires an independent estimate of the C–C bond length, free of strain. This intermediate situation is realized by stretching the second order bond, as in diatomic C_2 ($d_e = 1.312$ Å) by the same amount that transforms the ethylene bond (1.34 Å) into the first-order bond of ethane. This procedure is equivalent to removal of the screening couple from the internuclear region. By this argument the strain-free bond length is predicted as $d_0 = 1.51$ Å.

The calculation demonstrates that the concept of a strain-free bond has a clear electronic basis. It is important to note that in molecular mechanics the concepts of force constant and strain-free parameter are intimately linked. The calculated C–C pair, $k_e, d_0 = (4.88 \text{ Ncm}^{-1}, 1.51 \text{ Å})$, are very close to the best empirically optimized values of the most extended force fields. Although special strategies will be required to calculate specific k_e, d_0 pairs, the problem may be considered solved in general and the empirical estimates used with confidence. As a rule of thumb it is noted that, to first approximation,

$$d_0 = d(obs) - \Delta d/n \tag{6.6}$$

where Δd is the difference in length for bonds of integral order and n is the number of valence electrons per atom.

Non-bonded Interaction

The relationship between dissociation energies and bond orders is defined in equation (5.35) as:

$$D'_\nu = D'_1 + (\varepsilon/k_\nu)^2$$

with nuclear charge ε given by equation (5.14), or (5.19) for heteronuclear interactions, and screening parameter from equation (5.34), $k_\nu = 2^{(1-\nu/2)}$. From the relevant parameters for ethane: $d = 1.54$ Å, $d' = 0.8324$, $\varepsilon^2 = 0.1708$, $D'_1 = 0.4714$, one calculates:

$$D'(2) = 0.4717 + 0.3416 = 0.8133\,;\ D(2) = 609 \text{ kJmol}^{-1}$$

$$D'(3) = 0.4717 + 0.6442 = 1.1159\,;\ D(3) = 836 \text{ kJmol}^{-1}$$

It is noted that a fourth-order C–C bond is projected to have $d' > \tau$ and hence forbidden. Diatomic C_2 is shown to be of second order and therefore has a lone pair on each carbon atom.

The same relationship allows the calculation of zero-order (van der Waals) bonding parameters. By inverting the sign in equation (5.34) the dissociation energy of a C$\cdots$C van der Waals interaction follows as:

$$D'(0) = 0.4717 - 0.3416 = 0.1301\,;\; D(3) = 97 \text{ kJmol}^{-1}$$

corresponding to $d' \simeq 1.1$, $d_0 = 2.04$ Å, $k_r = 0.6$ Ncm^{-1}. The same calculation for H$\cdots$H, from $D'(1) = 0.2403$, $\varepsilon = 0.3340$, yields

$$D'(0) = 0.2403 - 0.2231 = 0.0172\,;\; D(3) = 31 \text{ kJmol}^{-1}$$

with $d'(0) = 1.33$, $d(0) = 1.02$ Å, $k_r = 4.5/(0.28^2 \times 301) = 0.19$ Ncm^{-1}.

Zero-order interaction in a heteroatomic pair is calculated in essentially the same way. The C$\cdots$H interaction serves to illustrate the method. Starting from the first-order interaction: $d = 1.08$ Å, $d' = 0.90$, $\delta\varepsilon = 0.1377$, $D' = 0.3983$. Hence

$$D'(0) = 0.3983 - 0.2754 = 0.1129\,;\; D(3) = 142 \text{ kJmol}^{-1}$$

$d'(0) = 1.06$, $d(0) = 1.27$ Å, $\Delta d = 0.19$ Å, $\Delta E = 0.8$ kJ, $k_r = 0.1$ Ncm^{-1}. These calculations are sensitive to small changes such as the choice of first-order parameter.

This approach can be used to model non-bonded interactions in molecular mechanics instead of using empirical potentials. The rule of thumb, equation (6.6) predicts effective strain-free zero-order bond lengths. As no interaction is possible at separations larger than $2 \times r_c$, the maximum $d_0(obs)$ for C$\cdots$C is 3.70 Å, and hence $d_0 = 3.70 - 0.28/4 \simeq 3.6$Å; d_0(C$\cdots$H)$\simeq 2.75 - 0.19/5 = 2.55$Å. The separation between non-bonded H atoms depends on the atom to which they are linked – the H ionization sphere is completely embedded within that of the larger atom. As a first approximation d_0(H$\cdots$H)$= d_0$(C$\cdots$H) is assumed.

Angle bending

Molecular-mechanics force fields distinguish between general and 1,3 non-bonded interactions. The obvious reason for this distinction is that the distance between ligands is affected when linked to the same central atom. Their final non-bonded separation depends, not only on ligand type, but also on the size of the central atom. In such a three-atom system the relevant parameters are the characteristic radius (r_c) of the central atom, together with the zero and first-order separations of the ligand pair. The general non-bonded distance (d_0) is stretched to measure

$$d_{1,3} = \frac{d_0}{d_1} \cdot r_c$$

For the ligand pairs C$\cdots$C, C$\cdots$H and H$\cdots$H on a central C atom the 1,3 distances are 2.45, 2.17 and 1.75 Å respectively. This effect explains the need of a special angle-bending parameter. The characteristic angles that correspond to these 1,3 separations are 106, 110 and 108° respectively, with angle-bending force constants estimated as $k_\theta = 0.6$ mdyneÅ^{-1}rad^{-1}. Most force fields assume a characteristic tetrahedral angle of 1.911 rad in all of these cases.

The Force Field

At the heart of molecular mechanics is a force field [89] and the ultimate force field should be fully transferable between all types of molecule. However, progress towards comprehensive force fields, such as the Universal Force Field of Casewit and Rappé [90], is invariably accompanied by an almost exponential increase in the number of parameters. The effort to reformulate molecular mechanics in terms of valence-bond concepts [91] reduces the number of formal parameters, but at the expense of almost the same number of hybridization parameters, which the authors [92] describe as follows:

> "The hybridizations used in this qualitative VB theory represent idealized hybridizations that are derived inductively from observed geometries."

The advantage is not obvious.

The complexity of problems addressed by molecular mechanics is such that multiparameter modelling is almost unavoidable. The best to hope for at this stage is to find a parameter set based on easily understood chemical concepts. The model outlined here is proposed in that spirit, although considerable refinement would be required before it translates into a useful tool.

It is based on the time-honoured concept of electronegativity, re-interpreted in terms of atomic ionization radii and the chemical potential of the valence state. The calculation [53] of these parameters by the fundamental Hartree-Fock-Slater model for non-hydrogen atoms, involves no empirical parameters or assumptions. It defines the valence state in terms of characteristic spheres to which a valence electron is confined at uniform charge density. Chemical bond formation occurs on the exchange of this valence charge density between atoms. The consequent polarization, when reduced to point-charge simulation, in dimensionless units, serves to describe all covalent dissociation energies as a function of interatomic distance. This function, significantly, applies specifically to interactions free of strain, defining parameters of fundamental importance in molecular mechanics.

To compensate for steric effects slight modification of ionization radii allows the derivation of observed dissociation energies, using the same general diatomic function for all covalent bonds, irrespective of bond order. Comparison of the modified characteristic radii and the fundamental ionization radii enables the calculation of harmonic force constants and defines a functional relationship between bonds of different order.

The bond-order function applies, not only to integral bond orders, but also to order zero, characteristic of all non-bonded interactions in a molecule. From these results it becomes possible, in principle, to define a force field, based on pairwise interaction, that should account for all structural and thermodynamic effects, apart from those related to orbital and spin angular momenta. The main purpose is not to produce yet another force field – the available products are more than adequate. What it does is to provide the much needed theoretical underpinning and reassurance that molecular mechanics is soundly based on first principles.

The force-field parameters as derived here have been used to model the structures of a series of differently branched aliphatic hydrocarbons [94].

6.4 Molecular Structure

The classical idea of molecular structure gained its entry into quantum theory on the basis of the Born–Oppenheimer approximation, albeit not as a non-classical concept. The B–O assumption makes a clear distinction between the mechanical behaviour of atomic nuclei and electrons, which obeys quantum laws only for the latter. Any attempt to retrieve chemical structure quantum-mechanically must therefore be based on the analysis of electron charge density. This procedure is supported by crystallographic theory and the assumption that X-rays are scattered on electrons. Extended to the scattering of neutrons it can finally be shown that the atomic distribution in crystalline solids is identical with molecular structures defined by X-ray diffraction.

6.4.1 Charge and Momentum Density

The electron density in an atom, molecule or crystal is described by a wave function which is subject to strict characteristic boundary conditions. As shown before (eqn. 5.3) an electron on a spherically symmetrical atom obeys the one-dimensional radial Helmholtz equation

$$(D^2 + k^2)R = 0$$

with $R(r) = \sum_k A_k \exp(ikr)$. For closely-spaced energy levels the wave vector, or momentum vector, k varies continuously and the coefficients A_k constitute a Fourier integral that represents the wave function in coordinate space,

$$\psi(q) = R(r) = \frac{1}{\sqrt{2\pi}} \int_{-\infty}^{\infty} A(k) e^{ikr} \mathrm{d}r \tag{6.7}$$

Likewise, an equivalent wave function is defined in momentum space

$$\varphi(k) = A(k) = \frac{1}{\sqrt{2\pi}} \int_{-\infty}^{\infty} R(r) e^{-ikr} \mathrm{d}r \tag{6.8}$$

The two functions $\psi(q)$ and $\varphi(k)$ are known as *Fourier transforms* of each other and they contain exactly the same information. By measuring the momentum density φ^2 it is therefore possible to determine the charge density $\rho(q) = \psi^2$.

Like the momentum variable, which in coordinate space is represented by an operator $p \leftarrow -i\hbar\partial/\partial q$, the position variable is represented in momentum space by the operator $q \leftarrow i\hbar\partial/\partial p$. The eigenfunctions of q are given by:

$$i\hbar\frac{\partial\varphi}{\partial p} = q_o\varphi$$

where the eigenvalue q_o is some constant value of the coordinate. The solution,

$$\varphi_q(p) = \exp\left(\frac{-ipq_o}{\hbar}\right)$$

defines the eigenfunctions of q in momentum space as *plane waves*. The roles of $\psi(x)$ and $\varphi(k)$ can be interchanged under the substitutions $x \leftrightarrow k$ and $i \rightarrow -i$. This substitution is known as a *conjugate transformation* and the variables x and k are said to form a conjugate pair of variables.

We note that the quantum-mechanical state of a photon, the counterpart of a particle in atomic systems, is described by a wave function in momentum space [15]–p.246. Electromagnetic waves, such as X-rays, that are scattered on an electron, are of this type. Taking the Fourier transform of such a scattered wave must therefore reveal the position of the scatterer.

The Scattering Process

A multitude of concepts such as X-ray, neutron and electron diffraction, X-ray crystallography, low-angle scattering, powder diffraction, scattering by non-crystalline and amorphous solids, all refer to the same physical phenomenon. Whereas X-rays and electrons are scattered by extranuclear charge clouds, more massive particles like neutrons and α-particles are scattered on atomic nuclei. In principle, all of these processes are of the same type, as described for X-rays below.

To estimate its *scattering cross section* an electron is considered as a charge e uniformly spread over a spherical surface of radius R. The energy stored in such a system, which constitutes an isolated conducting sphere, is calculated by simple electrostatics [95] as $E = e^2/8\pi\epsilon_0 R$ and equated with the rest energy of an electron of mass m_e to define the classical radius of the electron:

$$R = \frac{e^2}{8\pi\epsilon_0 m_e c^2} = 2.82 \times 10^{-15}\text{m}$$

In scattering theory this radius is interpreted as a maximum vibration amplitude. When set into vibration by a polarized X-ray plane wave of amplitude $\mathscr{E}$, the electron becomes a source of secondary scattered waves of amplitude

$$\mathscr{E}_s = \frac{R\mathscr{E}_0}{r}\cos 2\theta$$

for a scattering angle of 2θ at a distance r.

Scattering by an Atom

Since all electrons in an atom are not concentrated at a single point the waves scattered from different points within the charge cloud would generally be out of phase and interference occurs.

For a uniform charge distribution within a spherical atom the Fourier transform of the density has been shown (equation 5.6) to be of the form $\sin\alpha/\alpha$, for a wave of phase α in momentum space. From the Bragg equation (Figure 2.9), $\lambda = 2d\sin\theta$, it follows that electrons at a distance $d = \lambda/2\sin\theta$ apart, scatter in phase, *i.e.* with phase difference 2π. At a separation r the relative phase shift α, is given by:

$$\frac{\alpha}{2\pi} = \frac{r}{d} = \frac{2r\sin\theta}{\lambda}$$

The amplitude of a wave scattered on an electron at a distance r from the nucleus may therefore be defined as:

$$A' = A\frac{\sin\alpha}{\alpha}\,, \quad \text{where } \alpha = \frac{4\pi r\sin\theta}{\lambda}$$

and A depends on the function $u(r)\mathrm{d}r$, the number of electrons lying between the distances r and $r+\mathrm{d}r$ from the nucleus. Hence

$$A' = A_e\left[u(r)\mathrm{d}r\frac{\sin\alpha}{\alpha}\right]$$

where A_e is the amplitude scattered by an isolated electron. The resultant wave scattered by an atom is therefore given by

$$A_{at} = A_e\int_0^\infty u(r)\mathrm{d}r\frac{\sin\alpha}{\alpha},$$

where
$$A_e = \left(\frac{e^2}{8\pi\epsilon_0 m_e c^2}\right)\frac{\mathcal{E}_0}{r}$$

The function

$$f = \int_0^\infty u(r)\mathrm{d}r\frac{\sin(4\pi r\sin\theta/\lambda)}{4\pi r\sin\theta/\lambda}$$

is called the atomic scattering factor. It is usually tabulated as a function of $\sin\theta/\lambda$ and calculated by atomic Hartree-Fock, or related, methods. The atom scatters at any angle 2θ as though it were equivalent to f electrons. At small θ f approximates Z, the atomic number. For the sake of simplicity scattering by a single electron is considered as unity. The mean atomic scattering factor of the free C atom, calculated by HF methods, is shown in Figure 6.9.

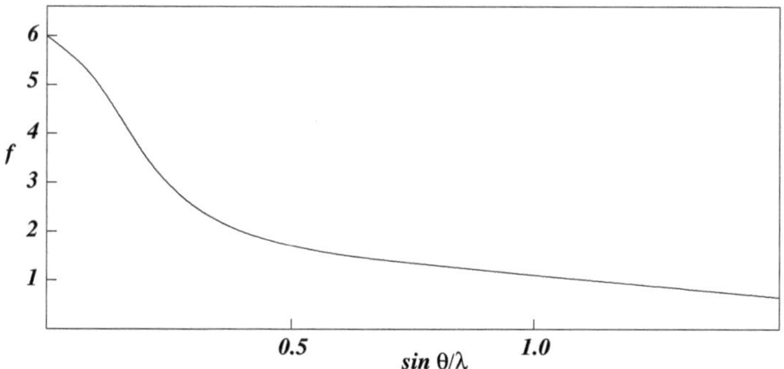

Figure 6.9: *Atomic scattering factor of C*

Macroscopic Samples

Interference between X-rays scattered at different atomic centres occurs in exactly the same way as for an atom. The scattered amplitude becomes a function of an atomic *distribution function.* In an amorphous fluid, a gas or non-crystalline solid the function is spherically symmetrical and the scattering independent of sample orientation. It only depends on the *radial* distribution of scattering centres (atoms).

Take the position of one atom as origin and denote the atomic density, in a given direction, at a distance r from the centre by the number of atoms between r and $\mathrm{d}r$, $\rho(r)$. The volume of the corresponding spherical shell is $4\pi r^2\mathrm{d}r$ and the number of atoms in the shell, $4\pi r^2\rho(r)\mathrm{d}r$, defines the radial distribution function. The intensity of scattered radiation from the sample is given by the Fourier transform

$$I(\theta) = 4\pi r^2 f \int_0^\infty \rho(r)\mathrm{d}r\frac{\sin\alpha}{\alpha} \quad , \quad \alpha = \frac{4\pi r\sin\theta}{\lambda}$$

The intensity of scatter can be measured experimentally and by taking the inverse transform, the radial distribution function is obtained as:

$$g(r) = 4\pi r^2\rho(r) \propto \int_0^\infty I(\theta)\frac{\sin\alpha}{\alpha}\mathrm{d}\theta$$

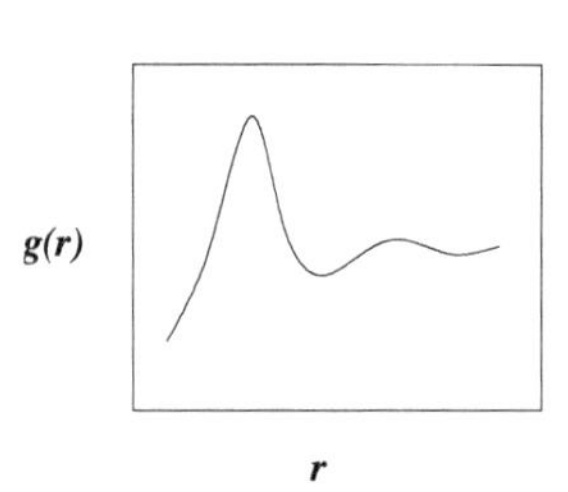

Figure 6.10: *The RDF of mercury*

The appearance of a typical RDF is shown in Figure 6.10. The maxima correspond to interatomic distances weighted by the atomic scattering factors.

Because of low concentration in the gas phase RDF peaks obtained by electron diffraction correspond exclusively to intramolecular interatomic distances. In principle, and practice, complete three-dimensional molecular structures can be derived, in special cases, by this method. In general however, there is no conformational information and three-dimensional structures can only be inferred by comparison of hypothetical models with the RDF.

6.4.2 Crystallographic Analysis

The electron density in a crystal is a three-dimensional periodic function. As the most general case consider the crystallization of a molecular compound. If, in the process, two identical molecules interact more efficiently for a specific mutual orientation, it is most likely that all molecules will adopt this same mutual disposition on solidification. The molecules fit together in three dimensions like wall-paper motifs in two dimensions. The resulting crystal can be described in terms of an elementary unit that repeats in three dimensions. This *unit cell* consists of a box that encloses, either a single molecule, or identical clusters of molecules.

All unit cells in a given crystal are identical in orientation and content. Each atom in the unit cell is related by *translational symmetry* to its counterparts in all unit cells. On connecting all such equivalent atoms by straight lines the three-dimensional grid defines individual unit cells geometrically, as shown two-dimensionally in Figure 6.11. In three dimensions the grid also

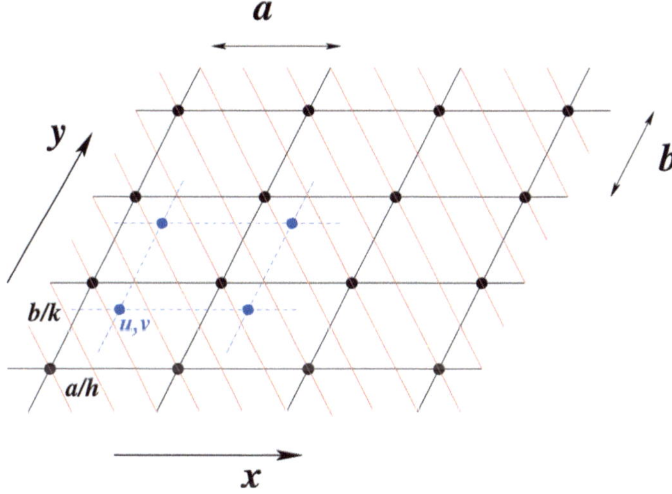

Figure 6.11: *2D crystallographic unit cells and lattices*

defines three sets of parallel planes, each of which contains all of the equivalent atoms. Whenever such a set of planes satisfies Bragg's equation with respect to an incoming X- beam, all of the equivalent atoms scatter in phase.

An alternative grid, shown in red, defines an alternative set of planes that also contains all equivalent atoms, and under different conditions, also satisfies Bragg's equation. Such planes, which may be constructed in an endless number of ways, all have one property in common – they make rational intercepts on the axes of the unit cell. The fractional intercepts are defined as a/h, b/k, c/l in terms of the unit cell constants and the integer *Miller indices, hkl.*

This construction is valid for each atom in the unit cell, and all atoms define identical sets of rational planes, but with an origin shift as shown by the blue lattice. All atoms in the unit cell therefore satisfy Bragg's equation for the same orientation of the crystal. They scatter together, but not in phase. The phase difference between the two sets of planes generated by atoms at fractional coordinates of 0,0,0 and $u = x/a$, $v = y/b$, $w = z/c$, in the unit cell, is calculated by noting that phase shifts of 2π are effected by coordinate shifts of a/h, b/k and c/l, respectively. Hence

$$\frac{\phi_a}{2\pi} = \frac{x}{a/h} \quad , \quad \frac{\phi_b}{2\pi} = \frac{y}{b/k} \quad , \quad \frac{\phi_c}{2\pi} = \frac{z}{c/l}$$

Thus $\phi_a = 2\pi h(x/a) = 2\pi hu$, *etc.*, for fractional coordinates u, v, w. The total phase difference is:

$$\begin{aligned} \phi &= \phi_a + \phi_b + \phi_c \\ &= 2\pi(hu + kv + lw) \end{aligned}$$

The one-dimensional Fourier transform of the density function becomes ($u = x/a$):

$$F(h) = \int_0^a \rho(u)e^{2\pi i hu}\mathrm{d}x \tag{6.9}$$

and the density, which is periodic in a, is written [4] as a one-dimensional Fourier sum of the form

$$\rho(\phi) = \sum_{n=-\infty}^{\infty} K_n e^{in\phi} = \sum_{n=-\infty}^{\infty} K_n e^{2\pi i nx/a}$$

[4]It was first pointed out by Fourier (1807) that every function in the closed interval $[-\pi, \pi]$ (*i.e.* $-\pi < x < \pi$) could be represented in the form

$$S = \tfrac{1}{2}a_0 + \sum_{n=1}^{\infty}(a_n \cos nx + b_n \sin nx)$$

If $f(x)$ is defined in the interval $[-L, L]$ of length $2L$,

$$f(x) = \tfrac{1}{2}a_0 + \sum_{n=0}^{\infty}[a_n \cos(n\pi x/L) + b_n \sin(n\pi x/L)]$$

which can be cast into the form

$$F(x) = \sum_{n=-\infty}^{\infty} K_n \exp[i(n\pi/L)x]$$

Substituted into (6.9):

$$F(h) = \int_0^a \left(\sum_{n=-\infty}^{\infty} K_n e^{2\pi i n u} \right) e^{2\pi i h u} \mathrm{d}x$$

For $h = n$, the integral

$$\int_0^a K_n e^{2\pi i(n-n)u} \mathrm{d}x = \int_0^a K_n e^0 \mathrm{d}x = K_n a$$

The Fourier coefficients are thereby defined as simply the sampling of the Fourier transform at the points $u = ha$ with the appropriate scale factor $1/a$. For $h \neq n$ the integral vanishes[5]:

$$\begin{aligned} \int_0^a K_n e^{2\pi i(h-n)u} \mathrm{d}x &= \left[\frac{a}{2\pi i(h-n)} \cdot e^{2\pi i(h-n)x/a} \right]_0^a \\ &= \left(\frac{a}{2\pi i(h-n)} \right) \left(e^{2\pi i(h-n)} - e^0 \right) \end{aligned}$$

For any h, $K_n = \frac{1}{a} F_h$, and

$$\rho(x) = \frac{1}{a} \sum_{h=-\infty}^{\infty} F_h e^{-2\pi i h u}$$

The term F_0 has the special meaning

$$F_0 = \int_0^a \rho(x) e^0 \mathrm{d}x = \sum Z$$

which shows that all electrons scatter in phase in the direction of the primary beam. The final results, extended to three dimensions are:

$$\rho(xyz) = \sum_h^{-\infty} \sum_k^{\rightarrow} \sum_l^{\infty} K_{hkl} e^{-2\pi i(hu+kv+lw)} \qquad (6.10)$$

$$F_{hkl} = \int_0^V f_j e^{2\pi i(hu+kv+lw)} \mathrm{d}V$$

The integral extends over all atoms in the unit cell. As in one dimension, it can be shown that $K_{hkl} = \frac{1}{V} F_{hkl}$.

[5]Since h and n are integers, $(h-n) = m$, the exponential term for $m \neq 0$, $e^{2\pi i m} = \cos 2\pi m + i \sin 2\pi m = 1 = e^0$

Structure Solution

Fourier representation of electron density suggests the possibility of direct structure analysis. If all *structure factors*, $F(hkl)$, are known, $\rho(xyz)$ can be computed at a large number of points in the unit cell and local maxima in the electron-density function are interpreted to occur at the atomic sites. A typical single-crystal diffraction pattern of the type used for measuring structure factor amplitudes is shown in Figure 6.12.

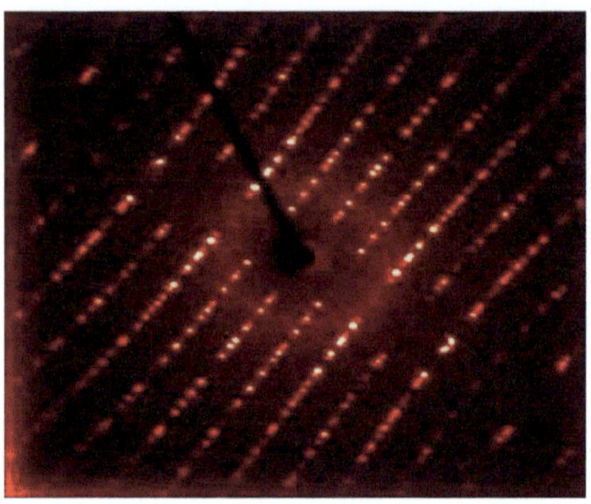

Figure 6.12: *A CCD X-ray diffraction pattern*

The structure factor, which is nothing but the wave function of the density, cannot be measured directly and the intensity of the diffracted wave $I = F^2(hkl)$, does not contain the phase information required for Fourier synthesis of the density.

If a centre of symmetry occurs at the origin of coordinates the pairing of structure factors, $F(hkl) = F(\bar{h}\bar{k}\bar{l})$, reduces the expression (6.10) to

$$\rho(xyz) = \frac{2}{V}\sum_h\sum_k\sum_l F(hkl)\cos 2\pi(hu + kv + lw)$$

The structure factor is now a real function, but a phase ambiguity of either 0 or 2π still exists. Without this information Fourier synthesis cannot be performed directly. Crystallographic analysis consists of strategies to overcome this *phase problem.*

An early strategy was to use measured intensities rather than structure factors as Fourier coefficients. Although the transform of I, known as a *Patterson function*, must, by definition contain the same information as the transform of F, it does not produce the density, but a weighted map of all vectors between density maxima in the unit cell[6]. Deconvolution of the Patterson function is facilitated by the presence of heavy atoms that dominate the scattering process so as to serve as a starting points for a series of structure refinement by Fourier synthesis.

In modern crystallography virtually all structure solutions are obtained by *direct methods.* These procedures are based on the fact that each set of hkl planes in a crystal extends over all atomic sites. The phases of all diffraction maxima must therefore be related in a unique, but not obvious, way. Limited success towards establishing this pattern has been achieved by the use of mathematical inequalities and statistical methods to identify groups of reflections in fixed phase relationship. On incorporating these into multi-solution numerical trial-and-error procedures tree structures of sufficient size to solve the complete phase problem can be constructed computationally. Software to solve even macromolecular crystal structures are now available.

6.4.3 Molecules and Crystals

The study of crystals, mainly through their interaction with electromagnetic radiation has developed into the modern science of crystallography. Four distinct variations of crystallography have developed in the hands of mathematicians, physicists, mineralogists and chemists. The four branches have the fundamentals in common, but each has developed its own flavour and applications. Chemical crystallography focusses on molecular structure as revealed by diffraction methods. It has many features in common with theoretical methods for the study of molecular structure.

Molecular structure is theoretically intimately related to electron-density distribution functions. In quantum-chemical analysis this density is synthesized as a molecular orbital, by a linear combination of real atomic orbitals, and minimized as a function of total energy. Crystallographically the unit cell density is represented by a Fourier sum over periodic electron wave functions

$$\rho(\boldsymbol{u}) = \frac{1}{V} \sum_{\boldsymbol{H}=-\infty}^{\infty} F(\boldsymbol{H}) e^{-2\pi i \boldsymbol{H}\boldsymbol{u}}$$

Both the LCAO-MO and the unit-cell density are represented by an infinite sum over a complete orthonormal set. In crystallography the number of

[6]Analogous to the RDF of amorphous materials.

terms is limited by the wavelength of the probe and hence the number of accessible diffraction maxima. Even at a tenfold over-determination of data to parameters, series-termination errors are known to give rise to complicating virtual effects. *Ab initio* basis sets are routinely smaller than the parameter set.

The basis functions, in both cases, are complex. This is a known complicating factor in the analysis of non-centrosymmetric crystal structures. In computational chemistry the problem is circumvented by using only real basis functions. The definition and limitation of such functions are discussed in section 2.6.3, but there is a more important mathematical factor that militates against this procedure:

> The orthonormal spherical harmonics from which the real functions derive are
>
> $$Y_l^{m_l}(\theta,\phi) = \left(\frac{(l-m_l)!}{(l+m_l)!}\frac{2l+1}{4\pi}\right)^{\frac{1}{2}} e^{-im_l\phi}P_l^{m_l}\cos(\theta)$$
>
> The importance of this formula is that each spherical harmonic is defined by specific values of the quantum numbers l and m_l. As an example, the solid spherical harmonics
>
> $$\mathscr{Y}_l^{m_l}(r) = r^l Y_l^{m_l}(\theta,\phi) \quad \text{for} \quad n=1 \quad \text{and} \quad m_l = \pm 1, \text{ are}$$
>
> $$\mathscr{Y}_1^{\pm 1} = \mp N(x \pm iy)$$
>
> A real function is obtained as the linear combination
>
> $$\begin{aligned} L^- = \mathscr{Y}_1^1 - \mathscr{Y}_1^{-1} &= N(-x-iy-x+iy) \\ &= -2Nx \end{aligned}$$
>
> while
>
> $$\begin{aligned} L^+ = \mathscr{Y}_1^1 + \mathscr{Y}_1^{-1} &= N(-x-iy+x-iy) \\ &= -2iNy \end{aligned}$$
>
> Based on these linear combinations two real functions, without sensible quantum numbers, are defined as
>
> $$p_x = 2Nx$$
>
> $$p_y = 2Ny$$
>
> and used in quantum chemistry instead of $\mathscr{Y}_1^1$ and $\mathscr{Y}_1^{-1}$. The argument goes that any linear combination of spherical harmonics is contained in the orthonormal set. Indeed,

$$\tfrac{1}{2}(L^+ + L^-) = -N(x+iy) = \mathscr{Y}_1^1$$
$$\tfrac{1}{2}(L^+ - L^-) = -N(x-iy) = \mathscr{Y}_1^{-1}$$

and the linear combination $L^-(real) + L^+(imaginary)$ has this property. However, $\frac{1}{2}(p_x+p_y) = N(x+y)$ is a real function outside the orthonormal set, despite the fact that p_x and p_y are orthogonal to each other. The use of such real functions is equivalent to treating all crystallographic structures as centrosymmetric.

The Fourier coefficients in crystallographic analysis are the measured structure factor amplitudes of diffraction maxima and correspond to the Fourier transform of the periodic density. Numerical solution of the phase problem enables the Fourier transformation that synthesizes the unit-cell electron-density function and hence the three-dimensional molecular structure. Quantum-chemical computations assume the molecular structure and calculate Fourier coefficients for a limited basis set to redefine the electron density.

6.4.4 Structural Formulae

Thousands of crystal structures have been analyzed by diffraction methods. Whenever covalence is the dominant chemical interaction, well-defined molecular units, held together by secondary forces such as van der Waals and/or hydrogen bonds, can be identified as the regular building blocks of the crystals. The geometrical features of such molecular units define the chemist's notion of structure. Still, there is no theory that defines molecular structure or electron density from first principles.

The challenge is to derive intramolecular connectivity from chemical composition. Once this has been achieved all further molecular properties can, in principle, be deduced. However, the first step is clearly impossible: As the number of atoms per molecule increases, so does the number of possible isomers, and there are no criteria for ranking the isomers. The only way to predict connectivity is from a historical record, such as a synthetic pathway, which, in turn, depends upon reaction conditions and the nature of primary reactants.

Organic chemists have devoted countless man-years to unravel the complexities of chemical synthesis, with remarkable success. The fruits of their labour, well beyond the scope of this book, confirm that, for any molecule, a logical precursor exists and by systematically working backwards a well-defined synthetic pathway to the final product can be mapped out. The inverse procedure, *i.e.* analysis by chemical degradation used to be the method of choice to establish the structural formula (*i.e.* connectivity) of an unknown compound. The empirical rules whereby molecular conformation was inferred from degradation studies reflect the way in which the conservation of orbital angular momentum dictates the relative orientation of sub-molecular fragments or radicals, (section 6.2).

The concepts, which are used in the design of synthetic routes, are based on the chemist's notion of electron density in molecular space. Most of these ideas have been arrived at by trial-and-error, or intuition, and refined by actual syntheses. An elaborate scheme, based on the concept of bond polarity, which is related to intrinsic differences in electronegativity, has grown into a book-keeping procedure to envisage electron flow during chemical reaction. An example of a polar bond occurs in chloromethane:

$$\mathrm{H_3C^{\delta+} \rightarrow Cl^{\delta-}}$$

To envision how chemical reaction arises from differences in polarity it is argued [96] that, since unlike charges attract, electron-rich sites in the functional groups of one molecule react with the electron-poor sites in the functional groups of another molecule. Bonds are made when the electron-rich reagent donates a pair of electrons to the electron-poor reagent. The movement of bonding electron pairs is followed by the use of curly arrows. The formalism is illustrated by some chemical reactions:

(1) $\mathrm{HO^- + H_3C{:}Br \longrightarrow HO{:}CH_3 + {:}Br^-}$

(2) $\mathrm{R_2C{=}CH_2 \xrightarrow{H^+} [R_2C^+{-}CH_3] \xrightarrow{Cl^-} R_2CCl{-}CH_3}$

(3) $\mathrm{B{:} + H{-}CR_2{-}CR_2X \longrightarrow R_2C{=}CR_2 + BH^+ + X^-}$

(1) The nucleophilic substitution reaction of hydroxide with bromomethane.

Although the synthesis rules are routinely formulated in terms of hybrid orbitals, they are in fact purely empirical, and not entirely logical. On the one hand an electrophile, H^+ is shown (2) to react with the pair of π-electrons of a C – C double bond to generate an intermediate carbocation. In the inverse process (3) a base (B:) is shown to attack through a C–H single bond without specifying the orbital that transfers the electron pair from C–H (double bond?) to C–C. This is an unnecessary complication as the empirical rules work well without the orbital baggage.

6.5 Emergent Structure

Like the exclusion principle (section 2.5.7) molecular structure has been recognized as an emergent property of matter [97], not accounted for by quantum theory.

It is quantum theory that shows how everything in the world is part of everything else, without boundaries and without isolated parts. However, it is most frequently an isolated system which is chosen for analysis. In chemistry it could be an isolated electron, an isolated atom, an isolated molecule, or an isolated crystal. It is easy to believe that this assumption would not introduce significant errors. Fact is that an isolated electron has no spin, an isolated atom has no size, and an isolated molecule has no shape. It follows that all of the important familiar properties of chemically significant entities arise through interaction with the environment.

Empirical force fields for the modelling of molecular shape are based on parameters measured in solid crystals, and not in isolated molecules. Although the method may therefore appear to simulate the shape of a single molecule, the reference structure is as observed in a crystal, surrounded by, and interacting with many neighbours. Force-field parameters are conditioned by the crystal environment and any attempt to derive them from first principles should take the influence of the environment into account.

The situation is the same for chemical bonds. Empirical facts about chemical bonds derive from measurements on material in the bulk. Chemical bonds are formed, either under crowded conditions or during high-energy collisions. In both instances there is a close encounter between activated species which implies previous interaction with external influences. The one thing not implied is that chemical bonds are generated spontaneously between isolated entities, either atoms or molecules. This means that in order to simulate the process of chemical binding or to understand the characteristics of chemical bonds, it is necessary to recognize the driving effect of the environment.

It has been argued (4.7.3) that spin is not a property of an electron, but a manifestation of the interaction between an electron and the vacuum. An applied magnetic field projects this spin into three-dimensional space as a two-level system.

Like electron spin, the valence state of an atom has no meaning in terms of free-atom wave functions. Like spin it could be added on by an *ad hoc* procedure, but this has never been achieved beyond the qualitative level. All conventional methods of quantitative quantum chemistry endeavour to simulate atomic behaviour in terms of free-atom functions.

In an environment of atoms in collision, interatomic contacts consist of interacting negative charge clouds. This environment for an atom is approximated by a uniform electrostatic field, which has a well-defined effect on the phases of wave functions for the electrons of the atom. It amounts to a complex phase (or gauge) transformation of the wave function:

$$\psi(x) \rightarrow e^{-i\xi(x)} \cdot \psi(x)$$

The effect is like that of a Faraday cage that confines the electron within a spherical surface. It is therefore no longer appropriate to formulate the wave function with infinite boundary conditions, assuming $\psi(r_0) \rightarrow 0$, for $r_0 < \infty$, instead.

This condition, which amounts to uniform compression of the atom, when simulated numerically, shifts the electronic energy to higher levels, and eventually leads to ionization. It means that environmental pressure activates the atom, promotes it into the valence state and prepares it for chemical reaction. The activation consists therein that sufficient energy is transferred to a valence electron to decouple it from the core. The wave function of such a freed electron (eqns 3.36, 5.31) remains constant within the ionization sphere.

The valence electron of a promoted atom readily interacts with other activated species in its vicinity to form chemical bonds. The mechanism is the same for all atoms, since the valence state always consists of a monopositive core, loosely associated with a valence electron, free to form new liaisons. Should the resulting bond be of the electron-pair covalent type, its properties, such as bond length and dissociation energy can be calculated directly by standard Heitler-London procedures, using valence-state wave functions (section 5.3.4).

The situation is further simplified by the zero kinetic energy of an electron in a stationary state. That leaves electrostatics as the only consideration in the calculation of bond properties, providing that the interatomic distance is known (section 5.3). The calculation is entirely classical. Molecular mechanics has the same classical basis. It works because of environmental gauge transformation of electronic wave functions.

6.5.1 Molecular Shape

There is no simple demonstration of how molecular shape, like spin or the atomic valence state, emerges as a consequence of environmental pressure, but there is the compelling argument that it never features in molecular physics, unless it is introduced by hand.

Apart from subtle exceptions, an isolated molecule differs from a molecule in a crystal in that the isolated molecule has no shape, whereas in a crystal it acquires shape, but loses its identity as an independent entity. This paradoxical situation is best understood through the famous Goldstone theorem, which for the present purpose is interpreted to state that any phase transition, or symmetry broken, is induced by a special interaction. When a molecule is introduced into an environment of other molecules of its own kind, a phase transition occurs as the molecule changes its ideal (gas) behaviour to suit the non-ideal conditions, created by the van der Waals interaction with its neighbours. An applied electric or magnetic field may induce another type of transformation due to polarization of the molecular charge density, which may cause alignment of the nuclei. When the field is switched off the inverse transformation happens and the structure disappears. The Faraday effect (6.2.3) is one example.

Another common phase change, from the gas to the liquid state, is caused by increased concentration and can be promoted by increased pressure or reduced temperature. It is mainly rotational symmetry that breaks down. It affects the heavier nuclei more than the electrons and short-range intermolecular interactions become important. For lack of better terminology these can be called polarization interactions and their effect is the disappearance of molecular units, and although the covalent interaction still predominates, it is no longer completely undirected. The arrangement of nuclei in space changes more slowly and transient structure appears. Below the transition point the symmetry is too low to identify molecules by their previous wave functions. The molecule is gaining structure, but losing identity. Interaction with the environment has now become too dominant to ignore when trying to identify individual molecules.

Transition from the liquid to the solid state happens as the vibrational motion of individual molecules spreads through the bulk of the material to reappear as lattice modes. The molecules are now coupled into a periodic array, but not necessarily in the same orientation. As crystals are cooled down further it is common for a whole series of phase transitions to occur as additional orientational forces become dominant and cause a lowering of the symmetry at each transition. The symmetry is often found to decrease from high-symmetry cubic forms through orthorhombic and monoclinic modifications, as ordering increases, ending up as triclinic crystals on approaching the zero point.

This general scheme of events, although valid for all materials, does not predict the same effects under the same conditions for all species. It depends quite critically on the complexity of the molecule. Large biomolecules or polymers have no gas-phase existence and even in solution, or the liquid state, may have a well-defined molecular shape, while small molecules like ammonia settle into a classical structure only at very low temperatures.

6.6 The Metaphysics

The transformation path from a structureless molecule to an entity with three-dimensional shape, embedded in a macroscopic holistic matrix, can conceptually be reversed, leading back to sub-atomic elements of matter. This retrogression starts from an isolated molecule as a collection of atoms held together by the electrodynamic interaction that conserves total electronic energy and orbital angular momentum. Without this interaction separated atoms consist of baryonic nuclei in electromagnetic equilibrium with negative charge clouds made up of electrons (leptons). In the ground state the electronic energy and o-a-m of an atom are both at a minimum as the lowest-energy electrons lack the o-a-m to further reduce thier energy by photon emission[7].

Baryons are held together in atomic nuclei by *strong* interaction and at least one further level of reduction at which protons and neutrons are separated, can be envisaged. Since a free neutron disintegrates spontaneously, ponderable matter can be considered as made up of protons, electrons and neutrinos. Each of these carry characteristic amounts of mass, charge and spin. The neutrino apparently has zero mass and charge.

The fundamental assumption in this work is that the three elementary units are indivisible wave structures in the vacuum. The quarks, assumed to be confined in a proton, are considered a manisfestation of the wave structure.

[7]During radioactive electron capture angular momentum is provided as a neutrino from the nucleus.

At this point the reduction has reached the level of philosophical discourse and shows remarkable similarity with the metaphysics of the 17th century mathematician and philosopher René Descartes. He argued that if the only thing that is clearly understandable in matter is mathematical proportions, then matter and spatial extension are the same. On this basis he rejected not only the idea of indivisible atoms but also of the void. Where there is space, there is by definition extension and, therefore matter, and hence the void is undefined.

This argument becomes more convincing within the theory of general relativity, which was unknown to Descartes. This theory equates the void with flat euclidean space-time and the appearance of matter with curved space (2.1.3). To understand what happens when space is curved it is instructive to consider the problem of covering a curved surface, like that of a sphere, with a flat sheet. The cover develops wrinkles that cannot be smoothed away. By analogy, elementary units of matter such as electron and proton are minimum persistent distortions of space with standing-wave structures, aspects of which are observed as elementary units of mass, charge and spin. The geometry of space-time curvature fixes the numerical values of the universal constants π, τ and e and by implication the structure and properties of material aggregates.

The philosophical appeal of curved space cosmology is that it allows a closed, rather than an infinite, universe. Experimental proof is in the observation of universal background radiation with a Planckian frequency distribution (Figure 2.5). Conjectural implications of closed space-time have been considered before [7, 62, 49].

Chapter 7

Chemical Change

The magic of chemistry comes with thrills and excitement, flashy reactions and fireworks, with colour and sound. It is not bonding and structure that grab the imagination, but spectacle and change. Here is the topic that tells the real story of chemistry. Chemical change, more than anything, happens in a crowded environment. Factors of importance are the state of aggregation, material concentration, temperature and pressure, collectively known as thermodynamic conditions. Students of chemistry, even at the elementary level, should be familiar with thermodynamic models of chemical reactivity. For a concise revision refer to [15]. A brief summary follows.

7.1 Thermodynamic Potentials

The fundamental assumption of thermodynamics is the conservation of energy, also during its conversion from one form into another. In chemical applications it is cumbersome to account for total energy in all its forms and the problem is avoided by focussing on differences in energy rather than absolute values. As the basis of calculation a convenient energy zero is arbitrarily defined and energy, relative to this state, is called the thermodynamic energy, consisting of three components:

$$U = TS - PV + \sum \mu_j N_j \tag{7.1}$$

or alternatively,

$$\Delta U = q + w + t$$

defining heat flux, mechanical work and matter flow. The parameters T, S, P, V, μ and N represent temperature, entropy, pressure, volume, chemical potential and mole number, respectively. By discounting one or more of these terms the zero point moves to a new level and alternative potential functions are defined, such as:

$$\begin{aligned}\text{Enthalpy} &: H = U + PV = TS + \sum \mu_j N_j \\ \text{Free energy} &: F = U - TS = -PV + \sum \mu_j N_j \\ \text{Free enthalpy} &: G = H - TS = \sum \mu_j N_j\end{aligned}$$

These formulae explain the common terminology for one-component closed systems:

$$\begin{aligned}H = q_P &\quad , \quad \text{heat content} \\ F = w_T &\quad , \quad \text{work function} \\ G/N = \mu &\quad , \quad \text{partial molar free enthalpy}\end{aligned}$$

The differential form of the potential expressions shows that chemical potential is defined by a first derivative of each potential:

$$\mu = \left(\frac{\partial U}{\partial N}\right)_{S,V} = \left(\frac{\partial H}{\partial N}\right)_{S,P} = \left(\frac{\partial F}{\partial N}\right)_{T,V} = \left(\frac{\partial G}{\partial N}\right)_{T,P}$$

Chemical reactivity, depending on the reaction conditions, can be described equally well in terms of any of these thermodynamic potentials and no effort will be made to differentiate between them in the following discussion.

7.2 Chemical Reactivity

It is possible to gain significant insight into chemical reactivity from a few simple principles, without getting involved with the abstract ideas of statistical thermodynamics.

A chemical reaction occurs as the material composition of a reaction mixture changes. Should this process happen spontaneously, chemical energy is released. Alternatively, supply of energy from an external source drives the chemical change. The energy produced during spontaneous change does not necessarily cause an increase in temperature as most of it may be dissipated as increased entropy. The course of a chemical reaction can therefore be followed by mapping changes in the energy of a system. As a general principle the propensity for chemical change in a mixture is considered to be a function of a potential-energy field, created by the mass ratios or amounts of substance in the mixture. This field is related to the quantum potential of the system and is said to reflect the chemical potential of the reaction mixture.

The chemical potential of an atom has been shown to depend on its electronegativity, or quantum potential, in the valence state. For a molecule, such a measure, although more difficult to estimate, still has the same meaning. In the case of a pure substance the chemical potential is an intensive property of the system, independent of the amount of material, but sensitive to thermodynamic changes in the environment. The quantity that varies with amount of substance is an energy, $u = n\mu$. The dimensionless variable n is conveniently expressed in moles of substance, defining μ as a molar energy. In a reaction mixture the chemical potential of each component $\mu_i = u/n_i$ is equivalent to its *partial molar energy.*

The observed energy of reaction depends on the thermodynamic conditions and the choice of zero point. The useful index in chemical reactions is therefore not the absolute value of the energy, but the change in energy during the course of the reaction. This change drives the reaction and is responsible for changing the supply of reactant by an amount Δn: $\Delta u = \mu\Delta n$. The quantity $\Delta u = \alpha$ is also known as the affinity of the reaction. In a reaction mixture consisting of several reactants and a number of reaction products, all of these components contribute to the affinity at any time during the course of thc reaction:

$$\alpha = \Delta u = \left(\sum_i n_i\mu_i\right)_{products} - \left(\sum_j n_j\mu_j\right)_{reactants}$$

Spontaneous chemical change occurs when $\Delta u < 0$ and ceases when $\Delta u = 0$. Chemical reaction therefore proceeds in the direction that minimizes the affinity and depends on the rate at which affinity changes.

Because of their variable thermodynamic state and concentration each reactant or product is characterized at any instant by an intrinsic *activity*, a_i, and the interplay between these activities defines the chemical action A, at that instant. The action changes at a rate proportional to A and to the change in affinity, as summarized by the linear homogeneous equation:

$$\mathrm{d}A = \beta A\mathrm{d}\alpha \quad \text{or} \quad \frac{\mathrm{d}A}{A} = \beta\mathrm{d}\alpha$$

Integration over the complete course of the reaction, from initial to final state,

$$\int_i^f \mathrm{d}(\ln A) = \beta\int_i^f \mathrm{d}\alpha$$

yields

$$\begin{aligned}\ln\left(\frac{A_f}{A_i}\right) &= \beta(\Delta u_f - \Delta u_i) \\ &= \beta\left(\sum_i \mu_i \Delta n_i\right)\end{aligned}$$

summed over all reactants and products. The bracket on the right represents an additive quantity and to satisfy the equation the bracket on the left must be a multiplicative function. At a given instant,

$$A = \prod_{i,j}\left\{\frac{(a_i)^{n_i}_{products}}{(a_j)^{n_j}_{reactants}}\right\} \tag{7.2}$$

Both chemical potential and affinity depend on the choice of a *standard state*. A convenient choice is $A_i = 1$. The action relative to the standard state ($^\ominus$) is given by:

$$\begin{aligned}\ln A_f &= \beta\left(\sum_i \mu_i \Delta n_i\right) \\ &= \beta\left(\Delta u_f - \Delta u^\ominus\right) \\ \Delta u_f &= \Delta u^\ominus + \frac{1}{\beta}\ln Q \end{aligned} \tag{7.3}$$

where $Q(= A_f)$ is known as the reaction quotient. For a single molecule eqn. (7.3) reduces to

$$\mu_i = \mu_i^\ominus + \frac{1}{\beta}\ln a_i \tag{7.4}$$

Activities, as yet undefined, must, by definition, be proportional to the concentration, or mole fraction, of each reactant in the mixture, *i.e.* $a_i = \gamma_i X_i$. The activity coefficient $\gamma_i \to 1$ as $X_i \to 1$.

Equations (7.3) and (7.4) are well-known thermodynamic expressions. A reaction reaches equilibrium as the affinity approaches zero, hence

$$Q = e^{-\beta\Delta u^\ominus} \equiv K\,,$$

known as the equilibrium constant. The elementary form of the equilibrium constant of the reaction:

$$a\mathrm{A} + b\mathrm{B} \leftrightharpoons c\mathrm{C} + d\mathrm{D}$$

is readily derived in terms of equilibrium molar concentrations as

$$K = \frac{[\mathrm{C}]^c[\mathrm{D}]^d}{[\mathrm{A}]^a[\mathrm{B}]^b}$$

7.3 The Boltzmann Distribution

The inverse argument shows how a chemical potential field imposes a particular energy distribution on units of matter.

Consider a quantized field with discrete energy levels, each occupied by a characteristic number of molecules. These numbers can be thought of as representing relative activities during the course of a hypothetical reaction that starts with n_i molecules at the initial energy level u_i and reaching equilibrium with n_j molecules of the final product at the level u_j. Intermediate levels are occupied by secondary products. As before

$$\ln\left(\frac{A_f}{A_i}\right) = -\beta\left(\Delta u_f - \Delta u_i\right) \tag{7.5}$$

in which the negative sign indicates that the number of particles increases with decreasing energy. Let a total of N molecules be spread over all available energy levels. Using (7.2) it follows that

$$A_f = n_j \Big/ \prod_{k\neq j}^{N} n_k \quad , \quad A_i = n_i \Big/ \prod_{l\neq i}^{N} n_l$$

$$\Delta u_f = u_j - \sum_{k\neq j}^{N} u_k \quad , \quad \Delta u_i = u_i - \sum_{l\neq i}^{N} u_l$$

It follows that

$$\frac{A_f}{A_i} = \frac{n_j}{n_i} \cdot \prod_{l\neq i}^{N} n_l \Big/ \prod_{k\neq j}^{N} n_k = \left(\frac{n_j}{n_i}\right)^2$$

$$\begin{aligned} \Delta u_f - \Delta u_i &= u_j - u_i - \sum_{k\neq j}^{N} u_k + \sum_{l\neq i}^{N} u_l \\ &= 2\left(u_j - u_i\right) \end{aligned}$$

Hence, by (7.5)

$$2\ln\left(\frac{n_j}{n_i}\right) = -2\beta(u_j - u_i)$$

which rearranges to

$$n_j = n_i e^{-\beta(u_j - u_i)}$$

recognized as the Boltzmann distribution for $\beta = 1/kT$. The total number of molecules at all energy levels

$$N = \sum_i n_i = n_0 \sum e^{-(u_i - u_0)/kT}$$

where n_0 and u_0 refer to the ground level.

Finally

$$\begin{aligned}\frac{n_i}{N} &= \frac{n_0 e^{-u_i/kT} \cdot e^{u_0/kT}}{n_0 \left(\sum_{i=0}^{\infty} e^{-u_i/kT} \cdot e^{u_0/kT}\right)} \\ &= \frac{e^{-u_i/kT}}{z} \end{aligned} \tag{7.6}$$

where z is the sum over levels, known as the molecular partition function of statistical thermodynamics,

$$z = \sum_j e^{-\beta\epsilon_j} \quad , \quad \beta = \frac{1}{kT}$$

The quantized energy ϵ_j can be of electronic, vibrational, rotational or translational type, readily calculated from the quantum laws of motion. In a macrosystem the sum over all the quantum states for the complete set of molecules, the sum over states defines the canonical partition function:

$$Z = \sum_{states} e^{-E_i/kT}$$

7.4 Entropy

Entropy production during chemical change has been interpreted [7] as the result of resistance, experienced by electrons, accelerated in the vacuum. The concept is illustrated by the initiation of chemical interaction in a sample of identical atoms subject to uniform compression. Reaction commences when the atoms, compacted into a symmetrical array, are further activated into the valence state as each atom releases an electron. The quantum potentials of individual atoms coalesce spontaneously into a common potential field of non-local intramolecular interaction. The redistribution of valence electrons from an atomic to a metallic stationary state lowers the potential energy, apparently without loss. However, the release of excess energy, amounting to $\Delta u = \mu_{val} - \mu_{met}$ per atom, into the environment, requires the acceleration of electronic charge from a state of rest, and is subject to *radiation damping* [99].

Emitted radiation carries off energy, momentum and angular momentum and so must influence the subsequent motion of the charged emitters. These reactive effects are usually considered of negligible importance and therefore neglected in most cases. The effect of radiation damping on elementary quantum transitions is immeasurably small and therefore ignored on account of Occam's razor. The laws of quantum mechanics are therefore considered to be strictly time-reversible. However, when charges at rest are suddenly accelerated for a short time the effect becomes appreciable. Redistribution of charge during chemical reaction represents situations of exactly this type.

The most general chemical reaction can be reduced to the process of making and breaking bonds through the rearrangement of valence electrons and atomic cores to minimize the electronic energy. This motion in molecular space is not frictionless and some energy is lost irretrievably to the radiation field. It is this universal friction that renders processes irreversible and creates the arrow of time. The time-irreversible second law of thermodynamics, like the exclusion principle, is thereby identified as an emergent property of macrosystems.

For book-keeping purposes the production of entropy during chemical change is considered as reducing the useful energy of the system by disorderly dispersion. In many cases this waste can be calculated statistically from the increase in disorder. To be in line with other thermodynamic state functions, any system is considered to be in some state of disorder at all temperatures above absolute zero, where entropy vanishes.

Thermodynamics is the workhorse of chemical engineering, but less important as a theory to elucidate the mechanism of chemical reactions.

7.5 Chemical Reaction

The quantum-mechanical formulation of the progress of a reaction such as

$$\mathrm{A} + \mathrm{B} \rightarrow \mathrm{C} + \mathrm{D}$$

starts [7] from a stationary product state $\psi_A.\psi_B$ of mixed reactants and proceeds via the entangled valence state ψ_{ABCD} towards the final product state $\psi_C.\psi_D$. There is no obvious mechanism for such events in terms of traditional quantum theory.

In Bohmian formalism it may be argued that the reaction system, considered closed, is described at all times by an equation $H\Psi = E\Psi$ in the time-dependent wave function $\Psi(A, B, C, D)$. The product states $\psi_A.\psi_B$ and $\psi_C.\psi_D$, as well as the valence state are special solutions of this equation under different boundary conditions. All rearrangements and transformations that determine the final outcome happen in the valence state. The valence state is not unique and is conditioned by thermodynamic factors. Under stormy conditions the reactants may be fragmented into smaller units, $A \rightarrow na_i$, $B \rightarrow mb_j$, *etc.* The number and nature of possible reaction products will depend critically on the degree of fragmentation. Fragmentation itself is brought about by electronic, vibrational and rotational transitions, with rates linked to the ambient conditions. The valence state may therefore be formulated in terms of variables, characteristic of either molecular fragments, atoms, nuclei, electrons and/or photons. It may be either holistic or partially holistic, with matching quantum potential. The extent of non-local interaction depends on the quantum potential and may be a factor limiting the extent of possible intramolecular rearrangement during chemical reaction. The traditional argument does not contemplate instantaneous transitions between states and intramolecular rearrangement becomes a complete mystery.

Real chemical reactions are violent affairs and not likely to proceed as smoothly as an ideal metalization. Reacting units have translational kinetic energy but only a fraction of activating encounters leads to binding. The final product is unlikely to incorporate all of the atoms promoted into a valence state; smaller fragments are more likely to separate before reaction has spread homogeneously through the entire system and all nuclei have moved into place. Formation of intermediate fragments represents relaxation to an energy well below the valence state and leads to cessation of further chemical interaction. In an atomic medium diatomic fragments are expected to be the major product. If these emerge in or near their valence states, further reaction may cause formation of oligomers. In complex reaction mixtures the course of reaction depends on the relative promotion status of the various constituents, which may be atoms, small molecules or ions. Although each individual reaction therefore has a specific course that depends on the composition of the reaction mixture and on environmental factors, it is of interest to identify the common principles that may influence reaction mechanisms.

It has been argued [7] that secondary interactions between primary fragments should normally result in a new non-local equilibrium situation involving all constituent atoms of an oligomer and their quantum potentials. However, this is not the soup of individually promoted atoms. It is more likely that interacting primary fragments would reach their own promotion state, which does not require sequestration of all cohesions established before. Any chemical process that occurs over a series of steps is thus predicted to yield diverse products dependent only on environmental conditions during each step. Certain fragments remain intact during rearrangement. These fragments, rather than their constituent atoms, contribute to the quantum potential of the whole and the shaping of a new product. The extent to which the molecular quantum potential dictates a robust three-dimensional shape depends on a wave function that remains localized on the molecule.

It is only the electronic wave function of an isolated free hydrogen atom that can conceivably be considered to extend indefinitely in a void. For any other situation, including the real world, local potential barriers must restrict it to a much smaller region. The more crowded the environment, the more closely is the wave function – and therefore the effect of the quantum potential – confined. Only when environmental crowding promotes the atom into its valence state does the wave function start penetrating into a larger region that covers the chemically interacting neighbourhood. As the primary reaction products separate, the total wave function factorizes into a product state

$$\psi_T = \phi_1^M \phi_2^M \cdots \phi_n^M . \tag{7.7}$$

Non-local connections between these molecular units are much feebler than within the molecules and vary with the state of aggregation. This conclusion seems to agree with conventional thinking in chemistry.

7.5.1 Atomic Reactions

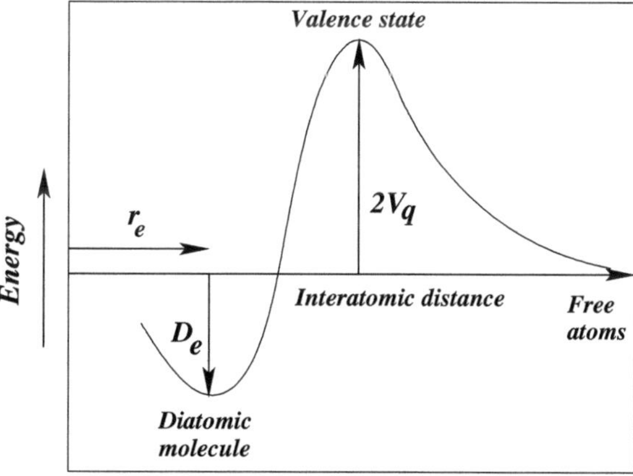

Figure 7.1: *Schematic drawing to illustrate promotion to the valence state and formation of a diatomic molecule.*

A characteristic degree (energy) of uniform compression is required to promote an atom into its valence state. In a compressed monatomic medium all atoms enter the valence state simultaneously. At this point the valence electrons have quantum potential energy only, but they are free to move away from their atomic cores. The barrier between actual atoms is never as uniform and impervious as in the simulation. Valence electrons gain kinetic energy and percolate into interatomic voids where they encounter other valence electrons with which they interact through the field effects of their quantum torque. In this fashion electrons become delocalized across the neighbourhood defined by those promoted atoms in close proximity; interacting via the quantum potential field. The reaction neighbourhood may consist of only two atoms that end up equally sharing the pair of valence electrons. This condition is the prototype of a covalent bond.

A schematic diagram to illustrate the course of reaction is shown in figure 7.1. The energy level of the valence state corresponds to the promotion of two atoms involved in the reaction. The energy of the system drops as the valence electrons spread out across the larger accessible space surrounding the atomic cores. Holistic interaction between the two valence electrons and the cores stabilizes the molecule by an amount D_e. According to the classical model this corresponds to an internuclear distance r_e.

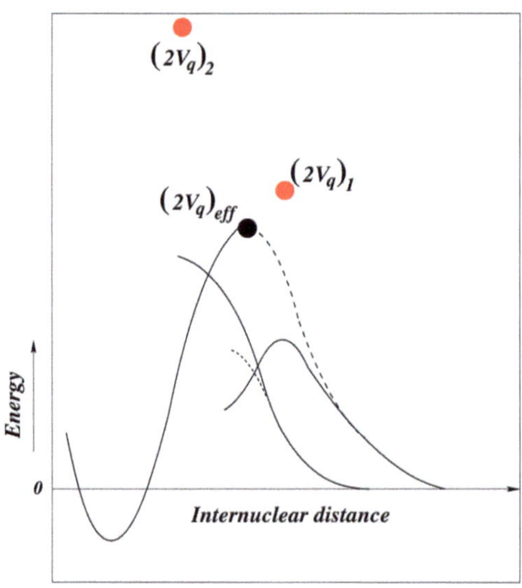

Figure 7.2: *Activation and interaction of a heteronuclear pair of atoms. Red dots indicate the activation levels of homonuclear diatomics. Because of the difference in polarity formation of the heteronuclear molecule is favoured.*

The idea of a chemical bond between two atoms in a molecule is akin to the classical model of a diatomic molecule, and its formation can be discussed along the same lines. The first assumption is that a pair of neighbouring atoms can be identified and isolated for study, well knowing that this action sacrifices all knowledge pertaining to overall intramolecular entanglement. The next approximation is to clamp the nuclei at classically variable coordinates. This approximation still allows freedom to study the electron density quantum-mechanically. However, in view of the nature of the valence state developed here there is nothing to gain by attempting all-electron calculations.

The predicted course of reaction between a heteronuclear pair of atoms is shown in Figure 7.2. Promotion is once more modeled with isotropic compression of both types of atom. The more electropositive atom (at the lower quantum potential) reaches its valence state first and valence density starts to migrate from the parent core and transfers to an atom of the second kind, still below its valence state. The partially charged atom is more readily compressible to its promotion state, as shown by the dotted line. When this modified atom of the second kind reaches its valence state two-way delocalization occurs and an electron-pair bond is established as before. It is notable how the effective activation barrier is lowered with respect to both homonuclear $(2V_q)_i$ barriers to reaction. The effective reaction profile is the sum of the two promotion curves of atoms 1 and 2, with charge transfer.

7.6 Chemical Kinetics

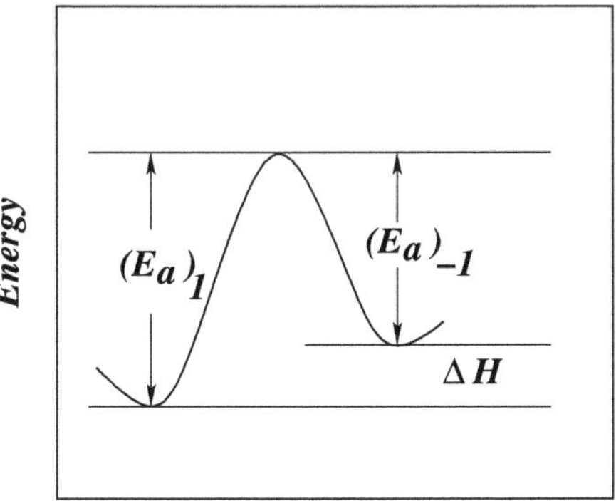

In a molar-scale diatomic reaction mixture ($R = kL$) the number of reactants in the valence state, at a given temperature, is given by an equation such as (7.6) with $u_i = (2V_q)_{eff} \equiv E_a$, called the *activation energy*. The mole fraction (concentration)

$$n_a/N = Ze^{E_a/RT} \tag{7.8}$$

of activated reactants determines the chemical action at temperature T. The well-known equation

$$k = Ae^{E_a/RT}$$

that defines reaction rate constant as a function of temperature, first obtained empirically by Arrhenius, clearly defines the same relationship as (7.8).

In the case of a complex reaction the rate constant is the product of rate constants for several elementary steps. The progress of such reactions is usually presented along a *reaction coordinate* – a hypothetical parameter that depends on all internuclear distances of relevance to the mechanism whereby reactants are converted into products.

The forward and reverse reactions

$$k_1 = A_1e^{-(E_a)_1/RT}$$

$$k_{-1} = A_{-1}e^{-(E_a)_{-1}/RT}$$

have different activation energies that define the thermodynamic equilibrium constant, $K = k_1/k_{-1} \propto e^{-\Delta H/RT}$.

All theories of chemical kinetics and reaction mechanisms are based on eqn. (7.8) by an estimation of the relevant partition function.

Chapter 8

The Central Science

8.1 Introduction

Chemistry, the link between earth and life sciences, is often considered as *The Central Science*. This description is equally appropriate in the reductionist hierarchy of knowledge, which ranges from philosophy through mathematics, physics, chemistry and biology, towards the behavioural sciences. The common principles that emerge in the periodic arrangement of matter, as a numerical function of either nucleons or electrons, in the nature of covalent interaction, in botanical phyllotaxis and in the observed gaps of the asteroid belt, suggest a central role for chemistry in an even more fundamental way. The golden ratio, Fibonacci numbers and Farey sequences feature prominently in all of these constructs. Should the same geometrical principle decide the planetary structure of the solar system and the structure of spiral galaxies, a universal self-similarity mediated by the symmetry of space-time could be inferred.

The parallel which was drawn by Nagaoka between the rings of Saturn and atomic structure is based on such self-similarity. Although the Saturnian ring system is stabilized by gravitation, with angular momentum, and the atom, which is stabilized electrodynamically, has no angular momentum, the structural difference is one of dimension only. In order to quench the orbital angular momentum, electronic rings are required to be spherical.

There is sufficient evidence [100] that galactic spirals and nautilus shells, as in Figure 8.1, have the same geometry as the golden logarithmic spiral. It is therefore only the structure of the solar system that prevents extrapolation from the sub-atomic to the galactic scale. However, knowing that [101] "the mean motions of a number of bodies in the solar system have been found to be commensurate to a remarkable degree, quite beyond the probability of

Figure 8.1: *Whirlpool galaxy M51 (Courtesy Hubble Heritage Team, ESA, NASA) and a fossil shell with the same spiral structure as the proposed distribution of matter in the solar system at the time of planet formation*

accidental commensurability", the planetary system should be re-examined for self-similar structure.

8.2 The Solar System

It has been known since 1772 that the planets orbit the sun at non-random distances, specified, in astronomical units[1], by the Titius-Bode law:

$$d = a + bc^n$$

in which $a = 0.4$, $b = 0.3$, $c = 2$ and $n = -\infty, 0, 1, 2, \ldots$ for the planets from Mercury to Uranus, including the asteroid belt, represented here by the minor planet Ceres, at $n = 3$. The formula overestimates the orbit of Neptune by 25%. This pseudo-geometrical progression indicates the possibility of a periodic fit of the orbital distances to a logarithmic spiral and/or some Farey sequence with Ford circles. On assigning a mean radius of unity to Neptune's

[1] $1\text{AU}= 1.496 \times 10^8\text{km}$

orbit, all other radii are predicted well by rational fractions thereof. The predicted distances are a considerable improvement over the Titius-Bode formula. The overall sequence separates into two sub-sequences at Ceres – the asteroid belt.

	Sol	Merc	V	E	M	C	J	S	U	N
	$\frac{0}{1}$	$\frac{1}{80}$	$\frac{1}{40}$	$\frac{1}{30}$	$\frac{1}{20}$	$\frac{1}{10}$	$\frac{1}{6}$	$\frac{1}{3}$	$\frac{2}{3}$	$\frac{1}{1}$
d	0	0.38	0.75	1.0	1.5	3.0	5.0	10.0	20.0	30.0
Obs		0.39	0.72	1.0	1.52	2.9	5.2	9.5	19.2	30.3

Sol	Merc	V	E	M	C
$\frac{0}{1}$	$\frac{1}{8}$	$\frac{1}{4}$	$\frac{1}{3}$	$\frac{1}{2}$	$\frac{1}{1}$

	C	J	S	U	N
$\frac{0}{1}$	$\frac{1}{10}$	$\frac{1}{6}$	$\frac{1}{3}$	$\frac{2}{3}$	$\frac{1}{1}$

This division corresponds to the division into rocky planets of the inner solar system and gaseous planets of the outer solar system, separated by the minor planets. By analogy, a larger sequence of solar satellites (planetoids, comets), of which the Pluto/Charon system is the first, is predicted to continue the regular progression beyond Neptune.

The orbits from Venus to Ceres are represented by the unimodular series $\mathscr{F}_4$. In the outer system the Ford circles of only Uranus and Neptune are tangent, but the likeness to Farey sequences in atomic systems is sufficient to support the self-similarity conjecture.

8.2.1 Spiral Structure

The generation of a golden spiral by the repeated construction of smaller golden rectangles through the removal of squares is demonstrated in Figure 8.2. The spiral is formed by circular segments inscribed in the successive squares. It simulates the structure of the objects shown in Figure 8.1.

Botanical growth proceeds by the appearance of florets at constant divergence angles, defined by Fibonacci fractions $(h/k) \times 360°$, along logarithmic spirals. Placing Neptune on the golden spiral at the remote corner of the golden rectangle and the other planets on their mean circular orbits, drawn to scale, the points of intersection with the spiral define a constant divergence

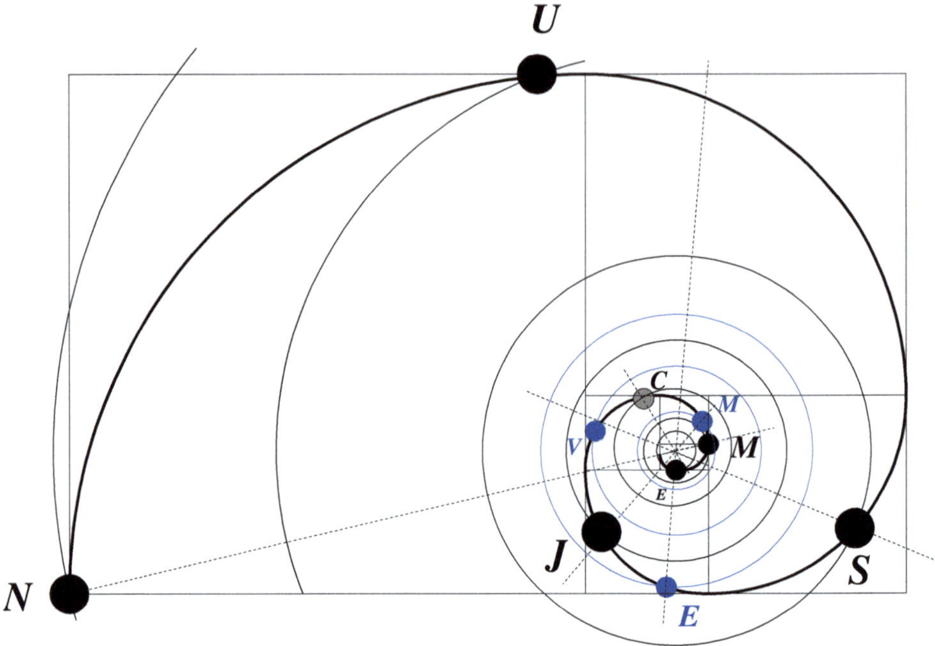

Figure 8.2: *Logarithmic spiral with superimposed mean planetary orbits. The circles in blue define the orbits of inner planets on a larger (self-similar) scale. The divergence angle of 108° causes those planets at angles of* $5 \times 108°$ *apart to lie on opposite sides of the spiral origin. These pairs are Neptune–Mars, Uranus–Earth, Saturn–Venus and Jupiter–Mercury. The hypothetical antipode of the asteroid belt, a second, unobserved group of unagglomerated fragments, has been swallowed up by the sun*

angle of 108°, with only few exceptions. The comparison between derived and observed orbital distances is shown in Table 8.1. The correspondence is sufficiently close to confirm that the original aggregation that produced the planets could have occured within the spiral arm of a condensing central core, which rotated within a nebular cloud – akin to the model proposed by Harold Urey [102] on chemical grounds. Images of rotating galaxies[2] provide an exact picture of the synthesis model proposed here.

The discrepancies in Table 8.1 exist because the derived values refer to

[2]A spiral galaxy, such as the famous Whirlpool M51 (Figure 8.1), resembles a vortex, most probably created by a central black hole.

Table 8.1: *The orbital distances (astronomical units) at perihelion, mean and aphelion of the planets, compared to scaled intercepts, with divergence angle of* 108°(3/10 × 360), *on a golden spiral*

Planet	Observed orbit (AU)			Derived
	P	*M*	*A*	
Mercury	0.31	0.39	0.47	0.34
Venus	0.72	0.72	0.73	0.6
Earth	0.98	1.0	1.02	1.0
Mars	1.39	1.52	1.66	1.7
Ceres		2.9		2.9
Jupiter	4.95	5.2	5.45	5.4
Saturn	8.99	9.5	10.06	9.7
Uranus	18.29	19.2	20.12	17.2
Neptune	29.8	30.0	30.3	30.0

the time of planet formation and observed values refer to the present. The fact that Uranus spins on an axis close to its orbital plane indicates that it suffered some encounter with a foreign body that must have modified its original orbit. Venus probably moved to a larger orbit because of losing atmospheric matter. Because of its proximity to the asteroid belt Mars could have gained sufficient mass, by coagulating a number of minor planets, and thereby reduce its orbital distance from the sun.

8.3 Chemical Science

After the necessary excursion into astronomy, enough evidence now exists to repeat that chemistry occupies a central position in a self-similar chain between sub-atomic matter and super-galactic structures. By virtue of this unique position in the natural sciences chemical theory should represent the link between physics, biology and the earth sciences. However, compared to the fundamental theories of physics, chemistry has no credible independent theory of its own.

The chemists of the 19th century discovered the Periodic Table of the Elements, but the underlying ideas were abandoned during the 20th century in favour of quantum physics. The early quantum theory of Bohr and Sommerfeld could be interpreted to be in line with the structure of the periodic table, but only with Pauli's postulate of an exclusion principle as an additional assumption. Later on the development of wave mechanics raised the expectation of a detailed prediction of the periodic table from first principles. Even today, chemistry texts pretend that this expectation has been fulfilled, but in two aspects the exercise has failed:

1. Quantum theory has still not been able to relate the exclusion principle to more basic concepts, accepting it as empirical fact, without which the periodic table remains a mystery.

2. The quantum theory predicts three transition series of 10 elements each as defined by the serial filling of d sub-levels. The actual series are observed to consists of 8 elements each.

8.3.1 Where did Chemistry go wrong?

The art of chemistry developed from the practice of alchemy, which was aimed at the production of certain potent agents, which in their pure form, were believed to transmute base metals into gold and hold the key to human immortality. The theories behind alchemy were closely linked to the occult and recognized sympathetic links between a special number of gods, and the same number of heavenly bodies, metals, minerals and the manipulations of alchemy. As a working hypothesis matter was considered made up of the three elements mercury, sulphur and salt, subject to the activity of spirits, such as fire.

The transformation of alchemy into a science happened during the Renaissance. What was needed was a rational understanding of matter, for which the foundations were laid by Galileo (1638) and Newton (1668) as the laws of motion, and by Boyle (1661) who rediscovered the conservation of matter during chemical change. Combination of these principles with Bernoulli's gas model produced the first convincing scientific theory of matter in the hands of Clausius and Maxwell ($\sim$ 1860). Known as the kinetic theory of ideal gases it succeeded to predict on the basis of a well-defined simple theoretical model most of the experimentally known properties of a gas. This achievement is the essence of a scientific theory. The assumptions are simple and clear, without any pretence of universal validity. The theory reached maturity on the basis of research in astronomy, mathematics, physics, chemistry and engineering, and still features prominently in all of these disciplines. In this sense it demonstrates a unity of knowledge in science. None of these disciplines is an island in an ocean of knowledge. They all share a common theory of understanding. It is only experimental data that are discipline specific.

In modern science there are three distinct theories of general applicability and although not always obvious they have a common root. All of current science can be understood on the hand of the theory of general relativity, quantum theory and the periodicity of matter. The first of these is commonly associated with physics, the latter with chemistry, and quantum theory features in different guises in both physics and chemistry. However, this distinction is completely arbitrary and artificial. It can be argued that the three theories separately find common ground in the structure of space-time, also known as the vacuum.

This conclusion opens up the intriguing possibility that all of science can be reduced to a single fundamental concept. The philosophy of reductionism makes exactly this assumption. It implies that the facts of biology can be reduced to the properties of chemical molecules, which in turn reduce to the atoms of physics, the nuclei of nuclear physics, to elementary particles and eventually to the symmetry of space-time or the vacuum.

Although the reductionist argument is of obvious validity, the inverse process of constructionism is impossible. This philosophy assumes that the properties of more complex systems can be predicted from those of a simpler one. By this logic theoretical chemists of the 20th century have persistently tried to reconstruct chemical behaviour from the fundamental equations of wave mechanics. To date the most powerful computers on the planet have failed consistently to reconstruct even the most fundamental property in all of chemistry, namely the structure of a molecule. Computations, known as quantum chemistry, all have to rely on the kick-start of an assumed molecular structure.

Even the layout of the periodic table of the elements cannot be derived from quantum theory without assuming an empirical concept, known as the Pauli exclusion principle. An alternative derivation (4.6.1) through number theory predicts the correct periodicity, without assuming the exclusion principle. In fact, the operation of an exclusion principle can be inferred from this periodic structure and reduced to a property of space, but it remains impossible to reconstruct or predict from more basic principles.

The exclusion principle is one example of an emergent property. It cannot be predicted, or even formulated, from all the known properties of a single electron, but emerges as an inevitable property of a world collective of electrons, or other fermions – particles with half a unit of spin angular momentum.

Perhaps the best-known emergent property in science is described by the second law of thermodynamics. A mystery lies therein that the laws of quantum mechanics that regulate the motion of sub-atomic particles are all time-reversible. The same is not true for macroscopic samples. Chemical systems have the tendency to approach a state of equilibrium spontaneously and irreversibly. Some factor, not anticipated by the fundamental laws, is clearly at work and the empirical law that correctly describes the behaviour of the system must be time-irreversible. The net effect is that energy appears to be lost in the course of spontaneous change and this is commonly ascribed to the production of entropy, which correlates with an increase in the orderly distribution of the energy. The cause of increased disorder is seldom addressed, but is evidently due to the friction that accompanies any macroscopic change.

In the case of a spontaneous quantum-mechanical change, often called a quantum jump, the effect is beyond detection and automatically eliminated by Occam's razor. Nonetheless, such a frictional effect is recognized in electrodynamics as radiation damping because of charge acceleration. Irreversibility of the second law may therefore be reduced to a primary cause, but not predicted by a primary law. Emergent effects are most common in biological systems. Cells have properties that cannot be predicted by molecular theory. The same applies when progressing to the higher levels of organization.

The unity of science appears indisputable. The way in which all facts hang together implies the reductionist principle. Failure of constructionism denies the priority of any branch of science over any other because of the unpredictability of emergent properties. The single discipline, normally not considered a science, but fundamental to all scientific facts, is mathematics. It now appears that science, at all levels, is intimately entangled with number theory.

8.3.2 Constructionism

The failure of constructionism in science needs to be re-examined in view of the demonstrated principles of self-similar cosmic symmetry. It has been argued that atoms, flowers and galaxies share the same design principles on different scales. In fact, the only similarity exists in the design, as if these entities have been constructed on the same scaled template, which is the local structure of space-time. Although this structure remains undetectable it seems to be faithfully reflected in the natural number system. Material entities at different levels can be related directly to one another. The strong interaction between nucleons, electromagnetic interaction between extranuclear electrons and gravitational interaction in the solar system produce uniquely different arrangements with a common design pattern. This is the pattern observed through the medium of number theory.

Emergent properties that differentiate between structures at different levels are generated by the unique interactions that operate at different levels of organization. The dominant mode of interaction between biological cells is a negligible factor in chemical systems and totally absent within atoms. Botanical phyllotaxis and the gravitational clumping in astronomical spirals, although regulated according to the same pattern[3], produce structures with vastly different properties. Reductionism refers to the recognition of common patterns whereas constructionism fails as an extrapolation between different modes of interaction. The structure of electromagnetic systems (atoms) cannot be predicted by the mechanics of gravitational systems (planetary rings), although they exhibit the same numerical regularities.

8.3.3 Emergent Chemical Properties

The properties that emerge in chemical studies include the exclusion principle, molecular structure and the second law of thermodynamics. Without these principles, not revealed by the laws of physics, there is no understanding of the properties of matter in the bulk. By way of example, the phenomena of optical activity and superconduction have never been fully explained by the laws of physics.

Most chemists would accept the general statement that optically active molecules are those without an alternating axis of symmetry, as adequate. This is perhaps a reliable diagnostic of natural optical activity, but it offers no explanation of the Faraday effect, which shows that achiral molecules become optically active in an applied magnetic field.

Starting from the emergent property of molecular structure the answer to this problem becomes obvious. Whereas brute-force quantum-chemical computations assume that observed molecular structure represents a minimum-energy configuration, it is noted that minimization of a scalar quantity such as energy can never produce a unique three-dimensional configuration, which is a vector quantity. The only molecular vector with the ability to dictate three-dimensional conformation is electronic orbital angular momentum. Minimization of kinetic energy is achieved by the optimal quenching of the orbital angular momentum, which dictates the relative orientation of inter-

[3]The common botanical divergence angle is $2\pi\tau^2$ radians (137.5°), compared to $(3/5)\pi \simeq \pi\tau$ of the planetary spiral. The botanical angle allows maximum radiation for each leave and is an area effect. Competition for intermediate material, along a semi-circle in the spiral arm, generates a divergence angle with linear dependence on τ.

acting molecular fragments or radicals, in symmetrical alignment. Situations of incomplete quenching leave residual angular momentum and also exhibit a lack of symmetry, or chirality. It is the magnetic moment associated with the residual angular momentum that couples to the magnetic field of polarized radiation. The Faraday effect occurs because of the resolution of angular momentum vectors in an external magnetic field.

In the second example BCS theory relates the appearance of a superconducting state to the breakdown of electromagnetic gauge symmetry by interaction with regular ionic lattice phonons and the creation of bosonic excitations. This theory cannot be extended to deal with high T_c ceramic superconductors and it correlates poorly with normal-state properties, such as the Hall effect, of known superconductors. It is therefore natural to look for alternative models that apply to all forms of superconductivity.

Such an alternative is suggested [62] by the grand periodic law as revealed by the distribution of prime numbers and the number theory of periodic systems. It applies equally to all atomic matter based on atomic number, mass number or neutron number. It suggests an idealized nuclear composition with protons and neutrons in the same ratio as the golden section and represents a configuration of special stability. Because of competing periodicities this composition is never realized. The compromise is formation of a high-spin layer of excess protons with a magnetic field that induces superconductivity in the regular structures of all known superconductors.

The time has come for chemistry to stop echoing the conclusions of physics and trying to reduce the emergent phenomena of the science to a deeper level of understanding. Only chemistry can probe the mechanisms of molecular formation and the effects of intermolecular interactions. Fundamental understanding of these issues, at the level of chemistry, rather than physics, will not only serve the chemist, but also improve the insight of biologists into the emergent phenomena of their discipline.

8.4 General Chemistry

In conclusion the simplification of complex theories, attempted here, will be put to the test by telling the story of chemistry in simple language.

Chemistry remains a mystery without understanding three basic attributes of matter: cohesion, structure and affinity, each of them shaped by an extremum principle. The principle of minimum energy regulates chemical cohesion, which results from the interaction between atoms, also known as chemical bonding. Minimization of angular momentum dictates the three-dimensional arrangement of atoms in chemical substances, which defines their structure. Affinity, which drives chemical reactions, reflects the differences in electronegativity and the electronic configuration of reactants. The driving force is the tendency to equalize electronegativity by redistribution of valence electrons. Both electron configuration and electronegativity are periodic properties of the elements.

Each of the three fundamental factors is responsible for the emergence of a special property of matter in the bulk, not predictable from the physics of microsystems. The energy minimum coincides with an extreme charge density, limited by the golden ratio, as a geometric property of space. This limitation is recognized as the empirically defined exclusion principle of statistical thermodynamics. Minimization of orbital angular momentum dictates the emergence of molecular shape. The affinity that drives a system towards electronic equilibrium, at the expense of the minimum-energy principle, creates a chemical potential field and the production of entropy as described by the second law of thermodynamics.

It is only against the background of these emergent properties that the manipulations of chemical practice in the laboratory can be properly appreciated. Some of the details are elaborated on in the following discussion.

8.4.1 Chemical Substance

The most direct way in which humans experience the world around them is through contact with material objects. Some of these are hard and brittle, others are soft and flexible and there seems to be a suitable material available for each and every possible use or purpose, ranging from glass, through putty, to fluids. A full description of the consistency and nature of any material or substance is said to be fixed by its *physical properties*. If these physical properties are the same for all samples of a given material, it defines a pure substance, which by definition, always has the same *chemical composition*. Chemical composition depends on the relative amounts of different chemical *elements*, locked up together in samples of the substance. If the same chemical elements are always found together in the same ratios, it is reasonable to conclude that the *atoms* of all elements in the sample are held in some specific pattern or arrangement. This arrangement will be secure and invariant if the interactions between all atoms in the sample are of a constant nature. These interactions are collectively known as *chemical bonding*.

To understand the essence of chemical bonding it is necessary to examine the possible ways in which a pair of atoms can either attract or repel each other. From the study of celestial bodies and their motion it has been inferred that an attraction, known as *gravity*, operates between any pair of massive objects. The same interaction should also occur between a pair of atoms, but because of their small masses the gravitational force between atoms is too feeble to account for chemical bonding.

8.4.2 Electromagnetism

An alternative type of interaction that operates between massive objects is caused by the *electromagnetic field.* Two well-known aspects of this field are known as *electricity* and *magnetism.* The latter is best known from the interaction between permanent magnets and from the response of compass needles to the magnetic field of the earth. At a deeper level of understanding magnetic effects are seen to result from the circulation of electric *charge.* Electricity occurs as either positive or negative charges. Experience shows that unlike electric charges attract each other whereas like charges repel. It is interesting to note that gravitational, electrical and magnetic interactions are all described by the same mathematical formula, known as an *inverse square law.* The squared quantity that features here is the distance, r, measured between the interacting points. In all cases, the force of interaction, F, is proportional to the inverse of r^2:

$$F = kr^{-2} \tag{8.1}$$

in which k is a *proportionality constant*[4].

Action at a distance is universally rejected as an absurdity. To account for interaction between objects not in physical contact, it has been necessary to consider such interaction as caused by *fields* of force, that spread through space and the vacuum. Analysis of their effects has shown that electric and magnetic fields always propagate together in the form of electromagnetic *waves*, that spread through the vacuum at constant speed. Such a wave consists of two components, in the form of electric and magnetic vectors that vibrate at right angles to each other and perpendicular to the direction of wave propagation.

When passing through a material medium the vibrations are impeded and hence become sensitive to the arrangement of matter in the medium. In many solid structures, especially crystals, a preferred direction of vibration that allows ready transit is selected by the waves. In the process a beam

---

[4]The constant k must be defined in such a way that the quantities on the two sides of the equation are expressed in the same units. Such *dimensional analysis* demands that k have the dimensions of a force times a distance squared. In terms of mass (M), length (L) and time (T), k has dimensions of ML^3T^{-2}.

of light is split into two rays, one of which uses the preferred direction for the vibration of electric vectors and the other for magnetic vibrations. The two rays move at different velocities through the medium. In some cases the secondary ray is completely absorbed by the medium and plane polarized light is transmitted through such material, known as a *polaroid.*

The theory of the electromagnetic field, as developed by James Clerk Maxwell, is arguably the most brilliant and useful result ever produced in science.

8.4.3 Relativistic Effects

A puzzling feature of electromagnetic radiation is the constant velocity of propagation in the vacuum, independent of the state of motion of both emitter and receptor. This constancy necessitated a complete reformulation of Galileo's model of relative motion and the development, in the hands of Lorentz, Einstein, Poincaré and others, of a special theory of relativity that embraced the motion of both mechanical objects and electromagnetic signals. The final conclusion was that the three-dimensional world is an illusion and that electromagnetic waves propagate in at least four dimensions of space-time. A by-product of the theory was to confirm the equivalence of matter and energy. Simple dimensional analysis shows that the proportionality factor must have the dimensions of a squared velocity, hence $E = mc^2$.

In its most general form Einstein's theory of general relativity, which also considers accelerated motion, led to a simple model of the gravitational field, by introducing the concept of curved space. Like an object that spontaneously rolls downhill on the surface of the earth it follows a similar gravitational potential path in curved space, caused by the presence of another massive object. There is no action at a distance, only response to the gradient of the gravitational potential-energy field. The field equations that led to this conclusion define a balance, or reciprocal relationship, between matter and curvature. Without matter space-time is flat, and without curvature space-time is void of matter and energy. Without this realization the nature of matter is incomprehensible.

The compelling conclusion, fundamental to the understanding of chemistry, is that units of matter and energy are no more than elementary distortions of the vacuum that appear when space is being curved. The nature of such distortions had already been recognized in electromagnetic theory, which ascribed the transmission of electromagnetic energy to wave motion. Without further qualification it remains a mystery, in terms of this model, how any transmitter manages to beam its energy to a single receptor by medium of a spherical wave. The wave spreads over the entire solid angle, but focusses its energy on a single point.

8.4.4 Interaction Theory

Closer scrutiny of the general wave equation provides an answer to the dilemma. As a second-order differential equation in the time variable, it has solutions in positive and negative time[5]. The negative time (or advanced) solutions are routinely rejected as physically impossible. This decision is based on prejudice rather than insight. Without evidence to accept only retarded (positive time) solutions as physically real, there is the possibility of response from a prospective receptor by means of advanced waves to establish one-to-one contact between emitter and absorber before transmission occurs.

Suppose the vacuum to be filled with a uniform radiation field, or waves. Interaction of this field with an emitter, in an excited state, causes modulation of the wave field that spreads in the form of a spherical retarded wave. On reaching a suitable absorber, at a lower energy level, the modulation stimulates a matching sympathetic response, which is returned as an advanced wave, that reaches the emitter at the very time of initial modulation. Such a superposition of advanced and retarded waves amounts to the creation of a standing wave between emitter and absorber. Emitter and absorber are now in contact and the transaction is completed on the transfer of excess energy to the absorber. The standing wave that persists for the duration of the transaction triggers the transfer in the form of a *photon*.

In terms of this model a photon is not a high-velocity particle. It is more appropriately described as a standing wave in collapse. When the handshake between emitter and absorber is established, their vibrations resonate to form a beat that disappears to the site of lower energy, which is thereby stimulated into a higher vibrational state, at a higher energy level. If the absorber (or emitter) is an electron at a quantized energy level on an atom, such interaction with a photon drives the electron to a higher (or lower) vibrational energy level, in a process that became known, misleadingly, as a quantum jump. The change in quantum level happens continuously. The transfer of energy happens at the speed of light without the involvement of a fast particle.

[5]Taking the square root of a squared quantity: $\sqrt{x^2} = \pm x$, leads to a sign ambiguity which cannot be ignored.

The wave packet that carries the energy vibrates in electromagnetic mode. Each linear vibration vector has two rotational components (compare Figure 6.3), which together determine the photon's state of polarization. The two rotational components define equal, but oppositely directed, angular momentum vectors, called right and left circular polarization, or spin. Photon spin, with a fixed magnitude of unity, becomes observable when the angular momentum is aligned in a magnetic environment.

8.4.5 Quantum Effects

The implied linear relationship between the vibrational state (frequency) of a photon and its energy requires the existence of another universal proportionality constant, such that:

$$E = h\nu \tag{8.2}$$

Dimensional analysis shows that Planck's constant, h, must have the same dimensions as angular momentum. The frequency of oscillation, associated with circular motion at angular velocity ω, becomes $\nu = \omega/2\pi$, and eqn 8.2 reduces to $E = h\omega/2\pi = \hbar\omega$, with Planck's constant in the form known as h-bar. It is instructive to note that the unit of angular momentum, or spin, carried by a photon has the magnitude of exactly $\hbar$.

Planck's constant was discovered as part of the solution to a nineteenth century conundrum in physics, known as the black-body problem. The challenge was to model the wavelength distribution of radiation emitted through the aperture in a closed cavity at various temperatures[6]. The standard equations of statistical thermodynamics failed to produce the observed spectrum, unless it was assumed that the energy of radiation with frequency ν was an integral multiple of an elementary energy quantum $h\nu$.

It is significant that in both cases Planck's constant appears in the specification of the dynamic variables of angular momentum and energy, associated with wave motion. The curious relationship between mass and energy that involves the velocity of a wave, seems to imply that the motion of mass points also has some wavelike quality. Only because Planck's constant is almost vanishingly small, dynamic variables of macroscopic systems appear to be continuous. However, when dealing with atomic or sub-atomic systems

[6] The isotropic microwave background radiation, observed in the vacuum, has the same wavelength distribution as a black body, which shows that the universe is closed, like a cavity, rather than open and expanding.

these variables must be treated as discrete and quantified by integers, known as quantun numbers.

As integers always appear in Nature associated with periodic systems, with waves as the most familiar example, it is almost axiomatic that atomic matter should be described by the mechanics of wave motion. Each of the mechanical variables, energy, momentum and angular momentum, is linked to a wave variable by Planck's constant: $E = h\nu = h/\tau$, $p = h/\lambda = h\bar{\nu}$, $L = h/2\pi$. A wave-mechanical formulation of any mechanical problem which can be modelled classically, can therefore be derived by substituting wave equivalents for dynamic variables. The resulting general equation for matter waves was first obtained by Erwin Schrödinger.

The understanding of chemical substances depends on understanding the behaviour of their elementary building blocks, which is achievable only through the medium of quantum theory. The complicating factor for chemists is the variety of different interpretations to quantum theory. If the only objective were the mathematical simulation of experimental measurements, there would be no need to understand the wave nature of matter. This approach tolerates incomprehensible concepts such as point particles with wave character, probability densities and molecules without extension. It creates an unbridgeable chasm between the practice and theory of chemistry. It is almost as bad as the practice of alchemy under the fanciful guidance of sympathetic magic, astrology and divine regulation. It is chemistry with Niels Bohr in the role of Hermes Trismegistos.

The remedy is not to attempt the reduction of chemistry to the one-particle solutions of quantum physics, without taking the emergent properties of chemical systems into account. Chemical reactions occur in crowded environments where the presence of matter in molar quantities is not without effect on the behaviour of the quantum objects that mediate the interactions. It is only against this background that quantum theory can begin to make a useful contribution to the understanding of chemical systems.

8.4.6 The Wave Mechanics

Schrödinger's equation is the wave-mechanical analogue of Hamilton's formulation of the classical laws of motion. Hamilton's function:

$$H = \frac{p^2}{2m} + V = E \tag{8.3}$$

defines the total energy as the sum of kinetic and potential energy contributions. When transformed into wave formalism, only the momentum acquires new meaning, as described by its wave equivalent, $p = h/\lambda$, which enters the general differential wave equation. On specification of the potential energy, the modified equation:

$$H\Psi = E\Psi \tag{8.4}$$

can be assembled directly for any chemical system of interest. In principle, the wave function Ψ, that solves the equation, contains all information pertaining to the dynamics of the system, and corresponding to allowed values of the energy, E. To arrive at equation 8.4 for an isolated hydrogen atom the potential energy has the simple form of equation 8.1 for the electrostatic potential energy of interaction between the positively charged proton and the negative electron.

The resulting equation can be solved by special techniques under the assumption that, because of the difference in mass, the proton remains stationary on the scale of the more rapid electronic motion. This assumption reduces the situation to a one-particle problem and the solutions specify the motion and energy of only the electron, relative to the static proton. The so-called *eigenvalues* of the electronic energy are, as expected for waves, described by a set of integers, known as quantum numbers. The principal quantum number, $n = 1, 2, 3\ldots$, defines the allowed energy levels of the electron, $l = 0, 1, \ldots, (n-1)$, defines the total angular momentum and $m_l = -l, \ldots, l$, the component of angular momentum in the direction of an effective magnetic field. The eigenfunctions, $\psi(n, l, m_l)$, corresponding to the eigenvalues $E(n, l, m_l)$, place no other restriction on the motion of the electron, which has the freedom to be at any distance from the proton.

Although wave equations are readily composed for more-electron atoms, they are impossible to solve in closed form. Approximate solutions for many-electron atoms are all based on the assumption that the same set of hydrogen-atom quantum numbers regulates their electronic configurations, subject to the effects of interelectronic repulsions. The wave functions are likewise assumed to be hydrogen-like, but modified by the increased nuclear charge. The method of solution is known as the self-consistent-field procedure.

8.4.7 The Chemical Environment

The SCF solutions of many-electron configurations on atoms, like the hydrogen solutions, are only valid for isolated atoms, and therefore inappropriate for the simulation of real chemical systems. Furthermore, the spherical symmetry of an isolated atom breaks down on formation of a molecule, but the molecular symmetry remains subject to the conservation of orbital angular momentum. This means that molecular conformation is dictated by the re-alignment of atomic o-a-m vectors and the electromagnetic interaction between neighbouring atoms.

The interaction depends on redistribution of electronic charge between an approaching pair of atoms. This charge density depends on atomic wave functions and the issue to decide is the way in which these eigenfunctions are modified in a crowded environment. The most obvious change is to the freedom of an electron to stray from the nucleus to infinity, without ionization, In the presence of other atoms, each atom is boxed in by the negative charge clouds of its immediate neighbours, so restricting its electrons to a finite sphere. On SCF simulation under this restriction, which amounts to uniform compression of the atom, all electronic energy levels are raised, but at different rates, inversely proportional to the quantum number l. An atom, as it eventually releases an electron on compression to its characteristic ionization radius, is said to be in its *valence state*. At this stage, sharing of valence electrons, considered to define the mechanism of covalent bonding, becomes possible.

8.4.8 Covalence

An electron in the valence state is confined to a sphere, defined by the ionization radius of the atom, and with electronic charge uniformly distributed. Such a charge density is correctly described by a wave function of constant amplitude within the sphere, and vanishing outside. The only parameter that differentiates between atoms of different type is the characteristic ionization radius, which is also a measure of the classical atomic property of *electronegativity*.

When the ionization spheres of two neighbouring atoms interpenetrate, their valence electrons become delocalized over a common volume, from where they interact equally with both atomic cores. The covalent interaction in the hydrogen molecule was modelled on the same assumption in the pioneering Heitler-London simulation, with the use of free-atom wave functions. By the use of valence-state functions this H-L procedure can be extended to model the covalent bond between any pair of atoms. The calculated values of interatomic distance and dissociation energy agree with experimentally measured values.

The generalized H-L procedure correctly simulates the properties of any covalent bond, also in complex molecules, providing that the number of valence electrons on an atom does not exceed the number of its first neighbours. Should this happen the excess valence density screens the repulsion between atomic cores, allowing them to move more closely together, and by increasing the extent of overlap, the bond strength is increased proportionately. The screening is said to increase the *bond order*.

The interaction between overlapping charge spheres can be simulated equally well on representing the equilibrium charge distribution by an equivalent set of electrostatic point charges. This method has the advantage of simulating bonds of all orders equally well, but the inability to predict bond lengths. Another advantage is that all covalent interactions can be represented by a single generalized covalence function, in dimensionless units, obtained by scaling interatomic distance and dissociation energy to effective ionization radii of unity. The cover diagram represents this general covalence function.

8.4.9 The Exclusion Principle

The generalized covalence diagram fits snugly into a golden rectangle, which is a geometrical construct with nearly magical properties. The two sides of the rectangle stand in the ratio of $1 : \Phi = 0.61803\cdots = \tau$, known as the golden section. The relationship, $\Phi = 1 + \tau = 1/\tau$, represents but one of its many intriguing qualities. The diagram shows that the minimum length of a purely covalent bond, in dimensionless units, is equal to τ and has a dissociation energy, in dimensionless diatomic units, of 2τ. The physical interpretation of this limit is that it corresponds to the maximum electron density in a covalent bond, known to be exactly two electrons. This conclusion agrees with the empirical result that two such electrons, at the same energy level, should have opposite spins. Other natural phenomena with a link to the golden ratio include the divergence angle of leaves on a stalk, the spiral stucture of nautilus shells and galaxies, the planetary distances from the sun, and many others. The only factor in common appears to be the curvature of space, thereby implicated to also underlie the exclusion principle.

The same principle will be encountered again in a discussion of the periodic properties of matter.

8.4.10 The Common Model

Despite the distinction, which is usually drawn between covalent, ionic, metallic and dispersion interactions, all of these are in fact of the same electromagnetic type. The popular notion that covalent interaction occurs by means of electron exchange between atoms has no physical basis. The common distinction between covalent and ionic contributions to an interaction reflects two different computational models, rather than different types of interaction.

All electromagnetic interactions are mediated by photons. During interaction that involves the participation of a source and a sink, energy is transmitted by the modulation of a standing wave, which generates an observable photon. As source and sink become balanced, transmission in either direction is equally probable. To maintain this state of balance the photon, which is passed back and forth, is no longer observable and therefore called a *virtual* photon. This exchange may be likened to the attraction between two rugby players running with a ball, which is passed back and forth between them. The optimal separation, which depends on the characteristics of the ball and the dexterity of the players, is dictated by the rules of the game. In physics, the rules of the game are defined by a field, in this case the electromagnetic field.

In the case of an ionic crystal the interacting units are readily identified as cations and anions, which exchange virtual photons. The ions are formed by the transfer of valence electrons to the more electronegative partners. Only a small fraction of the valence density remains in interstitial space. The resulting closed-shell ionic spheres are prevented from interpenetrating by the exclusion principle.

Metallic crystals are viewed as an array of cations bathed in a sea of valence electrons. In the absence of electronegative acceptors the valence density is concentrated preferentially on the interstitial sites between the cations to establish effective negative charge centres, with an overall interaction that resembles the ionic situation. In this case a larger percentage of the valence density is diffusely distributed between the major charge concentrations, leading to a substantial covalent contribution.

The feeblest type of chemical interaction occurs between neutral atoms, not in their valence state, and is typified by inert-gas crystals. In this case the atoms occur close-packed with a very small accumulation of charge on the interstitial sites. These charges are generated by mutual polarization of vibrating atomic charge spheres. Under high pressure an increased amount of valence density is squeezed into interstitial sites until a metal structure is formed.

Using the diamond structure as the prototype of covalent bonding it can be shown that the interaction between atomic cores and valence density, accumulated at interstitial sites, is consistent with both the conventional picture of covalence and with ionic-lattice type interactions.

8.4.11 Molecular Structure

The formation of chemical bonds reduces the total energy of activated species in contact. As all free atoms are spherically symmetrical there are no preferred directions for the formation of minimum-energy bonds. However, when interatomic interaction commences the electronic cloud, especially the valence shell, on each atom is polarized and a special direction in space is established. Electrons with non-zero quantum number l have orbital angular momentum, with a component $m_l\hbar$ in the polar direction. It means that the electron effectively rotates in a plane perpendicular to the polar axis. The only way to minimize this rotational energy is for the angular-momentum vector to align itself with another, equivalent vector, pointing in the opposite direction, also when a bond is formed. Any diatomic contact between reactants would then impose a constraint on the mutual orientation of the reactants to ensure that clockwise rotation on the one, is in balance with anti-clockwise rotation on the other.

In a complex molecule the orbital angular momentum vectors on each atom are required to maintain a fixed orientation with respect to those on all neighbouring ligands. The end result is a fairly rigid structure with well-defined three-dimensional geometry. The molecular structure that emerges in this situation is not predicted by atomic structure.

8.4.12 Electron Spin

The standard Schrödinger equation for an electron is solved by complex functions which cannot account for the experimentally observed phenomenon of electron spin. Part of the problem is that the wave equation 8.4 mixes a linear time parameter with a squared space parameter, whereas relativity theory demands that these parameters be of the same degree. In order to linearize both space and time parameters it is necessary to replace their complex coefficients by square matrices. The effect is that the eigenfunction solutions of the wave equation, modified in this way, are no longer complex numbers, but two-dimensinal vectors, known as *spinors*. This formulation implies that an electron carries intrinsic angular momentum, or spin, of $\pm\hbar/2$, in line with spectroscopic observation.

The way in which the spin factor modifies the wave-mechanical description of the hydrogen electron is by the introduction of an extra quantum number, $m_s = \pm\frac{1}{2}$. Electron spin is intimately linked to the exclusion principle, which can now be interpreted to require that two electrons on the same atom cannot have identical sets of quantum numbers n, l, m_l and m_s. This condition allows calculation of the maximum number of electrons on the energy levels defined by the principal quantum number n, as shown in Table 8.2. It is reasonable to expect that the electrons on atoms of high atomic number should have ground-state energies that increase in the same order, with increasing n. Atoms with atomic numbers 2, 10, 28 and 60 are

Table 8.2: *Allowed quantum numbers and electron counts for the first four wave-mechanical atomic energy levels*

n	l	m_l	m_s	Sub-total	Total	Grand total
1	0	0	$\pm\frac{1}{2}$	2	2	2
2	0	0	$\pm\frac{1}{2}$	2		
	1	$-1 \rightarrow 1$	$\pm\frac{1}{2}$	6	8	10
3	0	0	$\pm\frac{1}{2}$	2		
	1	$-1 \rightarrow 1$	$\pm\frac{1}{2}$	6		
	2	$-2 \rightarrow 2$	$\pm\frac{1}{2}$	10	18	28
4	0	0	$\pm\frac{1}{2}$	2		
	1	$-1 \rightarrow 1$	$\pm\frac{1}{2}$	6		
	2	$-2 \rightarrow 2$	$\pm\frac{1}{2}$	10		
	3	$-3 \rightarrow 3$	$\pm\frac{1}{2}$	14	32	60

predicted to have closed shells and to define the basis for the periodic classification of the elements. The prediction breaks down at atomic number 10. The periodic table of the elements bears little resemblance to the energy-level sequence indicated by Table 8.2, although several features, such as the appearance of fourteen lanthanide elements in sequence, and the grouping of the p-block elements, dovetail with the wave-mechanical model. A more important additional factor is clearly at work and once identified, it should reveal the logical structure of the periodic table.

8.4.13 Periodicity of Matter

The principle that governs the periodic properties of atomic matter is the composition of atoms, made up of integral numbers of discrete sub-atomic units – protons, neutrons and electrons. Each nuclide is an atom with a unique ratio of protons:neutrons, which defines a rational fraction. The numerical function that arranges rational fractions in enumerable order is known as a Farey sequence. A *simple* unimodular Farey sequence is obtained by arranging the fractions $(n/n+1)$ as a function of n. The set of k-modular sequences:

$$S_k = \frac{n}{n+k} \quad , \quad (n,k) = 0,1,2\ldots$$

contains all rational fractions, not necessarily in reduced form. The graph of this set, Figure 8.3, as a function of n generates a vivid summary of possible nuclide compositions. Related graphs are obtained by plotting the set as a function of either atomic number Z, neutron number N, or mass number $A = N + Z$.

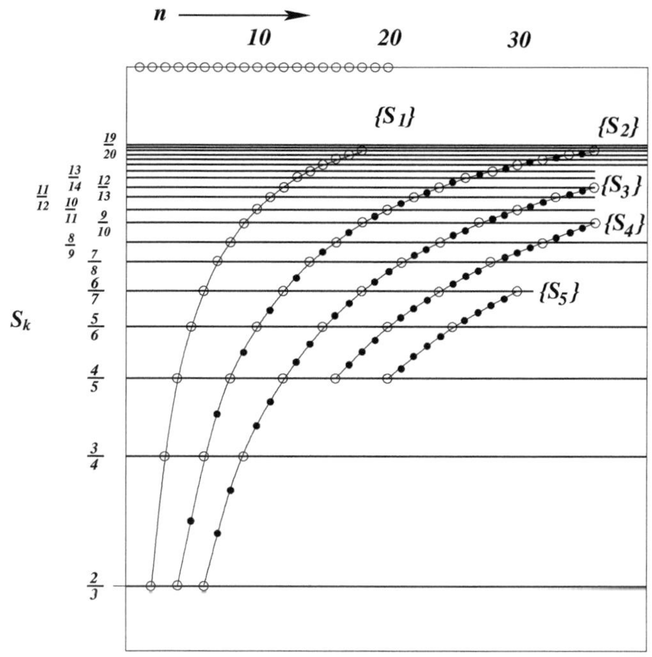

Figure 8.3: *The Simple k-modular Farey set*

The equivalence between $\{S_k\}$, the infinite Farey tree structure and the nuclide mapping is shown graphically in Figure 8.4. The stability of a nuclide depends on its neutron imbalance which is defined, either by the ratio Z/N or the relative neutron excess, $(N - Z)/Z$. When these factors are in balance, $Z^2 + NZ - N^2 = 0$, with the solution $Z = N(-1 \pm \sqrt{5})/2 = \tau N$. The minimum $(Z/N) = \tau$ and hence all stable nuclides are mapped by fractions larger than the golden mean.

To explore the periodic structure of the set $\{S_k\}$, and hence of the stable nuclides, it is convenient to represent each fraction h/k by its equivalent Ford circle of radius $r_F = 1/2k^2$, centred at coordinates h/k, r_F. Any unimodular pair of Ford circles are tangent to each other and to the x-axis. If the x-axis is identified with atomic numbers, touching spheres are interpreted to represent the geometric distribution of electrons in contiguous concentric shells. The predicted $\mathscr{F}_4$ shell structure of $2k^2$ electrons per shell is 2, 8, 8, 18, 18, 32, 32, *etc.*, with sub-shells defined by embedded circles, as 8=2+6,

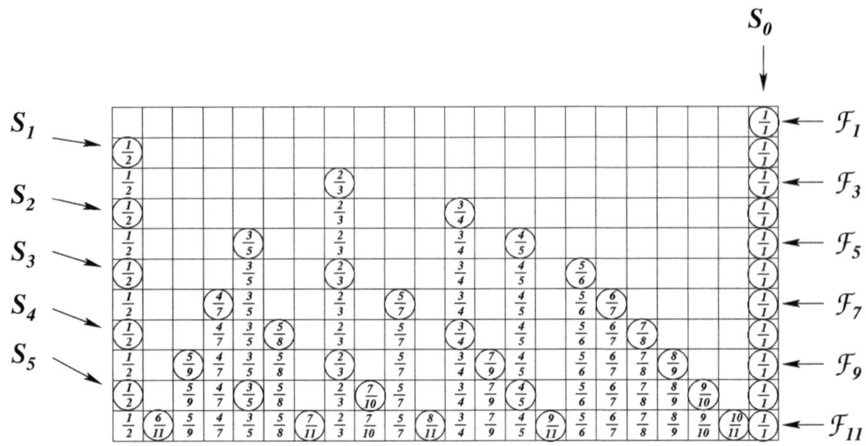

Figure 8.4: *Farey and Simple Farey sequences*

18=(2+8)+(2+6), 32=(2+6)+(6+2)+8+(2+6), in accord with the observed periodic Table 4.5(a). The convergence to τ follows the fractions of the unimodular series defined by Fibonacci numbers:

$$\begin{array}{ccccccccccccc} 1 & & 1 & & 2 & & 3 & & 5 & & 8 & \ldots \\ & \frac{1}{1} & & \frac{1}{2} & & \frac{2}{3} & & \frac{3}{5} & & \frac{5}{8} & & \ldots \end{array}$$

The derived Ford circles define the same periodicity as $\mathscr{F}_4$. An alternative convergence through Lucas fractions:

$$\frac{1}{1} \quad \frac{1}{2} \quad \frac{3}{5} \quad \frac{4}{7} \quad \ldots$$

proceeds to 0.5802..., from $\mathscr{F}_3$, with Ford circles that define the periodic sequence 2,8(18,32),50, shown in Table 5.4(b).

On a plot of Z/N *vs* Z for all stable nuclides the field of stability is outlined very well by a profile, defined by the special points of the periodic table derived from $\mathscr{F}_4$. Furthermore, hem lines that divide the 264 nuclides into 11 groups of 24 intersect the convergence line, $Z/N = \tau$, at most of the points that define the periodic function. If the hem lines are extended to intersect the line $Z/N = 0.58$, a different set of points are projected out and found to match the periodicity, derived from the wave-mechanical model.

The 264 nuclides consist of four groups with mass numbers $A = 4n$, $4n-1$, $4n-2$ and $4n-3$. The two even series consist of 81 members each and the odd series of 51 each. Consecutive members of each series differ by the equivalent of an α-particle, ^{4}He, *i.e.* 2 protons and two neutrons. The complete set is obtained in 52 steps of α-addition, interrupted at regular intervals by 11 steps of radioactive positron emission. The neutron excess increases at each interruption by four units, such that the heaviest member of the $4n$ series ^{208}Pb has a mass number of 4×52 and neutron excess, $n_x = 126 - 82 = 44 = 4 \times 11$. The atomic and neutron numbers of ^{208}Pb are both known to nuclear physicists as magic numbers A complete set of magic numbers are obtained by analysis of the neutron imbalance as a function of neutron number.

From a chemical point of view the most important result is that number theory predicts two alternative periodic classifications of the elements. One of these agrees with experimental observation and the other with a wave-mechanical model of the atom. The subtle differences must be ascribed to a constructionist error that neglects the role of the environment in the wave-mechanical analysis. It is inferred that the wave-mechanical model applies in empty space ($Z/N = 0.58$), compared to the result, observed in curved non-empty space, ($Z/N = \tau$). The fundamental difference between the two situations reduces to a difference in space-time curvature.

8.4.14 Nuclear Genesis

The notion that all nuclides were created by systematic addition of α particles implies that, without radioactive decay, the composition of heavy nuclides should converge to $Z/N = 1$, rather than τ. If, as generally assumed, nuclear synthesis happens in massive stars, the enhanced local curvature could conceivably sustain the favourable ratio of unity. Extrapolation of the hem-lines to $Z/N = 1$ (Figure 4.6) defines a periodic function that arises from the inversion of observed atomic electron configurations. This inversion is in line with the known effects of high pressure on electronic energy levels. The projected periodic structure could arise from a sub-level sequence such as $4f < 3d < 2p < 1s < \ldots$. Once ejected from the star into moderately curved space structural rearrangement of the nuclides inverts electronic configuration and increases neutron excess radioactively.

8.4.15 Reaction Theory

It is practically impossible to formulate a sensible theory of chemical reaction by treating the physical electron as a point particle. This assumption requires that valence electrons must be either in stationary balance with the positively charged atomic cores, or in rapid motion through the interior of the molecule. Crystallographic analysis rules out the first option. A stationary pair of electrons must scatter X-rays more effectively than a hydrogen atom, which in this case, as a bare proton, would be untetectable. Such a charge distribution is never observed. The second option, which implies constant radiation from accelerated charges, is equally unlikely[7].

Both of these observational problems are avoided by modelling an electron as a flexible standing-wave structure. Confined to an atom, electron waves form spherical shells around the nucleus, under electrostatic attraction. In order to approach the nucleus more closely the wave-like electron moves to a lower level of potential energy by releasing energy $h\nu$ in the form of a photon, which carries angular momentum $\hbar$. As the electron's o-a-m is reduced to zero, it cannot move any closer to the nucleus, and remains at its ground-state energy level, with the exchange of virtual photons. The number of electrons at this level is limited by the exclusion principle to a couple with paired spins.

Additional electrons are stacked around the inner core in a pattern, determined by the curved structure of space. The same pattern decides the optimum ratio of protons:neutrons for nuclides. The periodic distribution of extranuclear electrons and the periodic arrangement of nucleons therefore arise from the same periodic numerical function. The wave structure of the electronic configuration must obey Schrödinger's equation, providing that the potential energy is correctly specified. In the traditional formulation only the electrostatic potential is included and the eigenfunction solutions refer to an isolated (hydrogen) atom in empty, therefore flat, space. To allow for an environmental effect, a quantum potential, identified in Bohmian mechanics, must be taken into account, in addition to the classical potential. This problem has not been solved wave-mechanically, but number theory predicts, not only the correct periodicity for curved space, but also the Schrödinger solution for empty space. There are sufficient recognizable similarities between the two periodic functions to correlate the observed periodicity with wave-mechanical eigenvalues of energy and angular momentum.

Atoms do not interact spontaneously, unless the electrons on two atoms that come into contact are at different chemical potential energy levels. Such a difference could result from differences in polarizability, which may cause a redistribution of the overall charge density, involving both atoms. In most cases polarization is insufficient cause for chemical reaction, which normally requires activation by external factors. It could be due to energetic collision, thermal activation, high pressure or catalytic effects. The result is the

[7]Point-charge simulation of covalence treats an extended charge as mathematically equivalent to a set of hypothetical point charges.

same in all cases, but visualized best by the simulation of increased uniform compression, until a valence electron decouples from the core.

The effective ionization sphere has a characteristic value for each element and provides a direct measure of electronegativity, the basic parameter that quantifies all chemical interactions. The relative difference in electronegativity determines the extent and nature of valence-electron redistribution, which in turn differentiates between the major types of interaction, commonly known as ionic, covalent, metallic *etc.*

A large difference in electronegativity leads to complete transfer of valence electrons, from less to more electronegative sites on interacting species, with the creation of oppositely charged ions. For smaller differences, activated valence electrons are uniformly distributed among interstitial regions within a network of positive cores, held together by electromagnetic exchange of virtual photons with the regions of negative charge.

Well-defined products from the chaotic turmoil, which is a chemical reaction, result from a balance between external thermodynamic factors and the internal molecular parameters of chemical potential, electron density and angular momentum. Each of the molecular products, finally separated from the reaction mixture, is a new equilibrium system that balances these internal factors. The composition depends on the chemical potential, the connectivity is determined by electron density distribution and the shape depends on the alignment of vectors that quenches the orbital angular momentum. The chemical, or quantum, potential at an equilibrium level over the entire molecule, is a measure of the electronegativity of the molecule. This is the parameter that contributes to the activation barrier, should this molecule engage in further chemical activity. Molecular cohesion is a holistic function of the molecular quantum potential that involves all sub-molecular constituents on an equal basis. The practically useful concept of a chemical bond is undefined in such a holistic molecule.

The standard observation that thermal and photochemical electrocyclic and cycloaddition reactions always take place by opposite stereochemical pathways [96] is explained directly by the unit of angular momentum carried by a photon. Photochemical activation disturbs the balance between angular-momentum vectors and dictates a different molecular conformation.

Three factors decide the course of a chemical reaction: activation energy, energy gain and structure modification. The most important of these for the experimental chemist to control, is the structure. A number of empirical conformational rules enable structure design by molecular modelling. Most important is the preferred tetrahedral arrangement of ligands around saturated carbon atoms, with related rules for other elements. All of these rules reflect the conservation of orbital angular momentum. With structural changes identified, the energy of reaction is readily estimated as the net change over all bonds. This information is also accessible in the form of thermodynamic data. Despite favourable thermodynamics the feasibility of a reaction may be affected decisively by kinetic factors related to an activation barrier. Although the theoretical basis is well understood for diatomic interaction, activation energy is more difficult to estimate for reactions between complex molecules. Until methods for the estimation of the chemical potential for molecular valence states become available, experimental kinetic data remain the only guide towards barrier heights and catalysis.

The ultimate aim of practical chemistry and chemical engineering is custom design of materials with desirable physical, chemical and pharmaceutical properties. This is where theoretical chemistry can play an important role in future.

8.5 Chemical Cosmology

It would be wrong to create the impression that chemical science has no other purpose than the design and production of materials, which are useful in industry, agriculture, medicine, metallurgy, the arts, construction, electronics and many other spheres of human activity. It also shares the responsibilty, with other academic pursuits, to promote the unity of knowledge and understanding of the cosmic whole. At all levels of enquiry there emerge unique important concepts that contribute to a global understanding, when properly integrated. Apart from famous polymaths like Newton, who excelled as mathematician, physicist and alchemist, too few chemists of the modern era have explored the contributions that chemistry can make to cosmology.

A notable exception is Svante Arrhenius, whose eminently sensible suggestion [103], that new stars and planets arise from the debris of previous cycles, in an endless sequence, is never mentioned in modern cosmologies. Many other concepts of relevance are equally difficult to reconcile with the standard model of modern cosmology, which offers no explanation for the ubiquitous appearance of chiral structures, the reality of anti-matter and the periodic trends in the cosmic abundance and genesis of nuclides.

Significant new evidence is the self-similarity of sub-atomic, atomic, biological, planetary and galactic structures, all related to the golden section. The astronomical structures are assumed to trace out the shape of local space. The obvious conclusion is that all of space has a uniform non-zero characteristic curvature, conditioned by the universal constants π and τ. Space, in this sense, is to be interpreted as equivalent to the three-dimensional sub-space of the Robertson-Walker metric [104]. In standard cosmology this sub-space is assumed to be homogeneous and euclidean, in which form it can evidently not account for the observed self-similar (fractal) cosmic symmetry.

The popular notion that standard big-bang cosmology is universally accepted and complete, is mistaken. The state of the art was recently reviewed by the cosmologist Jayant Narlikar, who highlighted a number of speculative assumptions for which eperimental proof is either outstanding or conceptually impossible [105] – p. 494-509. The strongest argument in favour of the standard model is still based on the relative abundance of light nuclei. It is precisely on this issue that chemical evidence is most relevant.

8.5.1 Nuclear Synthesis

Figure 4.6 shows a plot of nuclear composition (proton:nuclear ratio) for the 264 stable nuclides, separated into 11 groups of 24. Cosmic (or solar) abundance of the nuclides has the same periodicity, as a function of mass number [62]. The hem lines that define the periodicity, intersect the line through the point of convergence, $Z/N \to \tau$, at the golden ratio, in eleven points that define the periodic table of the elements. The same lines, extrapolated to the convergence line, $Z/N \to 0.58$, for flat space-time, projects out the periodic function based on the wave-mechanical solution for an isolated hydrogen atom. It is inferred that the golden-ratio periodicity depends on the characteristic large-scale curvature of space-time.

Extrapolation of the hem lines to $Z/N = 1$ defines another recognizable periodic classification of the elements, inverse to the observed arrangement at $Z/N = \tau$. The inversion is interpreted in the sense that the wave-mechanical ground-state electronic configuration of the atoms, with sublevels $f < d < p < s$, is the opposite of the familiar $s < p < d < f$. This type of inversion is known to be effected under conditions of extremely high pressure [52]. It is inferred that such pressures occur in regions of high space-time curvature, such as the interior of massive stellar objects, a plausible site for nuclear synthesis.

In support of this conjecture it is noted that the known stable nuclides divide into four modular series of mass number, $A(\text{mod}4)$, each of which can be reconstructed by the addition of α-particles ($Z/N = 1$), starting from the neutron and antineutron combinations, n, nn^*, n^*, n^*n^* [62]. The scheme, like big-bang synthesis, depends on the availability of the light nuclides ^{2}H and ^{4}He. Unlike big-bang synthesis, all nuclides are proposed to be formed here in one equilibrium process, the only mechanism that explains periodic abundance.

The exact inverse of empty-space periodicity ($Z/N = 0.58$) is observed at $Z/N = 1.04$ $(1+\tau-0.58)$, by implication at infinite curvature. An attractive possibility identifies this infinity with a black-hole singularity, providing there is a mechanism to release the synthetic material.

The observation that both quasars and Seyfert galaxies appear to be exploding objects that release massive amounts of matter, ostensibly from nowhere [105] – p.343, in a mini-creation event, provides this mechanism. To complete the cycle it is only necessary to identify the black-hole singularity with the origin of the quasar emission. It becomes a viable possibility if the black hole and the quasar are on opposite sides of an interface between two regions of space-time.

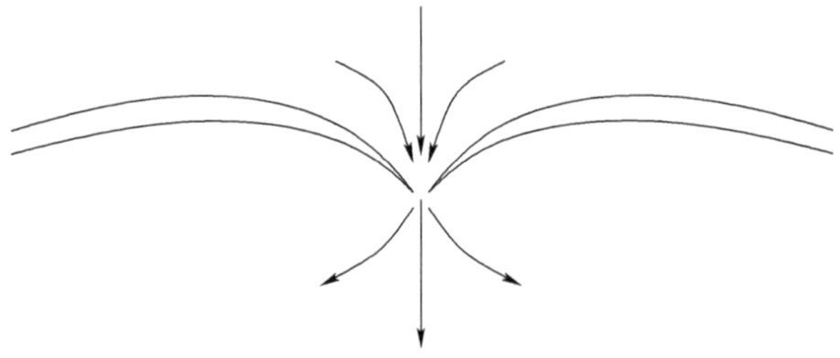

The inversion of atomic structure when forced through the black hole probably means that the disappearing matter emerges as anti-matter on the other side, and *vice versa.*

To complete the picture new matter gushes out into free space where electronic structure adapts to the low-curvature environment as in a phase transition, which renders some nuclides unstable against radioactive decay. The end product is 264 stable nuclides in the solar system.

8.5.2 Chirality of Space

Cosmic structure based on a vacuum interface has been proposed before [49, 7] as a device to rationalize quantum events. To avoid partitioning the universe into regions of opposite chirality the two sides of the interface are joined together with an involution. The one-dimensional analogue is a Möbius strip. Matter on opposite sides of the interface has mutually inverted chirality – matter and anti-matter – but transplantation along the double cover gradually interconverts the two chiral forms. The amounts of matter and anti-matter in such a universe are equal, as required by symmetry, but only one form is observed to predominate in any local environment. Because of the curvature, which is required to close the universe, space itself is chiral, as observed in the structure of the electromagnetic field. This property does not appear in a euclidean Robertson-Walker sub-space.

8.5.3 The Microwave Background

The major selling point of standard cosmology is the observed isotropic microwave background radiation, with black-body spectrum. In a closed universe it needs no explanation. Radiation, which accumulates in any closed cavity, tends, by definition, to an equilibrium wavelength distribution according to Planck's formula (Figure 2.5).

8.5.4 Spectroscopic Red Shifts

The second pillar of standard cosmology is the Doppler interpretation of galactic red shifts that leads to an expanding universe. The experimental observation is that light from distant galaxies has a different spectral composition compared to light emitted in the laboratory. In particular, the absorption lines in the spectra of distant galaxies are shifted to longer wavelengths at the red end of the spectrum. It is almost universally accepted that the red shifts, which correlate well with the estimated distance from the source, when interpreted as Doppler shifts, indicate that galaxies are moving apart. However, when analyzed statistically [105] – p.114, the effect "arises from the passage of light through a non-Euclidean spacetime. It does *not* arise from the Doppler effect, ...". In a generally curved space it is even simpler to demonstrate that a distant galaxy is at a different time coordinate and that the effect of the time difference (Δt) is mathematically equivalent to a Doppler shift.

The sensitivity of electronic configurations to gravitational fields offers an immediate explanation of the enormously different red shifts of light emitted by a quasar and by less massive objects, physically associated with the quasar. The furore [106] over the anomalous Fraunhofer lines of common metals in a quasar corona could also be defused by the conclusion that the electron configurations of elements within the quasar, and hence their spectroscopic properties, differ from their laboratory equivalents. The observed shifts are therefore not due to a fine-structure constant changing with time, but to the response of electronic energy levels to high pressure.

8.5.5 Conclusion

If cosmology is to be consistent with all of science it requires serious revision in order to come into line with chemical evidence pertaining to the periodicity of matter, the cosmic abundance of nuclides and the self-similar symmetry between objects large and small. All of these aspects, either refute, or remain neutral to the provisions of the big-bang model. A valid model, in line with the chemical facts, is a closed universe that resembles the double cover of a Möbius surface, identified as the vacuum. Extreme curvature punctures the interface to connect a material black hole with an anti-material quasar, or *vice versa*. Nuclear synthesis happens in this binary system and amounts to a recycling of matter between dying and new-born galaxies. The vortex structure that develops in the material falling into a black hole, traces out the same spiral shape as a developing solar system that grows along a vortex due to rotation in a nebular cloud, as long ago postulated by the philosopher Immanuel Kant. Being closed in space and time the search for the beginning or the end of the universe is as futile as searching for the beginning of a circle.

Bibliography

[1] A. Pais, *Inward Bound*, 1986, Clarendon Press, Oxford.

[2] A. Einstein, *On the electrodynamics of moving bodies*, in [3], p. 37-65.

[3] H.A. Lorentz, A. Einstein, H. Minkowski and H. Weyl, *The Principle of Relativity*, translated by W. Perrett and G.B. Jeffrey, 1952, Dover, NY.

[4] H.A. Lorentz, *Electromagnetic phenomena in a system moving with any velocity smaller than that of light*, in [3] p. 9-34.

[5] G.N. Lewis, *The Nature of Light*, Proc. Nat. Acad. Sci. U.S., 12 (1926) 22-29.

[6] A. Einstein, *The foundation of the general theory of relativity*, in [3], p. 111-164.

[7] J.C.A. Boeyens, *New Theories for Chemistry*, 2005, Elsevier, Amsterdam.

[8] A. Sommerfeld, *Atombau und Spektrallinien*, 4th ed., 1924, Vieweg, Braunschweig.

[9] W. Heisenberg, *The Physical Principles of the Quantum Theory*, English translation by C. Eckart and F.C Hoyt, Dover, 1949.

[10] H. Nagaoka, *Kinetics of a System of Particles illustrating the Line and Band Spectrum and the Phenomena of Radioactivity*, Phil. Mag. and J. Sci., 1904 (7) 445-455.

[11] H. Goldstein, *Classical Mechanics*, 2nd ed., 1980, Addison-Wesley, Reading, Mass.

[12] J. Kappraff, *Beyond Measure*, 2002, World Scientific, Singapore.

[13] B.L. van der Waerden, *Sources of Quantum Mechanics*, 1967, North-Holland, Amsterdam.

[14] M. Kaku and J. Thompson, *Beyond Einstein*, Oxford University Press, 1999, Oxford.

[15] J.C.A. Boeyens, *The Theories of Chemistry*, 2003, Elsevier, Amsterdam

[16] A. Kekulè, *Lehrbuch der Organischen Chemie*, 1861, Enke, Erlangen.

[17] J.C.A. Boeyens, *The Holistic Molecule*, in [18].

[18] J.C.A. Boeyens and J.F. Ogilvie (eds.), *Models, Mysteries and Magic of Molecules*, 2007, Springer.

[19] J.F Ogilvie<ogilvie@cecm.sfu.ca> e-mail communication, 26 Jan. 2008.

[20] W.G. Richards and D.L. Cooper, *Ab initio molecular orbital calculations for chemists*, 2nd ed., 1983, Clarendon Press, Oxford.

[21] D.R. Hartree, *The Wave Mechanics of an Atom with a Non-Coulomb Central Field.* Camb. Phil. Soc., 24 (1928) 89-132.

[22] A Bernthsen, *A Textbook of Organic Chemistry*, New edition, revised by J.J. Sudborough, 1922, Blackie, London.

[23] J. Kenyon and D.P. Young, *Retention of asymmetry during the Curtius and the Beckmann change,* J. Chem. Soc., 1941, 263-267.

[24] J. March, *Advanced Organic Chemistry. Reactions, Mechanisms, and Structure*, 4th ed., 1992, Wiley, New York.

[25] A. Einstein, *Zum Quantensatz von Sommerfeld und Epstein*, Verh. d. D. Physik. Gesell., 1917 (19) 82-92.

[26] K.R. Popper, *Quantum Theory and the Schism in Physics*, 1982, reprinted 1995, Routledge, London.

[27] W. Heisenberg, *The Representation of Nature in Contemporary Physics*, Daedalus, 1958 (87) 95-108.

[28] M. Gell-Mann in D. Huff and O. Prewett, *The Nature of the Physical Universe: 1976 Nobel Conference*, 1979, New York, p.29.

[29] J. von Neumann, *Mathematical Foundations of Quantum Mechanics*, English translation, Princeton University Press, 1955, page 325.

[30] J.S. Bell, *On the problem of hidden variables in quantum mechanics*, Rev. Mod. Phys., 1966 (38) 447-452.

[31] D. Finkelstein, *The logic of quantum physics*, Trans. N.Y. Acad. Sci., 1962/3 (25) 621-637.

[32] A. Rüger, *Atomism from cosmology: Erwin Schrödinger's work on wave mechanics and space-time structure*, Hist. Stud. Phil. Sci., 18 (1988) 377 - 401.

[33] I.N. Levine, *Quantum Chemistry*, 4th ed., 1991, Prentice-Hall, New Jersey.

[34] E. Schrödinger, *Collected Papers on Wave Mechanics*, translated by J.F. Shearer and W.M. Deans, 2nd. ed., 1978, Chelsea, NY.

[35] P. Jordan and O. Klein, *Zum Mehrkörperproblem der Quantentheorie*, Zeit. für Phys., 45 (1927) 751 - 765.

[36] P. Jordan and E. Wigner, *Über das Paulische Äquivalenzverbot*, Zeit. für Phys., 47 (1928) 631-651.

[37] E. Schrödinger, *Science Theory and Man*, 1957, Dover, London.

[38] E. Schrödinger, *Über die kräftefreie Bewegung in der relativistischer Quantenmechanik*, Sitzungsber. Preus. Akad. Wiss., (1930) 418-428.

[39] E. Madelung, *Quantentheorie in hydrodynamischer Form*, Z. für Physik, 1926 (40) 322-326.

[40] T. Takabayashi, *On the Formulation of Quantum Mechanics associated with Classical Pictures*, Prog. Theor. Phys., 1952 (8) 143-182.

[41] T. Takabayashi, *Remarks on the Formulation of Quantum Mechanics with Classical Pictures and on Relations between Linear Scalar Fields and Hydrodynamic Fields*, Prog. Theor. Phys., 1953 (9) 187-220.

[42] D. Bohm and J.P. Vigier, *Model of the Causal Interpretation of Quantum Theory in Terms of a Fluid with Irregular Fluctuations*, Phys. Rev., 1954 (96) 208-217.

[43] E. Schrödinger, *World Structure*, Nature, 1937 (140) 742-744.

[44] P.A.M. Dirac, *Classical theory of radiating electrons*, Proc. Roy. Soc. (London), 1938 (A167) 148 - 169.

[45] H.A. Lorentz, *Electromagnetic phenomena in a system moving with any velocity smaller than that of light*, Proc. Royal Acad., Amsterdam, 1904 (6) 809-831.

[46] A. Einstein, B. Podolsky and N. Rosen, *Can quantum-mechanical description of physical reality be considered complete?* Phys. Rev., 1935 (47) 777-780.

[47] J.A. Wheeler and R.P. Feynman, *Interaction with the Absorber as the Mechanism for Radiation*, Revs. Mod. Phys., 1945 (17) 157-181.

[48] E. Schrödinger, *Über eine bemerkenswerte Eigenschaft der Quantenbahnen eines einzelnen Elektrons.* Z. für Physik, 1922 (12) 13-23.

[49] J.C.A. Boeyens, *The geometry of quantum events.* Specul. Sci. Techn., 1992 (15) 192-210.

[50] J. Burdett, *Chemical Bonds: A Dialog*, 1997, Wiley, Chichester.

[51] B.H. Bransden and C.J. Joachain, *Physics of Atoms and Molecules*, 1983, Longman, London. p. 147.

[52] S. Goldman and C. Joslin, *Spectroscopic Properties of an Isotropically Compressed Hydrogen Atom*, J. Phys. Chem., 1992 (96) 6021-6027.

[53] J.C.A. Boeyens, *Ionization radii of compressed atoms*, J.C.S. Faraday Trans., 1994 (90) 3377-3381.

[54] H. Margenau and G.M. Murphy, *The Mathematics of Chemistry and Physics*, 1943, Van Nostrand, NY, p. 252.

[55] J.C.A. Boeyens, *Electron-pair bonds.*, S. Afr. J. Chem., 1980 (33) 14-20.

[56] N.C. Handy, *Density Functional Theory: An Alternative to Quantum Chemistry*, in [57].

[57] R. Broer, P.T.C. Aerts and P.S. Bagus, *New Challenges in Computational Quantum Chemistry*, 1994, University of Groningen.

[58] E. Clementi, G. Corongiu and L. Pisani, *New Challenges in Computational Chemistry*, in [57].

[59] J.C.A. Boeyens, *Molecular Mechanics and the Structure Hypothesis*, Structure and Bonding, 1985 (63) 65-101.

[60] J.C.A. Boeyens, *Electrostatic calculation of bond energy*, J. S. Afr. Chem. Inst., 1973 (26) 94-105.

[61] D.D. Ebbing and M.S. Wrighton, *General Chemistry*, 2nd ed., 1987, Houghton Mifflin, Boston.

[62] J.C.A. Boeyens and D.C. Levendis, *Number Theory and the Periodicity of Matter*, (2008), Springer.

[63] E.P. Batty-Pratt and T.J. Racey, *Geometric Model for Fundamental Particles*, Int. J. Theor. Phys., 1980 (19) 437-475.

[64] M. Wolff, *Beyond the Point Particle – A Wave Structure for the Electron*, Galilean Electrodynamics, 1995 (6) 83-91.

[65] J.C.A. Boeyens, *Angular Momentum in Chemistry*, Z. Naturforsch., 2007 (62b) 373-385.

[66] J.C.A. Boeyens and J. du Toit, *The theoretical basis of electronegativity*, Electronic J. Theor. Chem., 1997 (2) 296 - 301.

[67] R.G. Parr and R.G. Pearson, *Absolute Hardness, Companion Parameter to Absolute Electronegativity*, J Am. Chem. Soc., 1983 (105) 7512 - 7516.

[68] L. Pauling, *Nature of the Chemical Bond*, 3rd ed., 1960, Cornell University Press, Ithaca.

[69] J.C.A. Boeyens, *The Periodic Electronegativity Table*, Z. Naturforsch., 2008 (63b) 199-209.

[70] H. Eyring, J. Walter and G.E. Kimball, *Quantum Chemistry*, 1944, Wiley, NY.

[71] J.O. Hirschfelder, C.F. Curtiss and R.B. Bird, *Molecular Theory of Gases and Liquids*, 1954, Wiley, NY, p.237.

[72] J.C.A. Boeyens, *Multiple bonding as a screening phenomenon*, J. Crystallogr. Spectr. Res., 1982 (12) 245-254.

[73] J.C.A. Boeyens and D.J. Ledwidge, *A Bond-Order Function for Metal-Metal Bonds*, Inorg. Chem., 1983 (22) 3587-3589.

[74] C. Kittel, *Introduction to Solid State Physics*, 5th ed., 1976, Wiley, NY.

[75] R.C. Evans, *An Introduction to Crystal Chemistry*, 2nd ed., 1964, University Press, Cambridge.

[76] J.C.A. Boeyens and G. Gafner, *Direct Summation of Madelung Energies*, Acta. Cryst., 1968 (A25) 411-414.

[77] J.N. Israelachvili, *Intermolecular and Surface Forces*, 2nd. ed., 1992, Academic Press, London.

[78] D.R. Lide (ed.), *Handbook of Chemistry and Physics*, 82nd ed., 2001, CRC, Boca Raton.

[79] T.S. Koritsanszky and P. Coppens, *Chemical Applications of X-ray Charge-Density Analysis*, Chem. Rev.

[80] P.R. Holland, *The Quantum Theory of Motion*, 1993, University Press, Cambridge.

[81] F.A. Cotton and R.A. Walton, *Multiple Bonds between Metal Atoms*, 1982, Wiley, New York.

[82] J.C.A. Boeyens, *Quantum theory of molecular conformation*, C.R. Chimie 2005 (8) 1527-1534.

[83] S.F.A. Kettle, *Symmetry and Structure*, 2nd ed., 1995, Wiley, Chichester.

[84] H. Hart and R.D. Schuetz, *Organic Chemistry*, 5th ed., 1978, Houghton Mifflin, Boston.

[85] J.N. Murrell, *The Theory of the Electronic Spectra of Organic Molecules*, 1963, Methuen, London.

[86] J.C.A. Boeyens and S.D. Travlos, *A free-electron study of color vision*, J.Cryst.Spectr.Res., 1990 (20) 433-439.

[87] J.C.A. Boeyens, *Calculation of bond angles*, S. Afr. J. Chem., 1980 (33) 66-70.

[88] A.Y. Meyer, *Molecular mechanics and molecular shape.* Part VI., J. Mol. Struct. (Theochem), 1988 (179) 83-98.

[89] P. Comba, T.W Hambley, *Molecular Modeling of Inorganic Compounds*, 2nd ed., 2001, Wiley-VCH, Weinheim.

[90] A.K. Rappé, C.J. Casewit, K.S. Calwell, W.A. Goddard III, W.M. Skiff, *UFF, a full periodic table force field for molecular mechanics and molecular dynamics simulations.* J. Am. Chem. Soc., 1992 (114) 10024 - 10035

[91] D.M. Root, C.R. Landis, T. Cleveland, *Valence bond concepts applied to the molecular mechanics description of molecular shapes. 1. Application to nonhypervalent molecules of the P-block*, J. Am. Chem. Soc.,1993 (115) 4201 - 4209.

[92] C.R. Landis, T.K. Firman, T. Cleveland, D.M. Root, *Extending Molecular Mechanics Methods to the Descriptions of Transition Metal Complexes and Bond-Making and -Breaking Processes*, in [93], p 49 - 76.

[93] L. Banci, P. Comba (eds.), *Molecular Modeling and Dynamics of Bioinorganic Systems*, 1997, Kluwer, Netherlands.

[94] J.C.A. Boeyens, P. Comba and B. Martin, *Electronic Basis of Molecular Mechanics*, To be published.

[95] N.H. Frank, *Introduction to Electricity and Optics*, 2nd ed., 1950, McGraw-Hill, NY, p. 342.

[96] J. Mc Murray, *Organic Chemistry*, 1984, Brookes/Cole, Monterrey.

[97] J. C. A. Boeyens, *Environmental Factors in Molecular Modelling*, in [98], p. 99-103.

[98] W. Gans, A. Amann and J.C.A. Boeyens, *Fundamental Principles of Molecular Modeling*, 1996, Plenum, NY.

[99] J.D. Jackson, *Classical Electrodynamics*, 1962, Wiley, New York.

[100] M. Livio, *The Golden Ratio*, 2002, Review, London.

[101] G.E. Satterthwaite, *Encyclopedia of Astronomy*, 1970, Hamlyn, London.

[102] H.C. Urey, *The Planets: Their Origin and Development*, 1952, Yale University Press, New Haven.

[103] S.A. Arrhenius, *Worlds in the Making*, 1908, Harper, London.

[104] R. Adler, M. Bazin and M. Schiffer, *Introduction to General Relativity*, 1965, McGraw-Hill, N.Y.

[105] J.V. Narlikar, *An Introduction to Cosmology*, 3rd ed., 2002, University Press, Cambridge.

[106] *Blinding Flash*, New Scientist, 11 May (2002) 28-32.

Index

MIX
Papier aus verantwortungsvollen Quellen
Paper from responsible sources
FSC® C105338

If you have any concerns about our products,
you can contact us on
ProductSafety@springernature.com

In case Publisher is established outside the EU,
the EU authorized representative is:
Springer Nature Customer Service Center GmbH
Europaplatz 3, 69115 Heidelberg, Germany

Printed by Libri Plureos GmbH
in Hamburg, Germany